NEUROSCIENCE OF
PREFERENCE AND CHOICE

NEUROSCIENCE OF PREFERENCE AND CHOICE

COGNITIVE AND NEURAL MECHANISMS

Edited by

Raymond Dolan
Wellcome Trust Centre for Neuroimaging
University College London
London, UK

Tali Sharot
Wellcome Trust Centre for Neuroimaging
University College London
London, UK

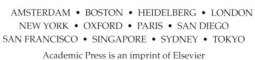

AMSTERDAM • BOSTON • HEIDELBERG • LONDON
NEW YORK • OXFORD • PARIS • SAN DIEGO
SAN FRANCISCO • SINGAPORE • SYDNEY • TOKYO
Academic Press is an imprint of Elsevier

Academic Press is an imprint of Elsevier
32 Jamestown Road, London NW1 7BY, UK
225 Wyman Street, Waltham, MA 02451, USA
525 B Street, Suite 1800, San Diego, CA 92101-4495, USA

First edition 2012

Notice
No responsibility is assumed by the publisher for any injury and/or damage to persons
or property as a matter of products liability, negligence or otherwise, or from any use or
operation of any methods, products, instructions or ideas contained in the material herein.
Because of rapid advances in the medical sciences, in particular, independent verification of
diagnoses and drug dosages should be made

British Library Cataloguing-in-Publication Data
A catalogue record for this book is available from the British Library

Library of Congress Cataloging-in-Publication Data
A catalog record for this book is available from the Library of Congress

ISBN: 978-0-12-381431-9

For information on all Academic Press publications
visit our website at: www.elsevierdirect.com

Typeset by MPS Limited, a Macmillan Company Chennai, India
www.macmillansolutions.com

Printed and bound in United States of America

12 13 14 15 16   10 9 8 7 6 5 4 3 2 1

Contents

II

CONTEXTUAL FACTORS

III

SOCIAL FACTORS

8. Social Factors and Preference Change

DANIEL CAMPBELL-MEIKLEJOHN AND CHRIS D. FRITH

9. Social and Emotional Factors in Decision-Making: Appraisal and Value

ELIZABETH A. PHELPS AND PETER SOKOL-HESSNER

IV

PERCEPTUAL FACTORS

10. Auditory Preferences and Aesthetics: Music, Voices, and Everyday Sounds

JOSH H. MCDERMOTT

V

IMPLICATIONS, APPLICATION AND FUTURE DIRECTION

Preface

How preferences are generated, and choices made, is the central focus of this book. Mindful that preferences is a subject of broad interest we asked the contributing authors to detail diverse approaches, appropriate to their core expertise – ranging from evolutionary biology to social psychology, psychophysics, cognitive neuroscience and economics. The contributions explore the neural and psychological processes that underlie decision-making, propose formal models of the associated mechanisms and illustrate their application to domains that include perception, social interaction and policy.

We see these various contributions as supporting two general claims. First, they argue that contrary to traditional decision-making theories, which assume choices are based on relatively steady preferences, preferences are in fact highly volatile and susceptible to the context in which the alternatives are presented. For example, the same object seems to be worth more if we are engaged in selling as opposed to buying it. Moreover, our preferences are modified by the mere act of choosing and altered by changing choice sets.

The observation that preferences are fluid invokes the idea that fluidity is embedded in the architecture of our brain. Part II of this book, in particular, is dedicated to this theme. Data from different scientific modalities, going from single cell recordings in non-human primates right through to brain imaging data in human adults, supports the notion that preferences are unstable and easily altered by context. One might assume that this instability in preferences is generated by an abundance or tyranny of choice in modern society. After all, unlike our ancestors, we can select from a near-infinite number of possibilities on where to live, who to marry, which profession to embark upon and how to spend our leisure time. However, by comparing valuation processes across non-human primates, human children, and human adults, Lakshminarayanan and Santos show that there are evolutionary conserved constraints on our preferences. Thus, like adults, children and non-human primates change their preferences based on the context in which the options are presented. Like humans, primates show loss aversion, post-choice re-evaluation, and the endowment effect.

The second theme to emerge is the commonality of the processes underlying preference generation. We suggest that regardless of whether we are selecting a musical tune, a perfume, or a new car – the brain uses similar computational principles to compute the value of our options, which are tracked by common neural systems. Part I of this book describes these systems and offers possible models for how decisions are taken. Part III then asks how these processes underlie social preferences and Part IV focuses on the perceptual processes that underlie preference generation.

The book concludes with a discussion of the implications of the research presented to policy, health and well-being. It highlights the societal importance of how understanding the neuroscience of preference and choice might help us make better decisions in our personal life and aid policy makers in enhancing the well-being of citizens.

Ray Dolan and Tali Sharot
Wellcome Trust Centre for Neuroimaging,
Institute of Neurology, University College London

Contributors

Andrew Caplin   Professor of Economics and Co-Director of the Center for Experimental Social Science, New York University

Daniel Campbell-Meiklejohn   Center for Functionally Integrative Neuroscience, Aarhus University and Wellcome Centre for Neuroimaging at University College London

Nick Chater   Behavioural Science Group, Warwick Business School, University of Warwick

Géraldine Coppin   Swiss Center for Affective Sciences, and Laboratory for the Study of Emotion Elicitation and Expression, University of Geneva

Ara Darzi   Division of Surgery, Imperial College London, London

Peter Dayan   Gatsby Computational Neuroscience Unit, UCL, London, UK

Raymond J. Dolan   Wellcome Trust Centre for Neuroimaging, Institute of Neurology University College London, London

Chris D. Frith   Center for Functionally Integrative Neuroscience, Aarhus University and Wellcome Centre for Neuroimaging at University College London

Paul W. Glimcher   Center for Neural Science, New York University

Lars Hall   Lund University Cognitive Science, Lund University

Petter Johansson   Division of Psychology and Language Sciences, University College London

Venkat Lakshminarayanan   Department of Psychology, Yale University, New Haven, USA

Hsin-I Liao   Department of Psychology, National Taiwan University, Taipei, Taiwan

Kenway Louie   Center for Neural Science, New York University

Benedetto De Martino   University College of London, IBN Department of Psychology, London and California Institute of Technology, Pasadena

Josh H. McDermott   Center for Neural Science, New York University, USA

Elizabeth A. Phelps   New York University, New York, USA

David Sander   Swiss Center for Affective Sciences, and Laboratory for the Study of Emotion Elicitation and Expression, University of Geneva

Laurie R. Santos   Department of Psychology, Yale University, New Haven, USA

Tali Sharot   Wellcome Trust Centre for Neuroimaging, University College London, London

Shinsuke Shimojo Division of Biology, Computation and Neural Systems, California Institute of Technology, Pasadena, CA, USA

Peter Sokol-Hessner California Institute for Technology, Pasadena, California, USA

Mkael Symmonds Wellcome Trust Centre for Neuroimaging, Institute of Neurology University College London, London

Ivo Vlaev Centre for Health Policy, Imperial College London, London

MECHANISMS

The Neurobiology of Preferences

Mkael Symmonds and Raymond J. Dolan

Wellcome Trust Centre for Neuroimaging, Institute of Neurology
University College London, London

OUTLINE

Neuroscience of Preference and Choice
DOI: 10.1016/B978-0-12-381431-9.00001-2

3

INTRODUCTION

Choice conjures the idea of a volitional or directed selection of desirable actions, motivated by internal likes and dislikes. Unsurprisingly, the conundrums raised by choice are center stage in philosophical and scientific debate since antiquity (Aristotle, 1998; MacPherson, 1968). Choice is generally envisaged as a conscious selection of alternatives, under an assumption that individuals possess "preferences," or a predilection to make certain types of choice in specific situations. This realization is often thought to reflect the expression of either acquired or "hard-wired" drives that are expressed in a biological substrate. Thus, an individual with a preference for chocolate over lemon is considered to possess an "internal machinery" capable of representing this enhanced valuation of chocolate relative to lemon, a valuation that provides the basis for a consistent and rational selection between these two goods.

Traditional economic thinking relates preferences to a statement about well-being. An agent who expresses a particular choice is considered to be maximizing their own subjective "utility" or welfare. Every decision or choice is considered an expression of preference and in the "revealed preference" framework (Samuelson, 1938), choice and preference are synonymous. While revealed choices are ultimately the dependent variable for classical economics, understanding the neurobiology of preferences necessitates that we entertain a range of mechanisms by which human choices are generated, expressed and influenced.

The Neuroscience of Decision-Making

The neuroscience of choice and preference dates back to the nineteenth century, with the emergence of the idea of functional specialization as a fundamental organizational principle of the brain. While phrenologists attributed behavioral characteristics to the contours of the scalp (Gall & Spurzheim, 1818), the observation of specific and consistent behavioral deficits following localized brain damage led to the development of neurology as a medical speciality (Broca, 1865; Jackson, 1873). The tradition of inferring function from structural and electrophysiological perturbations has remained powerful, enabling a mapping of both primary sensory and motor systems (Penfield & Boldrey, 1937), complex cognitive processes such as language (Head, 1920; Ojemann, 1978), memory (Scoville & Milner, 1957) and, more recently, a mapping of areas important in decision-making, strategy selection and learning (Bechara, Damasio, Damasio & Anderson, 1994; Shallice & Burgess, 1991).

The early twentieth century heralded the birth of behavioral psychology, pioneered by the classic findings of Pavlov, Skinner, Tolman, Hull and others (Schultz & Schultz, 2007). This tradition provided insights into core processes mediating learning and choice, but was restricted in scope by the primitive methods available to study concurrent neurobiological activity during decision-making. In the 1960s, Olds and colleagues produced a startling finding that stimulation of specific brain loci in animals imbued behavior with apparent hedonic value. For example, self-stimulation experiments in rats showed they were disposed to compulsively press a lever to the exclusion of other hedonic behavioral options. Thus, a dramatic form of preference could be driven by electrical stimulation of the rats' subcortical dopaminergic structures (Olds & Milner, 1954).

A refinement in neuroscientific techniques in the 1980s saw the emergence of single-unit recording methodologies. These revealed that activity of individual neurons in early visual areas could predict trial-by-trial choices of an animal in a random-dot motion discrimination experiment (Parker & Newsome, 1998). Such an approach provided the first direct link between neural activity at a single unit level and the expression of choice behavior. Subsequent studies have asked more sophisticated questions about the construction of value and choice in an economic framework. A key example is a report showing that a region of parietal cortex called LIP, expressed activity that correlated with the reward magnitude and probability (expected value) associated with an upcoming action (Platt & Glimcher, 1999).

The development of neuroimaging techniques, in particular functional magnetic resonance imaging (fMRI), has meant that questions related to choice and preference can now be addressed non-invasively in humans. Early neuroimaging studies of financial decision-making dissected out regions involved in processing monetary gain and loss (Elliott, Friston & Dolan, 2000; Thut, Schultz, Roelcke et al., 1997), as well as brain activation related to anticipation versus receipt of reward (Breiter, Aharon, Kahneman et al., 2001). More sophisticated studies have borrowed economically-inspired models of behavior to seek out a brain representation of key decision variables. Notable examples include a demonstration brain activity that tracks a Pascalian idea of a value representation constructed from a combination of amount and probability (Dreher, Kohn & Berman, 2006; Knutson, Taylor, Kaufman et al., 2005), to financial and ecological concepts of risk and uncertainty (Christopoulos, Tobler, Bossaerts et al., 2009; Mohr, Biele & Heekeren, 2010; Preuschoff, Bossaerts & Quartz, 2006), and the idea that anticipated temporal delay reduces the value of rewards (Kable & Glimcher, 2007; Pine, Seymour, Roiser et al., 2009).

The Nature of Preferences

Preferences can be thought of as biologically determined traits (Eysenck, 1990; Ebstein, 2006), but in reality they are dynamic and flexible, and indeed often inconsistent. The idea of preference is broad and diverse, encompassing a liking for different goods, a favoring of reward over punishment through to preferences for specific components of a decision ("decision variables"). The latter can encompass risk preference, impulsivity (the preference for delayed versus immediate goods) and social preferences ("other-regarding" preferences, altruism and fairness). Whether such preferences are convenient theoretical artefacts for classifying individual choice, or whether they relate to intrinsic biological processes and inter-individual differences is the subject of intense and wide-ranging programs of research. This breadth is evident in a facility to draw on human and animal psychology, neurobiology, economics, ecology and computational science, though a source of confusion can be each discipline's different terminology, theory, experimental technique and hypothetical assumptions that color the literature. There are many excellent reviews and recent descriptions of these conceptual edifices, along with attempts to bridge between them e.g. (Berns, Laibson & Loewenstein, 2007; Doya, 2008; Glimcher & Rustichini, 2004; Kenning & Plassmann, 2005; Sanfey, Loewenstein, McClure & Cohen, 2006). Our aim here is to discuss preferences from a biological perspective, acknowledging that our discussion will stray outside a boundary of what some would regard as preference.

We first survey examples of preference and choice that raise questions as to their biological implementation. We then discuss how strict economically-inspired definitions of preference might need to be loosened, or expanded, when probing the biological systems that drive choice behavior. Finally, we examine plausible mechanisms for a neural instantiation of preference and discuss how preference can be modulated by physiological, pharmacological or direct neural manipulation.

CONCEPTS OF CHOICE AND PREFERENCE

Revealed Preferences or Behavioral Biases?

"Revealed preference" is often assumed to imply that an individual *wants* the chosen outcome, but there are many scenarios where this is not necessarily the case. For example, the *status quo* bias (Samuelson & Zeckhauser, 1988), where individuals show a predilection to stick with a previously selected option, would not necessarily be thought of as an expression of an internal desire. Instead, it is more parsimonious to think of it as reflecting a biological intrinsic default response mode which needs to be overcome, usually requiring an expenditure of effort to recruit an alternative motor action program. Overcoming this default has been demonstrated to evoke enhanced activity in the subthalamic nucleus (Fleming, Thomas & Dolan, 2010), paralleling earlier findings from single-unit recordings where "incorrect" actions (saccadic eye movements in monkeys in a simple instrumental task) have been attributed to baseline (pre-stimulus) neuronal activity encoding a default motor action program (Lauwereyns, Watanabe, Coe & Hikosaka, 2002). These examples highlight a recurring dichotomy that different types of choice may be generated by entirely disparate mechanisms despite the end result (the apparent "revealed preference") being indistinguishable. Strictly speaking, the *status quo* bias *is* a preference to stay with the default rather than change, despite clearly not being driven by any desire to attain an internal goal or "utility maximize." Thus, we believe this example illustrates a necessity to invoke ideas beyond the typical remit of an economic definition of preference to fully understand the biological generators of choice.

Competing Valuation Systems

There are important examples where choices do not accord with internal wants. An addict may perform an action (e.g. taking drugs or alcohol) in the present despite expressing a desire to avoid doing this very action on a prior occasion. Explanations exist for this type of behavior that fit in with a classical concept of preference, such as state-dependent utility (Karni, Schmeidler & Vind, 1983), where physiologically-driven craving, at the point of consumption the by virtue of substance is truly valued more than disliked. Alternative explanations include the idea that choices alter depending upon the particular choice set available to the decision-maker (Koszegi & Rabin, 2007). However, we know that relapses are frequently triggered by environmental cues (e.g. seeing drug-related paraphernalia, watching others drinking alcohol, or the roguery of advertising) and this suggests an alternative possibility of a mistake, lapse or default action. Rather than being a momentary failure to execute a desired action, addicts seem to pursue a sequence of deliberate and complex actions to attain a goal that seems to be simultaneously craved but not wanted.

Young children and animals are notoriously impulsive, being unable to delay gratification of desire (Mischel et al., 1989). From an economic perspective they can be considered to express a high temporal discount rate (delayed reward is worth less than immediate reward, so 1 marshmallow now is preferred to waiting minutes for 2 marshmallows). However, the drives directing such behaviors do not always fall neatly in line with an economic model. In reverse-reward experiments, participants are simultaneously shown 2 options of different numerical magnitude (e.g. 5 versus 2 raisins). Whichever set of items is selected, the opposite set is actually received. So for a chimpanzee to obtain 5 raisins from the experimenter, they must select the smaller set of 2 raisins. Both primates (Boysen & Berntson, 1995) and preschool children (Carlson, Davis & Leach, 2005) are notoriously poor at learning to pick the smaller option, despite being visibly upset when they fail to get the larger reward.

Where is the preference in the above? Is it the chosen or unchosen option that chimps and children are distressed at not getting? The effect is not a failure in understanding the correct contingency or an inability to appreciate number. In fact, when edible items are replaced with inedible items, performance significantly improves (Boysen & Berntson, 1995; Schmitt & Fischer, 2011). Instead, the magnetic pull of a large pile of food interferes with the ability to preferentially select the smaller option. Similar examples from the animal literature include Hershberger's experiments (Hershberger, 1986), demonstrating that while chicks readily acquire an instrumental approach response to obtain food, they are totally unable to learn a reversed "move away" response to obtain the same outcome. These difficulties are thought to reflect conflict between Pavlovian (i.e. conditioned) and instrumental ("goal-directed") responses (see Chapters 2 and 5 by Dayan and De Martino). This speaks to the more general idea of multiple valuation systems that are under different degrees of control and amenable to different types of overt and covert influences.

Searching for a Hedonimeter

In contrast to the idea of multiple valuation systems is the concept of a unitary representation of value, corresponding to a classical idea of "utility." Utility theory is based on a set of axioms or rules about consistent behavior, and generates strong constraints upon allowable choice. However, there are multiple examples of behavioral biases that contradict the assumptions of utility theory, discussed at length elsewhere in this volume (e.g. the Allais paradox – see Chapter 5 by De Martino). Implicit in utility models of behavior is the idea that value is an integrated measure, constructed from different influences on choice and the product of interactions between internal preference and external information about a decision.

However, there is no biological necessity for there to be a single area, an internal homunculus or hedonimeter (Edgeworth, 1881) tracking or constructing implicit value. While sectors of the orbitofrontal cortex (OFC) have emerged as leading candidate regions for representing value, OFC lacks direct access to motor output networks and shares reciprocal connections with many other cortical and subcortical regions (e.g. the interconnected posterior parietal, cingulate and insular cortex and striatum) strongly associated with valuation (Cavada & Goldman-Rakic, 1989; Sesack, Deutch, Roth & Bunney, 1989; Shi & Cassell, 1998; Haber, 2003). Indeed, some neurobiological theories echo the revealed preference idea that value *is* choice, positing that valuable states of the world are simply states that fulfill prior expectations of actions, these expectations having been engendered by evolution to confer adaptive survival (Friston, 2010).

The Randomness of Choice

A major conundrum when thinking about neurobiological mechanisms in decision-making is the fact that choices are often noisy or stochastic. While "preference" to a traditional economist implies a consistent ordering of choices, real choices show a high variability, being probabilistic rather than deterministic (Herrnstein, 1974). This is accounted for in computational behavioral modeling by the addition of a stochastic choice generating function (e.g. logit-softmax or probit choice models). In practice, noise could arise prior to the valuation process, for example in primary or associative sensory cortex encoding stimulus properties as is evident in situations of sensory uncertainty (Knill and Pouget, 2004). Cortical neuronal firing rates are inherently stochastic, with one root of this randomness being variability in synaptic vesicular release (Faisal, Selen & Wolpert, 2008; Korn, Bausela, Charpier & Faber, 1993). Variability in spike trains often, but not always, conforms to a Poisson process (Tolhurst, Movshon & Dean, 1983; Maimon and Assad, 2009). Models of noisy evidence accumulation during the formation of a decision can thus account for a considerable degree of stochasticity seen in choice (Resulaj, Kiani, Wolpert & Shadlen, 2009).

In contrast to the idea that variability in behavior solely reflects noise in sensorimotor neurons, data from monkeys performing an depth-discrimination task ("near" or "far" random dot stereogram stimuli presented as a noisy stimulus) show that choice-related neuronal activity in disparity-sensitive neurons in area V2 during can be temporally decoupled from performance effects of sensory uncertainty on behavior (Nienborg & Cumming, 2009). While stimulus information in the first few hundred milliseconds of presentation predicts behavior, neuronal activity predictive of choice in these sensory neurons arises later. Moreover, this choice-related activity reflects a difference in sensitivity ("gain") for near and far stimuli, a change modulated by reward size. For large rewards, performance is better, but the correlation between V2 neuronal activity and choice is worse (i.e. a better coupling between stimulus-encoding properties of these neurons and behavior with large rewards at stake). This strongly suggests that "top-down" expectations influence a coupling between stimulus and choice in early sensory regions. In a similar vein, modulations in the degree of correlation between neuronal firing within areas also vary the stochasticity of a neural ensemble (Shadlen & Newsome, 1998).

Economists have dealt with this issue by proposing different models of stochastic preference. These models loosely map on to the many neurobiological loci where noise is generated. The valuation process itself could be prone to random error, akin to "random utility" models from economics, where noise is injected into the utility function (Hey & Orme, 1994), or motor selection itself could be variable (the analogous "trembling hand" models of choice (Harless & Camerer, 1994), where you know what you *want* to choose but imprecision in your executed actions leads to "mistakes"). Individual preferences themselves could be stochastically sampled from an underlying generative function (Loomes & Sugden, 1995). Other possibilities include the idea that potential outcomes from all possible choices influence the decision process. While choice is usually envisaged as a selection of an optimal action in accordance with one's own priors and preferences, there is good evidence that suboptimal outcomes (i.e. potentially irrelevant alternatives) impact on decisions. These irrelevant outcomes contribute to classic paradoxes of choice (Birnbaum, 2008), such as the impact of regret (Loomes & Sugden, 1982) and dynamically inconsistent strategic decisions (Symmonds, Bossaerts & Dolan, 2010a). Indeed counterfactual outcomes are represented in prefrontal cortex (Ursu & Carter, 2005) while fictive errors in economic learning tasks are seen in striatum (Lohrenz, McCabe, Camerer & Montague, 2007). Thus, choice variability could arise, in part, from a differential weighting of these suboptimal outcomes.

Thus, a range of plausible neurobiological mechanisms could contribute to variability in choice at a whole-organism level; noise at different levels of the decision process is not discrete. We note a dearth of evidence exploring the genesis of choice variability in humans. However, the mapping of such processes onto economic theories of stochastic preference could provide a fruitful source of testable hypotheses to better understand the whims and vagaries of choice, and moment-by-moment fluctuations in preference.

Different Types of Value?

If indeed there are multiple brain areas tracking value, this means that multiple valuation systems may be susceptible to distinct influences and potentially come into conflict. This raises deeper questions about what is actually meant by value and preference. The addiction example highlights a disparity between temporary cravings and long-term desires, a consideration that led Berridge (1996) to propose a distinction between "wanting" and "liking." While "liking" corresponds to hedonic pleasure (i.e. the experience of reward), "wanting" is more aligned with appetitive salience and motivation (i.e. anticipated reward). An extended account of this parsing of choice behavior involves a pharmacological division between these two components, with "wanting" dependent upon the dopaminergic system and "liking" on an endogenous opiod system. This in turn highlights another biological level within which preference can be generated – rather than being regionally encoded some aspects of preferences might be a property of a distributed pharmacological neuro-modulatory influence.

While providing a compelling heuristic, it turns out that a distinction between "wanting" and "liking" is not entirely clear cut. Tindell, Smith, Berridge and Aldridge (2009) has shown that salt-deprived rats can acquire preferences (i.e. instrumental choices) for previously disliked salt-giving levers in the absence of any new training or experience of the now beneficial salty outcome. This indicates that systems controling preference can also integrate information about physiological state, in the absence of information about hedonic pleasure or "liking," although physiological state may well have immediate access to a value-signaling system (as opposed to complex goals) given its central role in shaping adaptive fitness.

Dynamic Changes in Preference

A final consideration about preferences is their lability. Behavior often shows response shifts to exogenous and endogenous variables, from the typically economic (price, resource availability) to the biological and psychological (physiological state, arousal, emotion). Preference also changes over time, both in response to changes in environmental conditions (e.g. accumulation of wealth, reproductive success) and sources of unobserved heterogeneity (e.g. drifting of preferences over time). For biological success, it makes sense that preferences are amenable to "top-down" cognitive influences. For example, it may be appropriate in times of financial crisis to make a "risk-averse" choice and stick with a steady reliable source of income, while in more optimistic times "risk-seeking" or entrepreneurial choices may be of greater long-term benefit.

Many preferences are phylogenetically ancient. *Drosophila* flies can choose between competing cues to avoid punishment, a function supported by their dopaminergic system (Zhang, Guo, Peng et al., 2007). There is also a pervasive sensitivity to risk not only in humans (Kahneman & Tversky, 1979), but also in distantly related species including birds and fish (Real, Ott & Silverfine, 1982; Wunderle, Castro & Fetcher, 1987; Croy & Hughes, 1991; Bateson & Kacelnik, 1997). These "evolved" preferences suggest biological hard-wiring. Indeed, many of these species lack a highly developed prefrontal cortex, yet manifest preferences that are dynamically altered in response to complicated strategic motives. Flexible preferences mean that, even if a specific neural substrate is responsible for a general tendency to emit distinct behaviors, these regions necessarily need to receive information both from a hierarchy of "higher" (e.g. maintaining a strategy) and "lower-level" inputs (e.g. physiological or emotional state).

These behavioral switches link more broadly to the idea of "exploratory" behavior. A hesitancy to explore is used as a biological marker of trait anxiety in animal models (Hogg, 1996). Can we call exploration a preference? A "preference" to explore the unknown bucks the natural trend to stay with known rewarding actions (exploitation), and can lead to a different selection of actions (choices) depending on circumstance, a process poorly-captured in many economic models. Computationally, this explore–exploit dilemma is critical in learning theory, and there is evidence that exploratory behavior is instantiated in rostral anterior prefrontal cortex (Daw, O'Doherty, Dayan et al., 2006) and anterior cingulate cortex (ACC) (Procyk, Tanaka & Joseph, 2000). Thus an "exploratory" preference could reflect the function of a very different kind of neural process, more related to task-switching and behavioral flexibility (Ragozzino, 2007), perhaps an example of a top-down process that modulates classical preferences rather than a biological predilection in itself.

The Role of Neuroscientific Evidence

On the one hand, choice and preference can be considered the only relevant end-product of a "black-box" process (Gul and Pesendorfer, 2008). This is perfectly reasonable as an approach to decision-making as economic models are designed solely to predict choice, not to imply biological mechanisms. However, neuroscientists are primarily interested in mechanistic "proximate" questions – it is not sufficient to observe how cars move if you want to understand the internal workings of a combustion engine. The aims of neuroscience include understanding the biological underpinnings of behavior and beyond this a biophysically motivated model of choice. This in turn can yield insights about disease processes, reveal neural targets that determine pathological behavior, and even potentially "improve" an individuals' decision-making. As framed by Marr (1982), to address these questions one needs to specify computational goals, their algorithmic basis, as well as the biological hardware that implements these processes.

WHAT ARE THE POTENTIAL NEURAL MECHANISMS UNDERLYING PREFERENCE

What are the plausible mechanisms by which preferences are expressed in the brain? The most obvious is a direct neural instantiation of a hedonimeter (Edgeworth, 1881), a regionally specific encoding of intrinsic value. A more subtle question is to ask how "preference" is reflected within a single neural region. Alternatively, preferences may be determined by concentrations of neurotransmitters, a whole-brain property at the synaptic level. Preferences might be expressed at different time-points in the process of decision-making. Thus we can ask whether preferences are generated in early valuation or later motor response systems. Finally, rather than individual regions or systems in the brain being critical for generating preferences, we consider whether modulations in the interaction or communication within a network of regions holds the key to understanding inter-individual variability in choice.

Regionally-Specific Encoding of Intrinsic Value

Gottfried, O'Doherty and Dolan (2003) measured fMRI data during an appetitive olfactory conditioning paradigm, where arbitrary picture stimuli were paired with pleasant smells of different foods. Following learning, one of the olfactory stimuli was selectively devalued by consuming the associated food to satiety. This enabled a distinction between stimulus-bound neural responses predicting simply the type of odor (i.e. a "flavor" response) from responses that differentiated between valued versus devalued cues (i.e. an anticipated hedonic response that could reflect preference). While a network of regions in OFC, amygdala, piriform cortex and midbrain were activated during presentation of odors, only subregions of amygdala and OFC exhibited responses that paralleled the effects of selective satiety, showing decreased activation for sated compared to unsated odor cues. Of course this is not a causal demonstration of neural activity tied to preference, but exemplifies one of the earliest usages of neuroimaging to identify regions conforming to the internal representation of a decision variable. In this case, the fact that the OFC activation for the same physical stimulus changed in a manner that correlated with a behavioral manipulation of preference highlighted the OFC as a candidate region instantiating the expression of preferences. In the same study, other regions (for example, insula cortex and striatum) decreased their activity for the sated stimulus, but increased their activity for the alternative control stimulus. This observation highlights that an interpretation of findings is constrained by the purported models being tested. If, for example, one hypothesizes that the brain forms a relative ranking of value, rather than representing value on a cardinal scale, then the insula would fit this description. A rigorous specification of choice models can help marshal divergent findings and this is the essential advantage of model-based approaches, whether neuroeconomic or biophysically inspired.

Converging evidence for over a decade has identified the OFC as encoding reward value. Single neuron recordings have shown a correlation in firing of OFC neurons with preferences in monkeys (Tremblay & Schultz, 1999). Neuroimaging studies show that OFC responses are neither valence-specific, manifesting responses to both rewards and punishments (O'Doherty, Kringelbach, Rolls et al., 2001a), nor domain-specific showing responses to food and other primary reinforcers (Small, Zatorre, Dagher et al., 2001), money (Elliott et al., 2000) and social value (Rushworth, Behrens, Rudebeck & Walton, 2007). These responses also appropriately scale with increasing amounts of anticipated value (Elliott, Newman, Longe & Deakin, 2003), and actual behavioral preference (Plassmann, O'Doherty & Rangel, 2007).

Recent findings from direct neuronal recordings in monkeys show that a proportion of OFC neurons adapt to condition, manifesting similar ranges of response under different scales of outcome (Padoa-Schioppa, 2009). This suggests that the OFC not only responds to changes in value, but also integrates value with contextual and other goal-orientated information. This integration is potentially sophisticated and may pertain to outcomes that will be obtained at some point in the future, with subregions within OFC showing responses that track multiple components of goal-dependent action values in sequential decision-making tasks (Symmonds et al., 2010a). In different paradigms, OFC also encodes differences in stimulus value (FitzGerald, Seymour & Dolan, 2009), choice value (Boorman, Behrens, Woolrich & Rushworth, 2009) and counterfactual outcomes (Coricelli, Critchley, Joffily et al., 2005). These components are all important for the generation of choice. Nevertheless finding a region that maintains a stable, ordered, and scaled representation of value is not the same as finding the seat of preferences in the brain. OFC and similar regions could be representing value for other reasons; for instance, maintaining a representation of anticipated versus received rewards contingent on choice for learning purposes, or being an arbiter in providing control of behavioral responses by gating influences from other valuation regions. For instance, value-sensitive responses to reward feedback have been found in primary somatosensory (Pleger, Blankenburg, Ruff et al., 2008) and visual (Weil, Furl, Ruff et al., 2010) cortices in perceptual learning paradigms. Even if OFC performs computations of differences in value, perhaps being the generator of probabilistic choices for abstract value-based decisions rather than decisions based on comparisons of direct sensory evidence, it appears to be only one component of a network of value-sensitive regions. Moreover, it is plausible that "preference" or intrinsic value is imbued to decision values prior to this information arriving in OFC.

Multiple candidate regions that have been reported to express value-related responses include posterior parietal cortex (PPC) (Platt & Glimcher, 1999; Iyer, Lindner, Kagan & Andersen, 2010), posterior cingulate cortex (PCC) (McCoy & Platt, 2005), insula (Yacubian, Glascher, Schroeder et al., 2006), nucleus accumbens/ventral striatum (Elliott et al., 2003; Knutson et al., 2005), the ventral tegmental (VTA) and substantia nigra (SN) areas of midbrain (Tobler, Fiorillo & Schultz, 2005).

Other value-sensitive brain regions such as ACC appear to encode negatively-valenced components, such as cost or effort (Walton, Bannerman, Alterescu & Rushworth, 2003).

The range of findings support one hypothesis – that OFC and similar regions encode the (non-linear) sum of value-sensitive inputs in an excitatory manner that in turn positively correlates with desired stimuli and actions. Equally possible is that decrements in activity in a region could influence choice (perhaps by a reduction in tonic inhibition of action), as is evident in basal ganglia circuitry (Mink, 2003). In other words, even if there were a regionally specific encoding of preference, this does not mean that value necessarily corresponds to increases in neuronal activity. It is also possible that disparate pools of neurons within a region could exhibit very different relationships with stimuli or behavior, either individually influencing a target region or interacting within a region (e.g. by lateral inhibition [Blakemore & Tobin, 1972]). Single-cell recording studies in orbitofrontal cortex yield this mixed picture, revealing distinct populations of adjacent neurons that are value sensitive, some expressing increases and others decreases in firing rate with increments in subjective value (Kennerley, Dahmubed, Lara & Wallis, 2009).

A Change in Neural Sensitivity within a Brain Region

"Preference" describes a tendency to select a particular kind of action. The "sweet-toothed" choose sugary rather than savory foods; a "risk-seeking" person chooses uncertain options; an "impulsive" child fails to temper desire for immediate gratification while an "altruistic" individual takes account of others' welfare. Whilst these disparate tendencies may share common neural representations of value, economic models point to overall value as constructed from intrinsic subcomponents.

One can represent this process as:

$$Choice = f(Value) = f(\mu, \vartheta)$$

where μ are the stimulus properties or biological primitives relevant to behavior, and ϑ represents the parameters that govern how an individual treats or processes these properties. μ can relate to primary sensory properties of the stimulus (form, taste, smell, etc.), but equally can be a function of these sensory properties (i.e. $\mu(g(Stimulus))$). For example, a transformation of visual information presented in a gamble into decision variables such as magnitude and risk, or a non-linear weighting of more abstract stimulus feature such as reward probability. This framework partitions the decision-making process into four sub-components, each represented in the brain. Choice pertains to action, the end-product of a motoric network. Value relates to affective or hedonic anticipation or impact of those actions (as we discuss above), but μ and ϑ pertain to underlying decision variables. While μ is a function of the stimulus properties themselves and invariant with respect to behavioral predilections, ϑ describes how individuals treat each of these underlying properties (e.g. whether magnitude has a positive or negative impact on value). Thus, a search for the neural instantiation of preferences can be seen as a search for a representation of these underlying ϑ parameters.

One means of identification of such candidate areas would be to segregate areas representing μ (potentially processing these decision variables) from areas encoding ϑ (behavioral preference toward these decision variables). Lesion studies implicate both factors – for example the finding that patients with damage to their ventromedial prefrontal cortex (VMPFC) are less averse to uncertainty than healthy counterparts (Hsu, Bhatt, Adolphs et al., 2005; Shiv, Loewenstein & Bechara, 2005) could be interpreted as the emergence of normative economic behavior (a change in ϑ) or simply reflect a lack of sensitivity to this stimulus variable because of an inability to process uncertainty (no representation of μ). There have been a large cohort of studies demonstrating neural sensitivity to decision-relevant stimulus properties (μ), ranging from sensory attributes (O'Doherty, Rolls, Francis et al., 2001b) and reward identity (Chib, Rangel, Shimojo & O'Doherty, 2009), to abstract decision variables such as financial magnitude of reward (Elliott et al., 2000), probability (Hsu, Krajbich, Zhao & Camerer, 2009), expected value (Knutson et al., 2005), valence (Delgado, Nystrom, Fissell et al., 2000), and risk (Preuschoff et al., 2006). Other studies have attempted to identify regions correlating directly with preferences (ϑ). Structural studies have reported that prefrontal grey matter volume is inversely correlated with impulsivity (Bjork, Momenan & Hommer, 2009), and right prefrontal cortex lesions promote riskier choice (Clark, Manes, Antoun et al., 2003). Alternatively, functional observations include findings that activation in striatal (Tom, Fox, Trepel & Poldrack, 2007) and prefrontal (Gianotti, Knoch, Faber et al., 2009) regions correlate with loss and risk aversion, while lateral and medial OFC correlate positively and negatively with risk aversion, respectively (Tobler, O'Doherty, Dolan & Schultz, 2007).

When making a decision, brain areas governing preference need to interact with circuitry involved in valuation, a process we summarize above as $f(\mu,\vartheta)$. This influence could occur at an early stage of processing, during encoding of stimulus properties themselves. Both inferior frontal gyrus (IFG) and adjacent lateral prefrontal cortex express differential responses to high and low risk, with greater responses for high than low risk in risk-seekers, but an opposite response in risk-averse individuals (Christopoulos et al., 2009; Tobler, Christopoulos, O'Doherty et al., 2009). This expression of risk-sensitivity would be expected from areas extracting or constructing the subjective value of risk, making this cortical zone a good candidate locus for integration of different aspects of financial value with intrinsic economic preferences. This interaction between an observed behavioral preference and a regional expression of neural activity in response to a stimulus, context or decision trial has also been applied in studies of intertemporal and interpersonal ("social") choice. The effect of delay on behavior (preference for impulsive choice) is mirrored by the effect of delay on neural responses in ventral striatum, medial prefrontal and posterior cingulate cortex (Kable & Glimcher, 2007). In this task, striatal responses demonstrate an integration of two types of economic preference, for both magnitude (marginally decreasing utility) and delay (Pine et al., 2009).

Invetigations of moral judgements and social preferences implicate prefrontal and insular cortices (Hsu, Anen & Quartz, 2008; Koenigs, Young, Adolphs et al., 2007). For example, the Ultimatum Game is a stylized economic paradigm of social interaction used to investigate the neural basis of social preference. Here, a "proposer" might offer a fair 50:50 split of a $10 kitty to a "responder"– the crucial element is that the "responder" can elect to accept the offer, in which both parties receive the split as proposed, or reject the offer, in which case both parties get nothing. While the "rational" response in a one-shot anonymized version of this task is to accept any offer amount, even in these circumstances of no reciprocity or reputation-formation real-life responders typically reject offers of up to 30% of the kitty (Camerer, 2003). This punishes the proposer for an inequitable split despite a personal financial cost incurred by the responder, hence indexing the responder's trade-off between self- and other-regarding preferences. Here, the stimulus variable of interest (μ) is inequity itself, an objective measure of fairness. How such inequity influences the responder's choice to accept or reject is proposed to depend upon their "inequity-aversion" (ϑ) (Edgeworth, 1881; Fehr & Schmidt, 1999).

We recently employed a version of Ultimatum Game where participants thought they were responding to human proposers, but where the offers were actually experimentally controlled (Wright, Symmonds, Fleming & Dolan, 2011). This enabled a manipulation of inequity (i.e. the objective stimulus property μ), which linearly correlated with activity in posterior insula. Crucially we biased a subjective appreciation of fairness by altering the context of the decision – responders believed that offers were made by three socially different groups of subjects, offering fairer or less fair splits of the kitty. Identical offers from these different social groups induced differential neural responses to inequity in mid-insula, responses that paralleled a behavioral bias in choice. A different network of regions in precuneus, left prefrontal and temproparietal cortex reflected endogenous inequity-aversion across subjects (i.e. independent of the contextual manipulation), illustrating that even within the context of a specific task, preferences for the same stimulus feature can be expressed in different regions and modulated in a distinct manner.

Distributed Encoding of Value by Neurotransmitters

Midbrain dopaminergic neurons in the ventral tegmental area (VTA) and substantia nigra (SN) modulate their firing rate in response to rewarding stimuli (Tobler et al., 2005). Although dopamine was originally hypothesized to be a qualitative and quantitative hedonic signal (Yokel & Wise, 1978; Wise & Rompré, 1989), an aliquot of dopamine is not an intrinsic biological measure of utility, nor is dopamine release universally related to preference. A clear example is seen in genetically engineered rats with no endogenous dopamine synthesis but who still express relative preference for different types of reward (Cannon & Palmiter, 2003). However, dopamine does appear to play a central role in cost-benefit analysis (Phillips, Walton & Jhou, 2007). This effect appears to be modality-specific and dictated by the precise dopaminergic projection target, as dopamine depletion in ventral striatum reduces propensity for physical effort (Salamone, Correa, Farrar & Mingote, 2007), but not for time delay (Wakabayashi, Fields & Nicola, 2004). By contrast, systemic dopaminergic modulation does influence intertemporal decision-making (Wade, de Wit & Richards, 2000), an influence localized to orbital prefrontal cortex (and striatum) (Kheramin, Body, Ho et al., 2004; Pine, Shiner, Seymour & Dolan, 2010). However, there appear to be multiple dopamine-sensitive decision regions, as, for example, D1 blockade in ACC also reduces preference for expending effort for rats (Schweimer & Hauber, 2006).

Critically, dopamine neurons respond to surprising rewards (Schultz, Dayan & Montague, 1997), revealing a central role in associative learning. Although non-discriminative between reward types, dopaminergic firing in VTA does appear to reflect subjective (action) value with integrated responses to both delay and reward amount (Roesch, Calu & Schoenbaum, 2007). Dopaminergic neurons send diffuse projections to striatum (nigrostriatal pathway) and prefrontal cortex (mesocortical pathway) and thereby transmit a hedonic value or teaching signal to a variety of brain regions, for learning, stimulus evaluation, and directed action. Increasing striatal dopamine for instance does not elicit reward craving by itself, even in addicts, but does change a cocaine addicts' sensitivity to environmentally salient cues (Volkow, Wang, Telang et al., 2008).

Single-unit recording studies have shown that the tonic firing rates of dopaminergic midbrain neurons scale with reward uncertainty in risk-based decision-making tasks in primates (Fiorillo, Tobler & Schultz, 2003), although the interpretation of this finding is subject to dispute (Fiorillo, Tobler & Schultz, 2005; Niv, Duff & Dayan, 2005). These nuclei also receive inputs from areas such as the habenula (Matsumoto and Hikosaka, 2007), which responds to negatively-valenced stimuli and feedback (Ullsperger & Von Cramon, 2003). Hence, the VTA/SN could act as a hub to compare predicted action values and obtained outcomes. It is therefore possible that certain preferences could be engendered at this early stage, perhaps by differential responses to outcomes and inter-individual variability in the effect of precision of predictions on dopaminergic output.

There are substantial hormonal influences on behavior. For example, circulating hormones such as leptin and ghrelin act as satiety and hunger signals, reporting the status of body energy reserves (e.g. adipose tissue), energy requirements, and acute nutrient intake to hypothalamic and mid-brain targets in the central nervous system that regulate feeding behavior (Korotkova, Sergeeva, Eriksson et al., 2003). They also act on brain regions (in particular dopaminoceptive areas) implicated in human decision-making (Hommel, Trinko, Sears et al., 2006; Krügel, Schraft, Kittner et al., 2003). Metabolic state itself may thus directly affect the neural expression of preference, exemplified by findings that physiological state influence preferences for economic risk in humans (Symmonds, Emmanuel, Drew et al., 2010b).

Other neurochemical systems are implicated in preference, with both serotonergic depletion (Denk, Walton, Jennings et al., 2005) and NMDA antagonism (Floresco, Tse & Ghods-Sharifi, 2007) promoting impulsivity. Several neuromodulatory transmitters are purported to play a specific role in risky decision-making, including noradrenaline, serotonin, and dopamine (Rogers, Lancaster, Wakeley & Bhagwagar, 2004; Zeeb, Robbins & Winstanley, 2009). Clinically, Parkinson's disease, where nigrostriatal dopamine pathways are impoverished, can lead to disorders of decision-making (Cools, Barker, Sahakian & Robbins, 2003). Dopamine agonists, used to treat this disorder, can cause pathological gambling behavior as a side-effect of therapy (Gallagher, O'Sullivan, Evans et al., 2007). Additionally, manipulation of dopamine levels in rats disrupts decision-making under uncertainty in foraging tasks. Thus, administration of amphetamine (which augments dopamine release) increases preference for a risky choice, while the effects of amphetamine can be abolished by dopamine receptor blockade (St Onge and Floresco, 2008).

Finally, genetic polymorphisms affecting receptor function or expression are a suggested conduit for value and preference. For example, DRD4 polymorphisms modulate the incentive value of alcohol in alcoholics (MacKillop, Menges, McGeary & Lisman, 2007), while genetic polymorphisms affecting DRD2 receptor expression alter neuronal responses to food reward (Felsted, Ren, Chouinard-Decorte & Small, 2010). Although attributing specific whole-organism behavioral phenotypes or neural responses to genetic polymorphisms is highly contentious with results often contradictory and inconclusive (Hariri, Mattay, Tessitore et al., 2002; Schinka, Letsch & Crawford, 2002; Munafo, Clark, Moore et al., 2003; Kreek, Nielsen, Butelman & LaForge, 2005), nevertheless it is conceivable that alterations in receptor function in specific neural regions could bias valuation and action-selection process thereby engendering behavioral predilections.

Change in Neural Activity Coupled to Action

Action is supported by multiple brain regions and networks. Saccadic eye movements are driven by the frontal eye fields via superior colliculus, while limb movements are guided by motor efferents from Betz cells in primary motor cortex projecting to spinal anterior horn cells. Motor cortex receives inputs from thalamus, which in turn receives afferents from multiple areas including the supplementary motor area in premotor cortex, basal-ganglia and cerebellum (Houk & Wise, 1995), and interparietal sulcus, an associative region of cortex containing neurons with sensory receptive fields (Shadlen & Newsome, 1996).

This division between action and perception is evident in multiple neural processes. Interestingly, preferences for actions can be dissociated from preferences for stimuli, which echoes an idea that actions can in themselves create or establish preferences (Ariely & Norton, 2008; Sharot, De Martino & Dolan, 2009). For instance, OFC is predominantly associated with (learned) valuation of (reinforcing) stimuli, but there is evidence of a distinct process in ACC (which projects anatomically to premotor regions) of value encoding for reinforced actions, and ACC sulcus may be required for learning of action values (Kennerley, Walton, Behrens et al., 2006). Similar regions in ACC have also been shown to encode decision uncertainty (Behrens, Woolrich, Walton & Rushworth, 2007), and the ACC gyrus is involved in learning social values (Rushworth et al., 2007). Thus, it is possible that interactions between subregions of cingulate cortex might imbue taste for uncertainty or social preferences by differentially weighting action values represented in ACC.

ACC is also necessary for an appraisal of the energetic cost versus benefits of actions, and a preference to expend effort to achieve a goal. Take a situation where rats are required to make a decision between two alternatives, either to climb over a barrier to obtain a large food reward, or to expend low effort but receive a smaller food reward. ACC lesions alter the modal preference for expending effort to reap reward (Walton et al., 2003), an effect specific for energetic actions but not for non-effortful costs, such as a delay in obtaining reward (with the converse true for OFC) (Rudebeck, Walton, Smyth et al., 2006).

Neuroimaging studies have identified areas differentially responding to selected actions (e.g. the selection of risky versus safe options in a gambling task). Christopoulos et al. (2009) have parsed a risk-sensitive network into (right) IFG, a region whose activity promotes safe choices, from ventral striatum and ACC, engendering risky choice. Direct disruption of activity in this right lateral prefrontal region by repetitive transcranial magnetic stimulation biases choice towards risk-seeking (Knoch, Gianotti, Pascual-Leone et al., 2006b), while transcrianial anodal direct current stimulation, inducing activation in this area, promotes safer

choice (Fecteau, Knoch, Fregni et al., 2007). In the domain of intertemporal choice, McClure, Laibson, Loewenstein and Cohen (2004) have suggested an interaction between subcortical and cortical areas push and
pull choices towards impulsivity or patience.

More recently, studies have examined the interaction between preference and choice-related activity. Superior and inferior frontal gyrus
and OFC express greater or less activity prior to a risky choice in a manner that positively correlates with risk aversion (Engelmann & Tamir,
2009). A similar pattern is seen in PPC (Weber & Huettel, 2008), which
has polysynaptic links to basal ganglia and premotor regions (Tanne-
Gariepy, Rouiller & Boussaoud, 2002) and direct connections with insula
(Cavada & Goldman-Rakic, 1989). These choice-related activations also
relate to psychological measures such as harm avoidance in both ventral striatum (Matthews, Simmons, Lane & Paulus, 2004), and insula
(Paulus, Rogalsky, Simmons et al., 2003). In itself, this does not distinguish between modulation of stimulus evaluation and action selection, as risk-seekers will always tend to select a preferred riskier option
if the expected values of options are matched. However, this parallels
the interaction between brain areas governing preference and circuitry
involved in valuation (i.e. $f(\mu,\vartheta)$). Rather than envisaging this influence
at valuation or stimulus encoding, hypothesizing a modulation of choice-
related activity by preference postulates a direct effect of preference
on regional activity coupled with motor output, without a distillation
through a prefrontal or subcortical hedonometer. In other words, some
preferences could originate within the motor network, where biases arise
during a translation of sensory inputs into specific actions.

Change in Regional Connectivity and Modulation of Preference

The brain is a non-linear dynamical system, with dense intercon-
nections between and within brain areas. There has been an increasing
realization that connectivity between regions is a key component of a
functional neural architecture – this connectivity generates macroscopic
synchronization of neuronal pools (i.e. oscillatory activity) (Sporns,
Tononi & Edelman, 2000), and supports metastable representations that
change and switch at a behavioral timescale (Oullier & Kelso, 2006). This
connectivity itself could be the source of preferences in the brain – if parietal cortex or striatum quantifies specific decision variables, prefrontal
cortex integrates this information with context and incentive salience,
and premotor cortex plans actions, perhaps the true instantiation of preference lies in the strength of connections and mutual influence between
these regions.

Changes in behavior and associated neural responses in medial prefrontal cortex and striatum following knowledge acquisition have been

shown to be mediated by functional connections from dorsolateral pre-frontal cortex (DLPFC) (Li, Delgado & Phelps, 2011). Right DLPFC is also implicated in rejection of fair offers in the Ultimatum Game (Wright et al., 2011), and disrupting this region by transcranial magnetic stimulation induces a reversible increased tendency to accept normatively "unfair" offers (Knoch, Pascual-Leone, Meyer et al., 2006a). Dorsomedial prefrontal cortex (DMPFC) has been separately shown to express connectivity (correlation between time-series of inferred neuronal activity) with anterior insula and DLPFC during a risky decision-making task, with the relative strength of these functional connections dependent upon individual strategic preferences (Venkatraman, Payne, Bettman et al., 2009). Similarly, an interconnected network of DLPFC, IFG and VMPFC have been mapped out by functional connectivity analyses to highlight how DLPFC mediates restraint in choices about food (Hare, Camerer & Rangel, 2009). These studies indicate a network of regions involved in decision-making whose interactions, rather than intrinsic activity, dictate preference and choice.

CONCLUSION

The process of decision-making can be broken down into constituent parts, each supported by different neurobiological mechanisms. Preferences generated at each step are instantiated at a regional or network level, and controlled by multiple competing mechanisms. Consequently, there are multiple sites at which preferences can be endogenously or exogenously influenced. At one end of the spectrum is a demonstration that *in vivo* stimulation single of neurons in early sensory visual cortex can bias motion discrimination decisions in monkeys (Salzman, Britten & Newsome, 1990). At the other end, aberrant decision-making with altered behavioral preference is a feature of brain injury (Clark, Bechara, Damasio et al., 2008), Parkinson's disease (Gallagher et al., 2007), and can even be pharmacologically induced (e.g. by alcohol, amphetamines, and dopamine agonists) (Zeeb et al., 2009).

A coherent framework based upon neurobiological foundations allows arbitration between competing decision theories, and also reveals the sources of systematic biological influences on behavior. Neurobiological evidence thus furnishes process-based theories of choice (i.e. theories that describe choices as the product of a series of distinct computations in functionally or anatomically separable networks), and yields new hypotheses beyond the bounds of traditional economic dictum.

The various conceptual approaches to choice and preference, demonstrations of disparate influences on decision-making, and interactions between several different types of value, speak to different questions and levels of understanding of choice processes. For a biological model, we need to define the types of valuation systems we are considering, the network or areas involved in the associated computational processing, and the synaptic physiology that might underlie the process. In the same way that axiomatic economic models place well-defined constraints upon rational choice, biological parameters also place constraints on the plausible mechanisms that mediate the generation of actions and the preferences they express.

References

Ariely, D., & Norton, M. I. (2008). How actions create – not just reveal – preferences. *Trends In Cognitive Sciences, 12,* 13–16.

Aristotle (1998). The Nicomachean Ethics. In J. Ackrill & J. Urmson (Eds.), *Oxford world's classics.* Translated with an Introduction by David Ross. New York, NY: Oxford University Press.

Bateson, M., & Kacelnik, A. (1997). Starlings' preferences for predictable and unpredictable delays to food. *Animal Behavior, 53,* 1129–1142.

Bechara, A., Damasio, A. R., Damasio, H., & Anderson, S. W. (1994). Insensitivity to future consequences following damage to human prefrontal cortex. *Cognition, 50,* 7–15.

Behrens, T. E., Woolrich, M. W., Walton, M. E., & Rushworth, M. F. (2007). Learning the value of information in an uncertain world. *Nature Neuroscience, 10,* 1214–1221.

Berns, G. S., Laibson, D., & Loewenstein, G. (2007). Intertemporal choice – toward an integrative framework. *Trends In Cognitive Sciences, 11,* 482–488.

Berridge, K. C. (1996). Food reward: Brain substrates of wanting and liking. *Neuroscience And Biobehavioral Reviews, 20,* 1–25.

Birnbaum, M. H. (2008). New paradoxes of risky decision-making. *Psychological Review, 115,* 463–501.

Bjork, J. M., Momenan, R., & Hommer, D. W. (2009). Delay discounting correlates with proportional lateral frontal cortex volumes. *Biological Psychiatry, 65,* 710–713.

Blakemore, C., & Tobin, E. A. (1972). Lateral inhibition between orientation detectors in the cat's visual cortex. *Experimental Brain Research, 15,* 439–440.

Boorman, E. D., Behrens, T. E. J., Woolrich, M. W., & Rushworth, M. F. S. (2009). How green is the grass on the other side? Frontopolar cortex and the evidence in favor of alternative courses of action. *Neuron, 62,* 733–743.

Boysen, S. T., & Berntson, G. G. (1995). Responses to quantity: Perceptual versus cognitive mechanisms in chimpanzees (Pan troglodytes). *Journal of Experimental Psychology: Animal Behavior Processes, 21,* 82–86.

Breiter, H. C., Aharon, I., Kahneman, D., Dale, A., & Shizgal, P. (2001). Functional imaging of neural responses to expectancy and experience of monetary gains and losses. *Neuron, 30,* 619–639.

Broca, P. (1865). Sur le siège de la faculté du langage articulé. *Bulletins de la Société d'Anthropologie de Paris, 6,* 377–393.

Camerer, C. (2003). *Behavioral game theory: Experiments in strategic interaction.* Princeton, NJ: Princeton University Press.

Cannon, C. M., & Palmiter, R. D. (2003). Reward without dopamine. *Journal of Neuroscience, 23,* 10827–10831.

Carlson, S. M., Davis, A. C., & Leach, J. G. (2005). Less is more. *Psychological Science, 16,* 609–616.

Cavada, C., & Goldman-Rakic, P. (1989). Posterior parietal cortex in rhesus monkey: II. Evidence for segregated corticocortical networks linking sensory and limbic areas with the frontal lobe. *Journal of Comparative Neurology*, *287*, 422–445.

Chib, V. S., Rangel, A., Shimojo, S., & O'Doherty, J. P. (2009). Evidence for a common representation of decision values for dissimilar goods in human ventromedial prefrontal cortex. *Journal of Neuroscience*, *29*, 12315–12320.

Christopoulos, G., Tobler, P., Bossaerts, P., Dolan, R., & Schultz, W. (2009). Neural correlates of value, risk, and risk aversion contributing to decision-making under risk. *Journal of Neuroscience*, *29*, 12574–12583.

Clark, L., Manes, F., Antoun, N., Sahakian, B. J., & Robbins, T. W. (2003). The contributions of lesion laterality and lesion volume to decision-making impairment following frontal lobe damage. *Neuropsychologia*, *41*, 1474–1483.

Clark, L., Bechara, A., Damasio, H., Aitken, M., Sahakian, B., & Robbins, T. (2008). Differential effects of insular and ventromedial prefrontal cortex lesions on risky decision-making. *Brain*, *131*, 1311–1322.

Cools, R., Barker, R. A., Sahakian, B. J., & Robbins, T. W. (2003). L-Dopa medication remediates cognitive inflexibility, but increases impulsivity in patients with Parkinson's disease. *Neuropsychologia*, *41*, 1431–1441.

Coricelli, G., Critchley, H. D., Joffily, M., O'Doherty, J. P., Sirigu, A., & Dolan, R. J. (2005). Regret and its avoidance: A neuroimaging study of choice behavior. *Nature Neuroscience*, *8*, 1255–1262.

Croy, M. I., & Hughes, R. N. (1991). Effects of food supply, hunger, danger and competition on choice of foraging location by the fifteen-spined stickleback. *Spinachia spinachia* L. *Animal Behavior*, *42*, 131–139.

Daw, N. D., O'Doherty, J. P., Dayan, P., Seymour, B., & Dolan, R. J. (2006). Cortical substrates for exploratory decisions in humans. *Nature*, *441*, 876–879.

Delgado, M. R., Nystrom, L. E., Fissell, C., Noll, D., & Fiez, J. A. (2000). Tracking the hemodynamic responses to reward and punishment in the striatum. *Journal of Neurophysiology*, *84*, 3072–3077.

Denk, F., Walton, M., Jennings, K., Sharp, T., Rushworth, M., & Bannerman, D. (2005). Differential involvement of serotonin and dopamine systems in cost-benefit decisions about delay or effort. *Psychopharmacology*, *179*, 587–596.

Doya, K. (2008). Modulators of decision-making. *Nature Neuroscience*, *11*, 410–416.

Dreher, J. C., Kohn, P., & Berman, K. F. (2006). Neural coding of distinct statistical properties of reward information in humans. *Cerebral Cortex*, *16*, 561–573.

Ebstein, R. (2006). The molecular genetic architecture of human personality: Beyond self-report questionnaires. *Molecular Psychiatry*, *11*, 427–445.

Edgeworth, F. Y. (1881). *Mathematical psychics*. London: History of Economic Thought Books.

Elliott, R., Friston, K. J., & Dolan, R. J. (2000). Dissociable neural responses in human reward systems. *Journal of Neuroscience*, *20*, 6159–6165.

Elliott, R., Newman, J. L., Longe, O. A., & Deakin, J. (2003). Differential response patterns in the striatum and orbitofrontal cortex to financial reward in humans: A parametric functional magnetic resonance imaging study. *Journal of Neuroscience*, *23*, 303–307.

Engelmann, J. B., & Tamir, D. (2009). Individual differences in risk preference predict neural responses during financial decision-making. *Brain Research*, *1290*, 28–51.

Eysenck, H. J. (1990). Genetic and environmental contributions to individual differences: The three major dimensions of personality. *Journal of Personality*, *58*, 245–261.

Faisal, A. A., Selen, L. P. J., & Wolpert, D. M. (2008). Noise in the nervous system. *Nature Reviews Neuroscience*, *9*, 292–303.

Fecteau, S., Knoch, D., Fregni, F., Sultani, N., Boggio, P., & Pascual-Leone, A. (2007). Diminishing risk-taking behavior by modulating activity in the prefrontal cortex: A direct current stimulation study. *Journal of Neuroscience*, *27*, 12500–12505.

Fehr, E., & Schmidt, K. M. (1999). A theory of fairness, competition, and cooperation. *Quarterly Journal Of Economics, 114*, 817–868.

Felsted, J. A., Ren, X., Chouinard-Decorte, F., & Small, D. M. (2010). Genetically determined differences in brain response to a primary food reward. *Journal of Neuroscience, 30*, 2428–2432.

Fiorillo, C. D., Tobler, P. N., & Schultz, W. (2003). Discrete coding of reward probability and uncertainty by dopamine neurons. *Science, 299*, 1898–1902.

Fiorillo, C. D., Tobler, P. N., & Schultz, W. (2005). Evidence that the delay-period activity of dopamine neurons corresponds to reward uncertainty rather than backpropagating TD errors. *Behavioral and Brain Functions 1*, 7.

FitzGerald, T. H. B., Seymour, B., & Dolan, R. J. (2009). The role of human orbitofrontal cortex in value comparison for incommensurable objects. *Journal of Neuroscience, 29*, 8388–8395.

Fleming, S. M., Thomas, C. L., & Dolan, R. J. (2010). Overcoming status quo bias in the human brain. *Proceedings of the National Academy of Sciences of the United States of America, 107*, 6005–6009.

Floresco, S. B., Tse, M. T. L., & Ghods-Sharifi, S. (2007). Dopaminergic and glutamatergic regulation of effort-and delay-based decision-making. *Neuropsychopharmacology, 33*, 1966–1979.

Friston, K. (2010). The free-energy principle: A unified brain theory? *Nature Reviews Neuroscience, 11*, 127–138.

Gall, F. J., & Spurzheim, G. (1818). Anatomie et physiologie du système nerveux en général, et du cerveau en particulier: Avec des observations sur la possibilité de reconnoître plusieurs dispositions intellectuelles et morales de l'homme et des animaux par la configuration de leurs têtes: F. Schoell.

Gallagher, D. A., O'Sullivan, S. S., Evans, A. H., Lees, A. J., & Schrag, A. (2007). Pathological gambling in Parkinson's disease: Risk factors and differences from dopamine dysregulation. An analysis of published case series. *Movement Disorders, 22*, 1757–1763.

Gianotti, L. R. R., Knoch, D., Faber, P. L., Lehmann, D., Pascual-Marqui, R. D., & Diezi, C., et al. (2009). Tonic activity level in the right prefrontal cortex predicts individuals' risk taking. *Psychological Science, 20*, 33–38.

Glimcher, P. W., & Rustichini, A. (2004). Neuroeconomics: The consilience of brain and decision. *Science, 306*, 447–452.

Gottfried, J. A., O'Doherty, J., & Dolan, R. J. (2003). Encoding predictive reward value in human amygdala and orbitofrontal cortex. *Science, 301*, 1104–1107.

Gul, F., & Pesendorfer, W. (2008). The case for mindless economics. In A. Caplin & A. Shotter (Eds.), *Foundations of positive and normative economics.* Oxford University Press.

Haber, S. (2003). The primate basal ganglia: Parallel and integrative networks. *Journal of Chemical Neuroanatomy, 26*, 317–330.

Hare, T. A., Camerer, C. F., & Rangel, A. (2009). Self-control in decision-making involves modulation of the vmPFC valuation system. *Science, 324*, 646–648.

Hariri, A. R., Mattay, V. S., Tessitore, A., Kolachana, B., Fera, F., Goldman, D., et al. (2002). Serotonin transporter genetic variation and the response of the human amygdala. *Science, 297*, 400–403.

Harless, D. W., & Camerer, C. F. (1994). The predictive utility of generalized expected utility theories. *Econometrica, 62*, 1251–1289.

Head, H. (1920). Aphasia and kindred disorders of speech. *Brain, 43*, 87–165.

Herrnstein, R. J. (1974). Formal properties of the matching law. *Journal of the Experimental Analysis Behavior, 21*, 159–164.

Hershberger, W. A. (1986). An approach through the looking-glass. *Animal Learning and Behavior, 14*, 443–451.

Hey, J. D., & Orme, C. (1994). Investigating generalizations of expected utility theory using experimental data. *Econometrica, 62*, 1291–1326.

Hogg, S. (1996). A review of the validity and variability of the elevated plus-maze as an animal model of anxiety. *Pharmacology Biochemistry and Behavior, 54*, 21–30.

Hommel, J. D., Trinko, R., Sears, R. M., Georgescu, D., Liu, Z. W., Gao, X. B., et al. (2006). Leptin receptor signaling in midbrain dopamine neurons regulates feeding. *Neuron, 51*, 801–810.

Houk, J. C., & Wise, S. P. (1995). Distributed modular architectures linking basal ganglia, cerebellum, and cerebral cortex: Their role in planning and controlling action. *Cerebral Cortex, 2*, 95–110.

Hsu, M., Anen, C., & Quartz, S. R. (2008). The right and the good: Distributive justice and neural encoding of equity and efficiency. *Science, 320*, 1092–1095.

Hsu, M., Krajbich, I., Zhao, C., & Camerer, C. (2009). Neural response to reward anticipation under risk is nonlinear in probabilities. *Journal of Neuroscience, 29*, 2231.

Hsu, M., Bhatt, M., Adolphs, R., Tranel, D., & Camerer, C. F. (2005). Neural systems responding to degrees of uncertainty in human decision-making. *Science, 310*, 1680.

Iyer, A., Lindner, A., Kagan, I., & Andersen, R. A. (2010). Motor preparatory activity in posterior parietal cortex is modulated by subjective absolute value. *PLoS Biology, 8*, e1000444.

Jackson, J. H. (1873). On the anatomical and physiological localisation of movements in the brain. *The Lancet, 101*, 84–85.

Kable, J. W., & Glimcher, P. W. (2007). The neural correlates of subjective value during intertemporal choice. *Nature Neuroscience, 10*, 1625–1633.

Kahneman, D., & Tversky, A. (1979). Prospect theory: An analysis of decision under risk. *Econometrica, 47*, 263–291.

Karni, E., Schmeidler, D., & Vind, K. (1983). On state dependent preferences and subjective probabilities. *Econometrica, 51*, 1021–1031.

Kennerley, S. W., Walton, M. E., Behrens, T. E. J., Buckley, M. J., & Rushworth, M. F. S. (2006). Optimal decision-making and the anterior cingulate cortex. *Nature Neuroscience, 9*, 940–947.

Kennerley, S. W., Dahmubed, A. F., Lara, A. H., & Wallis, J. D. (2009). Neurons in the frontal lobe encode the value of multiple decision variables. *Journal of Cognitive Neuroscience, 21*, 1162–1178.

Kenning, P., & Plassmann, H. (2005). Neuroeconomics: An overview from an economic perspective. *Brain Research Bulletin, 67*, 343–354.

Kheramin, S., Body, S., Ho, M. Y., Velazquez-Martinez, D., Bradshaw, C., Szabadi, E., et al. (2004). Effects of orbital prefrontal cortex dopamine depletion on inter-temporal choice: A quantitative analysis. *Psychopharmacology, 175*, 206–214.

Knill, D. C., & Pouget, A. (2004). The Bayesian brain: The role of uncertainty in neural coding and computation. *Trends in Neurosciences, 27*, 712–719.

Knoch, D., Pascual-Leone, A., Meyer, K., Treyer, V., & Fehr, E. (2006). Diminishing reciprocal fairness by disrupting the right prefrontal cortex. *Science, 314*, 829–832.

Knoch, D., Gianotti, L. R. R., Pascual-Leone, A., Treyer, V., Regard, M., Hohmann, M., et al. (2006). Disruption of right prefrontal cortex by low-frequency repetitive transcranial magnetic stimulation induces risk-taking behavior. *Journal of Neuroscience, 26*, 6469–6472.

Knutson, B., Taylor, J., Kaufman, M., Peterson, R., & Glover, G. (2005). Distributed neural representation of expected value. *Journal of Neuroscience, 25*, 4806–4812.

Koenigs, M., Young, L., Adolphs, R., Tranel, D., Cushman, F., Hauser, M., et al. (2007). Damage to the prefrontal cortex increases utilitarian moral judgements. *Nature, 446*, 908–911.

Korn, H., Bausela, F., Charpier, S., & Faber, D. S. (1993). Synaptic noise and multiquantal release at dendritic synapses. *Journal of Neurophysiology, 70*, 1249–1254.

Korotkova, T. M., Sergeeva, O. A., Eriksson, K. S., Haas, H. L., & Brown, R. E. (2003). Excitation of ventral tegmental area dopaminergic and nondopaminergic neurons by orexins/hypocretins. *Journal of Neuroscience, 23*, 7–11.

Koszegi, B., & Rabin, M. (2007). Mistakes in choice-based welfare analysis. *American Economic Review, 97*, 477–481.

Kreek, M. J., Nielsen, D. A., Butelman, E. R., & LaForge, K. S. (2005). Genetic influences on impulsivity, risk taking, stress responsivity and vulnerability to drug abuse and addiction. *Nature Neuroscience, 8*, 1450–1457.

Krügel, U., Schraft, T., Kittner, H., Kiess, W., & Illes, P. (2003). Basal and feeding-evoked dopamine release in the rat nucleus accumbens is depressed by leptin. *European Journal of Pharmacology, 482*, 185–187.

Lauwereyns, J., Watanabe, K., Coe, B., & Hikosaka, O. (2002). A neural correlate of response bias in monkey caudate nucleus. *Nature, 418*, 413–417.

Li, J., Delgado, M. R., & Phelps, E. A. (2011). How instructed knowledge modulates the neural systems of reward learning. *Proceedings of the National Academy of Sciences of the United States of America, 108*, 55.

Lohrenz, T., McCabe, K., Camerer, C. F., & Montague, P. R. (2007). Neural signature of fictive learning signals in a sequential investment task. *Proceedings of the National Academy of Sciences of the United States of America, 104*, 9493–9498.

Loomes, G., & Sugden, R. (1982). Regret theory: An alternative theory of rational choice under uncertainty. *Economic Journal, 92*, 805–824.

Loomes, G., & Sugden, R. (1995). Incorporating a stochastic element into decision theories. *European Economic Review, 39*, 641–648.

MacKillop, J., Menges, D. P., McGeary, J. E., & Lisman, S. A. (2007). Effects of craving and DRD4 VNTR genotype on the relative value of alcohol: An initial human laboratory study. *Behavioral and Brain Functions, 3*, 11.

MacPherson, C. B. (1968). *Thomas hobbes, leviathan*. London, UK: Penguin Books.

Maimon, G., & Assad, J. A. (2009). Beyond poisson: Increased spike-time regularity across primate parietal cortex. *Neuron, 62*, 426–440.

Marr, D. (1982). *Vision*. New York: WH Freeman.

Matsumoto, M., & Hikosaka, O. (2007). Lateral habenula as a source of negative reward signals in dopamine neurons. *Nature, 447*, 1111–1115.

Matthews, S. C., Simmons, A. N., Lane, S. D., & Paulus, M. P. (2004). Selective activation of the nucleus accumbens during risk-taking decision-making. *NeuroReport, 15*, 2123–2127.

McClure, S. M., Laibson, D. I., Loewenstein, G., & Cohen, J. D. (2004). Separate neural systems value immediate and delayed monetary rewards. *Science, 306*, 503.

McCoy, A. N., & Platt, M. L. (2005). Risk-sensitive neurons in macaque posterior cingulate cortex. *Nature Neuroscience, 8*, 1220–1227.

Mink, J. W. (2003). The basal ganglia and involuntary movements: Impaired inhibition of competing motor patterns. *Archives of Neurology, 60*, 1365–1368.

Mischel, W., Shoda, Y., & Rodriguez, M. (1989). Delay of gratification in children. *Science, 244*, 933–938.

Mohr, P., Biele, G., & Heekeren, H. (2010). Neural processing of risk. *Journal of Neuroscience, 30*, 6613–6619.

Munafo, M. R., Clark, T. G., Moore, L. R., Payne, E., Walton, R., & Flint, J. (2003). Genetic polymorphisms and personality in healthy adults: A systematic review and meta-analysis. *Molecular Psychiatry, 8*, 471–484.

Nienborg, H., & Cumming, B. G. (2009). Decision-related activity in sensory neurons reflects more than a neuron's causal effect. *Nature, 459*, 89–92.

Niv, Y., Duff, M. O., & Dayan, P. (2005). Dopamine, uncertainty and TD learning. *Behavioral and Brain Functions, 1*, 1–9.

O'Doherty, J., Kringelbach, M. L., Rolls, E. T., Hornak, J., & Andrews, C. (2001). Abstract reward and punishment representations in the human orbitofrontal cortex. *Nature Neuroscience, 4*, 95–102.

O'Doherty, J., Rolls, E., Francis, S., Bowtell, R., & McGlone, F. (2001). Representation of pleasant and aversive taste in the human brain. *Journal of Neurophysiology, 85*, 1315–1321.

Ojemann, G. A. (1978). Organization of short-term verbal memory in language areas of human cortex: Evidence from electrical stimulation. *Brain and Language, 5*, 331–340.

Olds, J., & Milner, P. (1954). Positive reinforcement produced by electrical stimulation of septal area and other regions of rat brain. *Journal Of Comparative and Physiological Psychology, 47*, 419–427.

Oullier, O., & Kelso, J. A. S. (2006). Neuroeconomics and the metastable brain. *Trends in Cognitive Sciences, 10*, 353–354.

Padoa-Schioppa, C. (2009). Range-adapting representation of economic value in the orbitofrontal cortex. *Journal of Neuroscience, 29*, 14004–14014.

Parker, A., & Newsome, W. (1998). Sense and the single neuron: Probing the physiology of perception. *Annual Review of Neuroscience, 21*, 227–277.

Paulus, M. P., Rogalsky, C., Simmons, A., Feinstein, J. S., & Stein, M. B. (2003). Increased activation in the right insula during risk-taking decision-making is related to harm avoidance and neuroticism. *NeuroImage, 19*, 1439–1448.

Penfield, W., & Boldrey, E. (1937). Somatic motor and sensory representation in the cerebral cortex of man as studied by electrical stimulation. *Brain, 60*, 389–443.

Phillips, P. E. M., Walton, M. E., & Jhou, T. C. (2007). Calculating utility: Preclinical evidence for cost–benefit analysis by mesolimbic dopamine. *Psychopharmacology, 191*, 483–495.

Pine, A., Seymour, B., Roiser, J. P., Bossaerts, P., Friston, K. J., Curran, H. V., et al. (2009). Encoding of marginal utility across time in the human brain. *Journal of Neuroscience, 29*, 9575–9581.

Pine, A., Shiner, T., Seymour, B., & Dolan, R. J. (2010). Dopamine, time, and impulsivity in humans. *Journal of Neuroscience, 30*, 8888–8896.

Plassmann, H., O'Doherty, J., & Rangel, A. (2007). Orbitofrontal cortex encodes willingness to pay in everyday economic transactions. *Journal of Neuroscience, 27*, 9984–9988.

Platt, M. L., & Glimcher, P. W. (1999). Neural correlates of decision variables in parietal cortex. *Nature, 400*, 233–238.

Pleger, B., Blankenburg, F., Ruff, C. C., Driver, J., & Dolan, R. J. (2008). Reward facilitates tactile judgments and modulates hemodynamic responses in human primary somatosensory cortex. *Journal of Neuroscience, 28*, 8161–8168.

Preuschoff, K., Bossaerts, P., & Quartz, S. R. (2006). Neural differentiation of expected reward and risk in human subcortical structures. *Neuron, 51*, 381–390.

Procyk, E., Tanaka, Y., & Joseph, J. P. (2000). Anterior cingulate activity during routine and non-routine sequential behaviors in macaques. *Nature Neuroscience, 3*, 502–508.

Ragozzino, M. E. (2007). The contribution of the medial prefrontal cortex, orbitofrontal cortex, and dorsomedial striatum to behavioral flexibility. *Annals of the New York Academy Science USA, 1121*, 355–375.

Real, L., Ott, J., & Silverfine, E. (1982). On the tradeoff between the mean and the variance in foraging: Effect of spatial distribution and color preference. *Ecology, 63*, 1617–1623.

Resulaj, A., Kiani, R., Wolpert, D. M., & Shadlen, M. N. (2009). Changes of mind in decision-making. *Nature, 461*, 263–266.

Roesch, M. R., Calu, D. J., & Schoenbaum, G. (2007). Dopamine neurons encode the better option in rats deciding between differently delayed or sized rewards. *Nature Neuroscience, 10*, 1615–1624.

Rogers, R. D., Lancaster, M., Wakeley, J., & Bhagwagar, Z. (2004). Effects of beta-adrenoceptor blockade on components of human decision-making. *Psychopharmacology, 172*, 157–164.

Rudebeck, P. H., Walton, M. E., Smyth, A. N., Bannerman, D. M., & Rushworth, M. F. S. (2006). Separate neural pathways process different decision costs. *Nature Neuroscience, 9*, 1161–1168.

Rushworth, M., Behrens, T., Rudebeck, P., & Walton, M. (2007). Contrasting roles for cingulate and orbitofrontal cortex in decisions and social behavior. *Trends In Cognitive Sciences, 11*, 168–176.

Salamone, J. D., Correa, M., Farrar, A., & Mingote, S. M. (2007). Effort-related functions of nucleus accumbens dopamine and associated forebrain circuits. *Psychopharmacology, 191*, 461–482.

Salzman, C. D., Britten, K. H., & Newsome, W. T. (1990). Cortical microstimulation influences perceptual judgements of motion direction. *Nature, 346*, 174–177.

Samuelson, P. A. (1938). A note on the pure theory of consumer's behavior. *Economica, 5*, 61–71.

Samuelson, W., & Zeckhauser, R. (1988). Status quo bias in decision-making. *Journal of Risk and Uncertainty, 1*, 7–59.

Sanfey, A. G., Loewenstein, G., McClure, S. M., & Cohen, J. D. (2006). Neuroeconomics: Cross-currents in research on decision-making. *Trends in Cognitive Sciences, 10*, 108–116.

Schinka, J., Letsch, E., & Crawford, F. (2002). DRD4 and novelty seeking: Results of meta analyses. *American Journal of Medical Genetics, 114*, 643–648.

Schmitt, V., & Fischer, J. (2011). Representational format determines numerical competence in monkeys. *Nature Communications, 2*, 257.

Schultz, D. P., & Schultz, S. E. (2007). *A history of modern psychology.* Wadsworth Publishing Co.

Schultz, W., Dayan, P., & Montague, P. R. (1997). A neural substrate of prediction and reward. *Science, 275*, 1593–1599.

Schweimer, J., & Hauber, W. (2006). Dopamine D1 receptors in the anterior cingulate cortex regulate effort-based decision-making. *Learning & Memory, 13*, 777–782.

Scoville, W. B., & Milner, B. (1957). Loss of recent memory after bilateral hippocampal lesions. *Journal of Neurology, Neurosurgery & Psychiatry, 20*, 11.

Sesack, S., Deutch, A., Roth, R., & Bunney, B. (1989). Topographical organization of the efferent projections of the medial prefrontal cortex in the rat: An anterograde tract-tracing study with Phaseolus vulgaris leucoagglutinin. *Journal of Comparative Neurology, 290*, 13–242.

Shadlen, M. N., & Newsome, W. T. (1996). Motion perception: Seeing and deciding. *Proceedings of the National Academy of Sciences of the United States of America, 93*, 628–633.

Shadlen, M. N., & Newsome, W. T. (1998). The variable discharge of cortical neurons: Implications for connectivity, computation, and information coding. *Journal of Neuroscience, 18*, 3870–3896.

Shallice, T., & Burgess, P. (1991). Deficits in strategy application following frontal lobe damage in man. *Brain, 114*, 727–741.

Sharot, T., De Martino, B., & Dolan, R. J. (2009). How choice reveals and shapes expected hedonic outcome. *Journal of Neuroscience, 29*, 3760–3765.

Shi, C., & Cassell, M. (1998). Cortical, thalamic, and amygdaloid connections of the anterior and posterior insular cortices. *Journal of Comparative Neurology, 399*, 440–468.

Shiv, B., Loewenstein, G., Bechara, A., Damasio, H., & Damasio, A. R. (2005). Investment behavior and the negative side of emotion. *Psychological Science, 16*, 435–439.

Small, D. M., Zatorre, R. J., Dagher, A., Evans, A. C., & Jones-Gotman, M. (2001). Changes in brain activity related to eating chocolate. *Brain, 124*, 1720–1733.

Sporns, O., Tononi, G., & Edelman, G. M. (2000). Connectivity and complexity: The relationship between neuroanatomy and brain dynamics. *Neural Networks, 13*, 909–922.

St Onge, J. R., & Floresco, S. B. (2008). Dopaminergic modulation of risk-based decision-making. *Neuropsychopharmacology, 34*, 681–697.

Symmonds, M., Bossaerts, P., & Dolan, R. (2010). A behavioral and neural evaluation of prospective decision-making under risk. *Journal of Neuroscience, 30*, 14380–14389.

Symmonds, M., Emmanuel, J., Drew, M., Batterham, R., & Dolan, R. (2010). Metabolic state alters economic decision-making under risk in humans. *PLoS ONE, 5*, e11090.

Tanne-Gariepy, J., Rouiller, E., & Boussaoud, D. (2002). Parietal inputs to dorsal versus ventral premotor areas in the macaque monkey: Evidence for largely segregated visuomotor pathways. *Experimental Brain Research, 145,* 91–103.

Thut, G., Schultz, W., Roelcke, U., Nienhusmeier, M., Missimer, J., Maguire, R. P., et al. (1997). Activation of the human brain by monetary reward. *NeuroReport, 8,* 1225–1228.

Tindell, A. J., Smith, K. S., Berridge, K. C., & Aldridge, J. W. (2009). Dynamic computation of incentive salience: "Wanting" what was never "liked". *Journal of Neuroscience, 29,* 12220–12228.

Tobler, P. N., Fiorillo, C. D., & Schultz, W. (2005). Adaptive coding of reward value by dopamine neurons. *Science, 307,* 1642–1645.

Tobler, P. N., O'Doherty, J. P., Dolan, R. J., & Schultz, W. (2007). Reward value coding distinct from risk attitude-related uncertainty coding in human reward systems. *Journal of Neurophysiology, 97,* 1621–1632.

Tobler, P., Christopoulos, G., O'Doherty, J., Dolan, R., & Schultz, W. (2009). Risk-dependent reward value signal in human prefrontal cortex. *Proceedings of the National Academy of Sciences of the United States of America, 106,* 7185–7190.

Tolhurst, D., Movshon, J., & Dean, A. (1983). The statistical reliability of signals in single neurons in cat and monkey visual cortex. *Vision Research, 23,* 775–785.

Tom, S. M., Fox, C. R., Trepel, C., & Poldrack, R. A. (2007). The neural basis of loss aversion in decision-making under risk. *Science, 315,* 515–518.

Tremblay, L., & Schultz, W. (1999). Relative reward preference in primate orbitofrontal cortex. *Nature, 398,* 704–708.

Ullsperger, M., & Von Cramon, D. Y. (2003). Error monitoring using external feedback: Specific roles of the habenular complex, the reward system, and the cingulate motor area revealed by functional magnetic resonance imaging. *Journal of Neuroscience, 23,* 4308–4314.

Ursu, S., & Carter, C. S. (2005). Outcome representations, counterfactual comparisons and the human orbitofrontal cortex: Implications for neuroimaging studies of decision-making. *Brain Research Cognitive Brain Research, 23,* 51–60.

Venkatraman, V., Payne, J., Bettman, J., Luce, M., & Huettel, S. (2009). Separate neural mechanisms underlie choices and strategic preferences in risky decision-making. *Neuron, 62,* 593–602.

Volkow, N. D., Wang, G. J., Telang, F., Fowler, J. S., Logan, J., Childress, A. R., et al. (2008). Dopamine increases in striatum do not elicit craving in cocaine abusers unless they are coupled with cocaine cues. *NeuroImage, 39,* 1266–1273.

Wade, T. R., de Wit, H., & Richards, J. B. (2000). Effects of dopaminergic drugs on delayed reward as a measure of impulsive behavior in rats. *Psychopharmacology, 150,* 90–101.

Wakabayashi, K. T., Fields, H. L., & Nicola, S. M. (2004). Dissociation of the role of nucleus accumbens dopamine in responding to reward-predictive cues and waiting for reward. *Behavioural Brain Research, 154,* 19–30.

Walton, M. E., Bannerman, D. M., Alterescu, K., & Rushworth, M. F. S. (2003). Functional specialization within medial frontal cortex of the anterior cingulate for evaluating effort-related decisions. *Journal of Neuroscience, 23,* 6475–6490.

Weber, B. J., & Huettel, S. A. (2008). The neural substrates of probabilistic and intertemporal decision-making. *Brain Research, 1234,* 104–115.

Weil, R. S., Furl, N., Ruff, C. C., Symmonds, M., Flandin, G., Dolan, R. J., et al. (2010). Rewarding feedback after correct visual discriminations has both general and specific influences on visual cortex. *Journal of Neurophysiology, 104,* 1746–1757.

Wise, R. A., & Rompré, P. P. (1989). Brain dopamine and reward. *Annual Review of Psychology, 40,* 191–225.

Wright, N. D., Symmonds, M., Fleming, S. M., & Dolan, R. J. (2011). Neural segregation of objective and contextual aspects of fairness. *Journal of Neuroscience, 31,* 5244–5252.

Wunderle, J. M., Jr., Castro, M. S., & Fetcher, N. (1987). Risk-averse foraging by bananaquits on negative energy budgets. *Behavioral Ecology and Sociobiology, 21,* 249–255.

Yacubian, J., Glascher, J., Schroeder, K., Sommer, T., Braus, D. F., & Buchel, C. (2006). Dissociable systems for gain- and loss-related value predictions and errors of prediction in the human brain. *Journal of Neuroscience, 26,* 9530–9537.

Yokel, R. A., & Wise, R. A. (1978). Amphetamine-type reinforcement by dopaminergic agonists in the rat. *Psychopharmacology, 58,* 289–296.

Zeeb, F. D., Robbins, T. W., & Winstanley, C. A. (2009). Serotonergic and dopaminergic modulation of gambling behavior as assessed using a novel rat gambling task. *Neuropsychopharmacology, 34,* 2329–2343.

Zhang, K., Guo, J. Z., Peng, Y., Xi, W., & Guo, A. (2007). Dopamine-mushroom body circuit regulates saliency-based decision-making in Drosophila. *Science, 316,* 1901–1904.

2

Models of Value and Choice

Peter Dayan

Gatsby Computational Neuroscience Unit, UCL, London, UK

INTRODUCTION

The relationship between value and choice seems very simple. We should choose the things we value and value the things we choose. Indeed, one of the most beautiful results in economics goes exactly along these lines – if only our choices between possible actions were to satisfy some simple, intuitive, prerequisites, such as being transitive, then these actions could be arranged along a single axis of preference, and could be endowed with values that could be treated as governing choice. Although psychologists or behaviorally-inclined economists might go no further than asking people to report their subjective values in one of a variety of

33

ways, neuroscientists could search for postulated forms of such a value function in the activity of neurons, or their blood flow surrogates.

Unfortunately, life is not so simple. A subject's choices fail to satisfy any set of intuitive axioms, instead exhibiting the wide range of anomalies explored in this book and elsewhere. When subjects can be persuaded to express values, not only do these also show anomalies and inconsistencies, but also their anomalies are not quite the same as those of the choices with which they would be associated. Thus, subjects will subscribe to values that are not consistent with their expressible preferences.

There have been many interesting approaches to save part of the bacon of value and/or the relationship between value and choice by appealing to external or internal factors. The former (such things as the radical asymmetry in the information state between the experimenter and the subjects) are covered in other chapters in this book. In this chapter, we will show that even if there was a Platonic utility function governing idealized values of outcomes, rationalizable complexities of the internal architecture of control imply a range of apparent inconsistencies.

We start by noting three separate systems that have been mooted as being involved. Two (called model-based and model-free – Daw, Niv & Dayan, 2005; Dickinson & Balleine, 2002) are instrumental; a third is Pavlovian. A fourth, episodic, system has also been suggested (Lengyel & Dayan, 2007), but enjoys less empirical support and we will not discuss it here.

Crudely speaking (and we will see later why this is too crude in the current context), for instrumental systems, the relationship in the environment between choices and affectively important outcomes such as rewards and punishments plays a key role in determining choice. Subjects repeat actions that lead to high value outcomes (rewards), and avoid ones that lead to low value outcomes (punishments).

Conversely, for the Pavlovian system, choices are determined by predictions of these outcomes, irrespective of the actual environmental relationship between choices and outcomes (Dickinson, 1980; Mackintosh, 1983). Thus, for instance, animals cannot help but approach (rather than run away from) a source of food, even if the experimenter has cruelly arranged things in a looking-glass world so that the approach appears to make the food recede, whereas retreating would make the food more accessible (Hershberger, 1986).

Rather than starting from choices, models of all three systems start from values, from which choices are derived. Consistent with this, value signals have duly been found in a swathe of neural systems including, amongst others, the orbitofrontal cortex, the amygdala, the striatum, and the dopaminergic neuromodulatory system that innervates various of these loci (see Morrison & Salzman, 2010; Niv, 2009; O'Doherty, 2004,

2007; Samejima, Ueda, Doya & Kimura, 2005; Schultz, 2002; Wallis & Kennerley, 2010, and references therein). In fact, the mechanisms turning these values into choices, is less clear.

Complexities in the relationship between value and choice are two central sources of anomalies. First, the different systems can disagree about their values (and thus the choices they would make) (Dickinson & Balleine, 2002). However, choice is, almost by definition, unitary. Thus, if the values produced by the different systems differ, then the ultimate behavior will clearly have to fail to follow all of them. Nevertheless, we will argue that it is adaptive to have multiple systems (even at the expense of value-choice inconsistencies), since they offer different "sweet-spots" in the trade-off between adaptivity and adaptability.

The second source of anomalies is the Pavlovian system itself. As described above, it has the property that choices are determined directly by predictions, without any regard for their appropriateness. Thus, for example, given the chance, the subjects in the looking-glass world would clearly exhibit a preference for food, but nevertheless emit actions that are equally clearly inconsistent with this preference.

Having discussed the individual systems and their properties, we then consider issues that arise from their interaction. For instance, we have interpreted various findings as suggesting that Pavlovian mechanisms may interfere with model-based instrumental evaluation (Dayan & Huys, 2008). Equally, if one system controls behavior, then it can prevent other systems from being able to gain sufficient experience to acquire a full set of values and associated preferences.

We conclude the chapter with some general remarks about the naivety of the original expectation for a simple mapping between value and choice for subjects suffering from limited computational power in an unknown and changing world.

REINFORCEMENT LEARNING

Reinforcement learning (Sutton & Barto, 1998) formalizes the interaction between a subject and its environment. In the simplest cases, the environment comprises a set of possible states $\chi = \{x\}$, actions $\mathcal{A} = \{a\}$ and outcomes $\mathcal{O} = \{o\}$. We will also consider that the subject is in an internal motivational state $m \in \mathcal{M}$, such as hunger or thirst, although sometimes, as we discuss later, we will concatenate external and internal states to make a single state of the world (also called x).

In human experiments, states are often like the stages of a task, and are typically signaled by cues such as lights and tones. A state could also be a location in a maze. Actions involve such things as picking a stimulus or pressing a button.[1] The environment specifies a set of rules governing the transitions between states depending on the action chosen. This is often treated as a Markov chain, characterized by a transition matrix T_{xy} (a) that specifies the probability of moving from state x to state y given action a (Puterman, 2005; Sutton & Barto, 1998). We will treat the outcome $o(x,a)$ as being a deterministic function of the state x and action a; it is a simple generalization to make the outcomes also be probabilistic.

In this context, the subject's choices comprise a policy π. We consider so-called closed-loop policies, which specify the probability $\pi(x,a,m)$ of taking action a at state x in motivational state m. The question we have to address is what this policy is intended to achieve. As mentioned in the introduction, our strategy in this chapter is to start from the premise that there is an ideal, Platonic, utility function $u(o,m)$, which depends on the outcome o and the subject's motivational state m, and derive the consequences for this of the complexities of the architecture of control.

If the subject only had to make a single choice, then we could write down so-called action-values or Q-values (Watkins, 1989) as:

$$Q(x,a,m) = u(o(x,a),m)$$

and choose for an optimal policy the action with the largest Q-value. Formally, this implies that:

$$\pi(x,a,m) = \begin{cases} 1 & \text{if } Q(x,a,m) > Q(x,a',m), \forall a' \neq a \\ 0 & \text{otherwise} \end{cases} \tag{1}$$

In general, we can quantify the value of state x as the average utility available from that state, following policy π. This can be written as:

$$V^{\pi}(x,m) = E[u(o(x,a),m)]_{a \sim \pi} = E[Q(x,a,m)]_{a \sim \pi} \tag{2}$$

where $E[\ ]$ is the expectation operator, here taken over any randomness in the policy π.

There are at least two complicating factors that corrupt the simple policy in Eqn. 1. The first one, which underlies the multiplicity of instrumental controllers, is that subjects often have to make many choices at a sequence of states, and can receive many outcomes, rather than just one. How should the agent value the whole sequence of outcomes $\{o(x_t,a_t)\}$ unfolding over time? This has subtle complexities – for instance, the outcomes could be imperfectly substitutable; for instance, if the act of

---

[1] Here, we will only consider discrete choices, and not worry about other factors such as the latency or vigor with which those choices are made (Niv, Daw, Joel & Dayan, 2007).

consuming one morsel of food makes the next morsel either less attractive (e.g., because of satiety), or more attractive (as with salted peanuts).

For the present, we define values and action-values in terms of discounted sums of future utilities under a fixed motivational state, as in:

$$\sum_{t=0}^{\infty} \gamma^t u(o_t(x_t, a_t), m)$$

where $0 \leq \gamma < 1$ is called a discount factor which down-weights distant rewards compared to proximal ones.[2]

Starting at state x, we therefore seek the policy π_t, which can now, in principle, depend on time t, that maximizes the expected value:

$$V^{\pi}(x_0, m) = \left[\sum_{t=0}^{\infty} \gamma^t u(o_t(x_t, a_t), m) \right]_{a_t \sim \pi_t; \, x_{t+1} \sim T_{x_t x_{t+1}}(a_t)} \quad (3)$$

averaging over the actions and transitions. For convenience, we sometimes write:

$$o_t = o_t(x_t, a_t) \text{ and } u_t = u(o_t(x_t, a_t), m)$$

Similarly, the action-value of state x_0 and action a is defined as:

$$Q^{\pi}(x_0, a, m) = E\left[u(o(x, a), m) + \gamma \sum_{t=0}^{\infty} \gamma^t u_{t+1} \right] \quad (4)$$

where now the expectation is over all the transitions, but only over the policy π after the first step, since action a is taken initially. The extra factor of γ in the second term of the sum comes from the fact that we reset the iterator t to start at $t = 0$.

In principle, it is straightforward to find the optimal policy. We can do this by expanding the whole search tree of possible state that the agent might successively occupy, and summing over the resulting paths down the tree (each of which defines a sequence of states and outcomes) weighted by their respective probabilities. The trouble is that a tree typically grows exponentially with the number of layers considered, making this extremely difficult.

A key idea from the field of dynamic programming is that, since the environment is Markovian, there turns out to be a state-independent policy $\pi(x,a,m)$ that maximizes Eqn. 3 simultaneously for all starting states

---

[2] In fact, discounting in humans and animals is typically not exponential, as in Eqn. 3, but rather hyperbolic (Ainslie, 2001). This has an important effect, and has been the subject of substantial empirical research; however, we will not consider it further here.

x (Sutton & Barto, 1998). This licenses a variety of algorithmic methods that are more efficient than tree expansion and that can be used to find the optimal policy (Puterman, 2005). However, their neural realization may still be problematic.

The second problem surrounding the use of Eqns 1 and 3 is learning. Subjects must typically acquire knowledge of the transition and outcome structure of the environment instead of it being provided directly. However, learning comes at a potentially high price – both the possibility of getting a catastrophically bad outcome, and having to try (i.e., explore) actions to determine the statistics of their transitions, even when early attempts to use those actions led to poor outcomes. We will see that the need for learning inspires the use of both model-based and model-free control. We will also interpret the idiosyncrasies of the Pavlovian system as coming from evolutionary pre-programming of generally appropriate choices. These are couched as very strong prior expectations, which therefore obviate certain forms of learning. This will be the sense in which (evolutionarily) adaptive choice trumps adaptability.

Having defined the environment and optimal behavior, we can now turn to the three systems for turning value into choice. As mentioned above, we will assume that they share the same utility function $u(o,m)$. However, the way that they calculate values and determine choices is different.

CONTROLLERS

We defined instrumental control in terms of its sensitivity to the relationship in the environment between actions and outcomes, reflected in action values $Q^\pi(x,a,m)$. Finding the policy π that optimizes Eqn. 3 clearly depends on this sensitivity. Model-based and model-free controls are ways of doing this, which differ in the information about the environment they use and the computations they perform. The Pavlovian controller depends on the values of states $V^\pi(x,m)$, but with a predetermined mapping from value to action, i.e., a fixed choice of policy $\pi(x,a,m)$ that depends on these values. It is the fixedness that makes Pavlovian control insensitive to action-outcome contingencies.

We describe the controllers only briefly; they are portrayed in detail in our terms in Daw et al. (2005); Dayan (2008); Dayan and Daw (2008); Dayan, Niv, Seymour and Daw (2006).

Model-based Control

The model-based controller is a realization of Tolman's ideas about cognitive mapping (Tolman, 1949). It uses experience in the environment to learn a model of the transition distribution, outcomes and motivationally-sensitive utilities. We do not discuss the learning rules here, but they are straightforward forms of the Rescorla–Wagner or delta rule (Rescorla & Wagner, 1972; Widrow & Hoff, 1960). Furthermore, the model-based controller infers choices by performing a computation akin to a more or less direct implementation of Eqn. 3; for instance, building and evaluating the search decision tree to work out the optimal course of action.

The signature of model-based control is its motivational sensitivity (Dickinson & Balleine, 2002). That is, since it infers the best action by expanding the sum in Eqn. 3, it assesses the value of the expected outcomes in a way that is sensitive to the motivational state. Consider the case that during learning the subject is hungry, making the experienced utilities $u(o, \text{hunger})$ for food large but then, before the test, the subject is allowed to eat to satiety. Provided that the internal model of the utility $u(o, \text{satiety})$ is apprised of the effect of satiety, thus according the outcome a low utility, the subject will not be willing to make choices to achieve that outcome.

This apparent goal-directedness of the choices of the model-based system shows that they are perfectly aligned with the Platonic utility function. There have been substantial investigations into the neurobiological substrate of this controller in rats, monkeys and humans, leaving particular roles for the orbitofrontal cortex (OFC) (Padoa-Schioppa, 2007; Valentin, Dickinson & O'Doherty, 2007; Wallis & Kennerley, 2010), the prelimbic cortex (Killcross & Coutureau, 2003), the basolateral nucleus of the amygdala (Balleine, 2005; Cardinal, Parkinson, Hall & Everitt, 2002), and dorsomedial regions in the striatum (or anterior caudate in humans) (Balleine, 2005; Balleine & O'Doherty, 2010). In particular, the orbitofrontal cortex, a key region for many chapters in this book, perhaps in association with the basolateral nucleus of the amygdala, appears to realize the motivationally-sensitive internal model of the utility. This is one piece of evidence that choice can indeed be based on internal utilities.

However, there are limits to this motivational sensitivity. For instance, in Chapter 3, Sharot discusses the fact that when occupying one motivational state (say hunger, or calmness), we are bad at estimating the utility of outcomes in a different motivational state (say thirst or rage).[3] If the OFC is indeed responsible for the motivational estimation, and model-based control dominates, this could reflect the inputs it receives associated with bodily state. These could pin the OFC to reporting aspects of the here and now, and make it hard to be affectively predictive. Thus, even the model-based system can express choices that would be

---

[3] Other animals may be even worse at this (Suddendorf & Corballis, 1997), though note the interesting exception of Western scrub-jays (Correia, Dickinson & Clayton, 2007).

inconsistent with its own values (in this case future), depending on what sensitivities these values have.

One could imagine a more sophisticated model-based method according to which the motivational state m_t is itself considered to be time-varying, and to be encompassed within the state of the environment $x_t \Rightarrow (x_t, m_t)$. If the model-based controller was capable of predicting the effect of actions (and outcomes) on the motivational state, then it would even be possible to solve problems of non-substitutability (for example, that eating food at one time will make later foods less appealing). Again, although there is no direct evidence, it seems unlikely that the model-based system is capable of solving such estimation problems in general.

Model-based control is statistically efficient, but computationally prohibitive, as discussed in Daw et al. (2005). The computational cost of expressing and searching the tree for the optimal action (or indeed other ways of solving for the optimal policy) is the close companion of the motivational flexibility with which this endows the subject. We have suggested that the consequence of this cost is inaccuracy or uncertainty, for instance, arising from interference in the working memory concerned.

Model-free Control

In response to the computational complexity of model-based control, reinforcement learning (RL) has developed a number of methods of model-free control (Sutton, 1988; Sutton & Barto, 1998). The underlying idea is the simple one of self-consistency.

Consider the case of Eqn. 3 in which the policy is deterministic (taking action $\pi(x,m)$ at state x), and the transitions are deterministic, so x_{t+1} can be written as a function of x_t and $\pi(x_t, m)$.

Then, we can write:

$$V^\pi(x_0, m) = \sum_{t=0}^{\infty} \gamma^t u_t = u_0 + \gamma \sum_{t=0}^{\infty} \gamma^t u_{t+1}$$

However, if the values are correct, then the right hand term is just $\gamma V^\pi(x_1, m)$, starting from the state $x_1 = s(x_0, \pi(x_0, m))$, making the whole expression:

$$= u_0 + \gamma V^\pi(x_1, m)$$

Any discrepancy or inconsistency:

$$\delta_0 = u_0 + \gamma V^\pi(x_1, m) - V^\pi(x_0, m) \tag{5}$$

between the two sides of this equation can be used to improve the estimate of the value of $V^\pi(x_0, m)$.

Critically, one can use such discrepancies to learn estimates of the values $V^{\pi}(x,m)$ just from experience of the environment. That is, they can be learned from samples of states x_t, actions a_t, and utilities $u_t = u(o_t(x_t,a_t),m)$. This can be done without building a model of any aspect of the environment.

The quantity in Eqn. 3 is just the value of being at a particular state x_0 in the environment, given the policy. By itself, this is not enough to determine what choices to make at x_0. However, it turns out to be just as straightforward to acquire the state-action value $Q^{\pi}(x_0,a,m)$ as in Eqn. 4, using one of two main algorithms, Q-learning (Watkins, 1989) and SARSA (Rummery & Niranjan, 1994). In more exotic schemes, it is also possible to use the prediction error in Eqn. 5 to train a form of state–response (SR) mapping that indicates the optimal choice (Barto, Sutton & Anderson, 1983).

Note that the model-free controller thus seeks to learn the same quantity that the model-based controller estimates, from the same experience.[4] However, it does so by caching and then recalling the results of experience rather than building and searching the tree of possibilities. Thus, the model-free controller does not even represent the outcomes o_t that underlie the utilities, and is therefore certainly not in any position to change the estimate of its values if the motivational state changes. Consider, for instance, the case that after a subject has been taught to press a lever to get some cheese, the cheese is poisoned, so it is no longer worth eating. The model-free system would learn the utility of pressing the lever, but would not have the informational wherewithal to realize that this utility had changed when the cheese had been poisoned. Thus it would continue to insist upon pressing the lever. This is an example of motivational insensitivity.

Such insensitivity is a popular signature of habitual control (Dickinson & Balleine, 2002), exhibited by rats and humans alike (Tricomi, Balleine & O'Doherty, 2009), and has been used to implicate dorsal striatum (Balleine, 2005; Balleine & O'Doherty, 2010; Killcross & Coutureau, 2003; Tricomi et al., 2009), and, at least in the case of rewards, dopaminergic neuromodulation (Barto, 1995; Montague, Dayan & Sejnowski, 1996; Suri & Schultz, 1999) in the operation of the habit system. There is even evidence in the activity of dopamine neurons for the operation of Q-learning in rats

---

[4] Note, however, that the model-based controller learns the transition structure T_{xy} (a) of the environment separately from the utilities, whereas the model-free controller does not. Thus, for instance, in latent learning experiments such as those pioneered by Tolman (1949), the model-based controller can uniquely take advantage of trials which only provide information about the transitions in the absence of utilities (Gläscher et al., 2010).

(Roesch, Calu & Schoenbaum, 2007) (albeit SARSA in macaques; Morris, Nevet, Arkadir et al., 2006). In terms of this chapter, model-free control is clearly capable of expressing preferences that do not fit with its utility function. In the context of explaining the behavior and impact of dopamine in appetitive contexts, Berridge (2004, 2007) distinguishes the concept of *liking*, which one could perhaps consider in terms of Platonic utility, from *wanting*, which we would locate with the model-free values. It appears that if the liking can be influenced by opioids, then wanting is at least also influenced by drugs that manipulate dopamine (Berridge, Robinson & Aldridge, 2009; Peciña, 2008; Peciña & Berridge, 2005).

Note, however, that it is possible for model-free control to be sensitive to the motivational state. That is, provided that there is a representation of the motivational state m, and each state x and outcome is experienced in state m, then the prediction error of Eqn. 5 can be used to train predictions $V^{\pi}(x,m)$ that are conditional on m. However, they cannot generalize as flexibly as model-based control, which can change the choice at one state x based on the current motivational appetitiveness of the outcome available at another state, even if it had not previously experienced state x under the current motivational state.

The model-free controller has the opposite characteristics to the model-based controller, being statistically inefficient, but computationally trivial (Daw et al., 2005). The statistical inefficiency arises from the form of the prediction error in Eqn. 5. The value $V^{\pi}(x_0,m)$ is criticized partly on the basis of the value $V^{\pi}(x_1,m)$. However, at the outset of learning, $V^{\pi}(x_1,m)$ is itself inaccurate, and therefore makes for an initially noisy learning signal. This is called the problem of bootstrapping. On the other hand, the action values or SR policies that it acquires can be used without any need to build or search a tree of possibilities, and is therefore not susceptible to computational inaccuracy or noise.

Pavlovian Control

Pavlovian control is also based on predictions of affectively important outcomes such as rewards and punishments. However, rather than determining the choices that would lead to the acquisition or avoidance of these outcomes, it expresses a set of hard-wired preparatory and consummatory choices. Preparatory choices are largely sensitive to the net valence of the prediction, involving such actions as approach and engagement with appetitive outcomes, and withdrawal and inhibition from aversive ones, apparently mediated partly by the action of dopamine and serotonin, respectively (see Boureau & Dayan, 2011, for an extensive review); the nucleus accumbens also plays a critical role (Faure, Reynolds, Richard & Berridge, 2008; Reynolds & Berridge, 2001, 2002). Indeed, the architecture of the striatum, with direct and indirect pathways involved in execution and inhibition and dopamine D1 and D2 receptors, respectively, is one obvious substrate for these biases (Alexander & Crutcher, 1990; Frank, Seeberger & O'Reilly, 2004; Frank & Claus, 2006; Smith & Villalba, 2008). Consummatory choices, which are expressed when the outcomes are actually present, are sensitive to the nature of the outcomes themselves.

The affective predictions that underlie Pavlovian choices are very similar to those that underlie instrumental choices. Thus, we might expect there to be model-based and model-free predictions, with the same characteristic sensitivities to motivational states that we discussed above. However, the motivational status of Pavlovian values has been harder to assess with the precision of the methods that have so clearly distinguished goal-directed and habitual systems (Dickinson & Balleine, 2002), and so this is currently not clear.

The critical characteristic of Pavlovian choices is that they are insensitive to their effects on the delivery of the outcomes that they themselves are intended to predict, crudely depending on $V^\pi(x,m)$ rather than $Q^\pi(x,a,m)$. This is what leads to problems in the looking glass world mentioned in the introduction (Hershberger, 1986). It has also been quantified in the form of negative auto-maintenance or omission schedules, in which subjects continue to emit preparatory Pavlovian responses based on predictions of getting an outcome in cases in which those responses themselves are deemed by the experimenter to lead to the omission of those outcomes (Sheffield, 1965; Williams & Williams, 1969). Closer to the themes of this book, Pavlovian influences tilt in the same direction as such neuroeconomic anomalies as framing and loss aversion (Bushong, King, Camerer & Rangel, 2010; Dayan et al., 2006; Dayan & Seymour, 2008; De Martino, Kumaran, Seymour & Dolan, 2006; Kahneman & Frederick, 2007).

This Pavlovian misbehavior (Dayan et al., 2006) might seem bizarre from the perspective of instrumental control. However, it can also be seen as offering evolutionary pre-specified choices that are broadly appropriate as a default across environments. There is a huge advantage for a subject not to have to learn for itself what actions to take, particularly in the case of threats (Boureau & Dayan, 2011; Dayan & Huys, 2009). Indeed, the periacqueductal grey (Blanchard & Blanchard, 1988) offers an exquisite topography of species-typical (Bolles, 1970) defensive actions, including fighting, fleeing, and freezing.

In terms of the argument in this chapter, Pavlovian choices are based on the Platonic utilities of outcomes, and indeed are often consistent with the acquisition or avoidance of those outcomes in a way that is appropriate to those utilities. However, this consistency is irrelevant at the moment that the choices are made, and so choice and utility can be quite independent.

COMBINATION

We have described three systems that are involved in making choices. Even in the case that they share a single, Platonic, utility function for outcomes, the choices they express can be quite different. The model-based controller comes closest to being Platonically appropriate, although we noted that even it may be incapable of assessing higher order effects associated with the sequences of outcomes. The choices of the model-free controller can depart from current utilities because it has learned or cached a set of values that may no longer be correct. Pavlovian choices, though determined over the course of evolution to be appropriate, can turn out to be instrumentally catastrophic in any given experimental domain.

Net Choice

The most important issue remaining is to consider how choice emerges from the collection of controllers. Unfortunately, the mechanisms of this are not yet very clear. In building models of behavior, and relating them to fMRI BOLD, one approach is to generate net Q-values, as in:

$$Q^{\text{net}}(x,a,m) = \alpha^{\text{MB}}Q^{\text{MB}}(x,a,m) + \alpha^{\text{MF}}Q^{\text{MF}}(x,a,m) + \alpha^{\text{Pav}}Q^{\text{Pav}}(x,a,m)$$

$$(6)$$

using the superscripts "MB" for model-based, "MF" for model-free and "Pav" for Pavlovian. Here, we consider preparatory Pavlovian actions:

$$Q^{\text{Pav}}(x,a,m) = \begin{cases} V^{\pi^{\text{net}}}(x,m) & \text{if } a \text{ is the Pavlovian associate of } V^{\pi^{\text{net}}}(x,m) \\ 0 & \text{otherwise} \end{cases}$$

as being chosen by, and with the strength of, the valence of the prediction $V^{\pi^{\text{net}}}(x,m)$.

In Eqn. 6, $0 \leq \alpha \leq 1$, $\alpha^{\text{MB}} + \alpha^{\text{MF}} + \alpha^{\text{Pav}} = 1$, are the relative weights of each controller, and the net policy is governed by:

$$\pi^{\text{net}}(x,a,m) = \frac{e^{\beta Q^{\text{net}}(x,a,m)}}{\sum_{a'} e^{\beta Q^{\text{net}}(x,a',m)}}$$

where $\beta > 0$ is a parameter that allows for stochasticity in choice. If β is very large, then this policy strongly favors the action with the largest net Q-value. If β is near 0, then all actions are chosen nearly equally often. In fact, except for rather unusual utility functions (Todorov, 2009) or in a competitive, game theoretic, context (Fudenberg & Tirole, 1994), it is not usually appropriate to have anything other than a deterministic policy (we comment on exploration below). Nevertheless, it has proved impossible to find deterministic hidden variables that underlie the apparent stochasticity of animal and human behavior. Furthermore, models of the neural implementation of choice derive such stochasticity from internal noise processes associated with neural activity (Bogacz, Brown, Moehlis et al., 2006; Usher & McClelland, 2001). Of course, any residual apparent randomness becomes a further discrepancy between Platonic utility and expressed choice.

If Eqn. 6 is indeed the appropriate way to construct a net Q-value for each action, it becomes critical to understand how the relative weights α^* are determined. When considering just model-based and model-free controls, Daw et al. (2005), suggested that α^{MB} and α^{MF} be set according to the relative uncertainties of each system. Indeed, it is relatively straightforward to compute the uncertainty of the model-based system, although much more difficult to do this for the model-free system (Dearden, Friedman & Russell, 1998). This sufficed to account for the key behavioral observations about the transition of motivational sensitivity of behavior over the course of experience (Dickinson & Balleine, 2002; Holland, 2004). Given few trials, the statistical efficiency of model-based control out-competes the computational efficiency of model-free control, and so behavior is motivationally sensitive, and thus goal-directed. However, after more extensive training, the computational uncertainty induced by model-based inference overwhelms the effects of model-free bootstrapping, and so behavior becomes model-free or habitual and motivationally insensitive.

The progressive switch from model-based to model-free control has been examined directly in a latent-learning paradigm (Tolman, 1949). Here, experience that targeted only the model-based controller was provided in a first set of trials, such that only it could make a sensible suggestion in the second set (Gläscher, Daw, Dayan & O'Doherty, 2010). However, behavior switched from being model-based to model-free much faster than would have been expected from pure uncertainty.

The manner in which a task is presented could also potentially bias the use of one or other system. For instance, in many experimental paradigms in economics, human subjects are presented with explicit information about probabilities and outcomes. They then exhibit a variety of choice biases associated, for instance, with the prospect theory (Kahneman, 2003), such as over-weighing small probability outcomes. However, if the options are presented experientially, as they would be to a rat or a macaque, human subjects show opposite biases such as underweighting low probability events (Hertwig, Barron, Weber & Erev, 2004; Hertwig & Erev, 2009; Weber, Shafir & Blais, 2004). It could be that the two different methods of presentation bias the weighting towards model-based and model-free systems, respectively.

It is not quite so obvious how to assess the uncertainty associated with Pavlovian control, since it seems impossible to provide a sensible weighting for an evolutionarily-specified prior. Thus, at present, we have little choice but to use the ad-hoc procedure of fitting α^{Pav} to observed behavior.

Interaction

There is also evidence that the systems interact in various more complicated ways. For instance, the operations that the model-based system executes to build and evaluate a search tree can be considered as internally-directed actions, rather like the way cognitive control is envisioned (Hazy, Frank & O'Reilly, 2007). The potential regress – what determines the choice of these actions – has not been completely addressed. Further, one might expect that model-based search in deep trees would have to be truncated as the number of leaves grow too large. In computational terms, the value that the model-based system should use at an internally truncated leaf is the model-free estimate (Samuel, 1959; Sutton & Barto, 1998), but there is, as yet, no evidence if this actually happens.

There is some evidence for three relevant interactions. First, it has been mooted that Pavlovian processes associated with approach and withdrawal might influence model-based inference, by favoring paths in the search tree associated with positive values, and pruning paths associated with negative values. If this happened on the basis of predicted immediate utilities $u(o_t,m)$, then this could have a severe effect on an accurate estimation in certain contexts, even if it was in general a useful and appropriate heuristic method. Dayan and Huys (2008) suggested exactly this possibility as a way of linking lowered serotonin levels in depression to the phenomenon of depressive realism.

Second, it has long been suggested that the model-based system might train the model-free system by generating fictitious samples of behavior (Sutton, 1990). The characterization of hippocampal replay (Foster & Wilson, 2006; Gupta, van der Meer, Touretzky & Redish, 2010), in which recent past experience is recreated during sleep or resting wakefulness is consistent with this; although its impact on behavior is not clear. The model-based system might also influence the prediction errors that determine model-free learning (Daw, Gershman, Seymour et al., 2011).

Along related lines, consider the case that the model-based system is in (self-)confident control, but is incorrect; for instance, by expressing a poorly matching prior. By virtue of controlling the subject's experience, the model-free system might not be able to acquire appropriate values, and so could be permanently suppressed.

Finally, each system could manipulate the course of behavior to ameliorate what it sees as other systems perversions of its attempts to perform optimization on its Platonic utilities. This has perhaps been most extensively investigated in the case of hyperbolic discounting (Ainslie, 2001). This notoriously leads to preference reversals, implying that a subject at one time cannot rely on its future self to hold fast to a policy that it currently considers optimal. Thus, in order to deny the future self the chance to fail, the current self might optimally engage in a form of (potentially expensive) binding behavior, as in the story of Ulysses and the Sirens (Elster, 1979). The instrumental systems might try to prevent malign influences of the Pavlovian system in just the same manner (Dayan et al., 2006). This could happen either explicitly, for instance, if the model-based system included Pavlovian effects in its model of the environment, or implicitly, by observing the successes and failures of strategies that limit Pavlovian misbehavior.

DISCUSSION

The intent of this chapter was to point out that, even if there is a single, Platonic, utility function, the needs of control in an unknown, uncertain and changing world are such that we can expect a great deal of substantial divergence between: (1) the choices that could be computed to be consistent with the utility assuming perfect knowledge and unlimited computational time; and (2) the choices that any practical system will be able to generate. Computational heuristics (such as the model-free controller) and statistical heuristics (such as the Pavlovian controller) can offer excellent, if suboptimal, performance. Indeed, the field of reinforcement learning has addressed the complexities by inventing just such controllers, each of which works well in a certain regime of knowledge and computation time, and also by hard-wiring certain options. Thus, multiplicity is a rationalizable way of addressing the complexities mentioned, but can lead to clashes between Platonic utility and choice. Further, model-free and Pavlovian choices can themselves be inconsistent with their own utilities.

There are additional factors making it hard to square value and choice, notably uncertainty, stemming from initial ignorance and ongoing change. In response to this, subjects have to solve the complex problems associated with the trade-off between *exploitation*, which involves performing as well as possibly using, just existing knowledge, and *exploration*, which reduces uncertainty in order potentially to be able to perform better in the future. This trade-off is particularly sensitive to prior expectations about the environment, which can themselves be established by prior experience in partially related environments. This issue has been studied, for instance, in the context of controllability and learned helplessness (Huys & Dayan, 2009). Thus, it might be very hard to understand whether the resulting choices are consistent with exploration that is appropriate to the Platonic utility. The model-free controller cannot even build a sophisticated model of its own uncertainty, and so might resort to Pavlovian heuristics such as optimistic initial evaluation (if insufficiently-visited states are, by default, considered to be attractive, it is appropriate to plan to sample them). There is certainly evidence for such optimism (Wittmann, Daw, Seymour & Dolan, 2008), although its model-based or model-free status is unclear.

We chose the case of a single, Platonic, utility function in order to make the point about the partial independence of value and choice most starkly. However, it is certainly possible that there are distinct utility functions associated with different systems, and that there is therefore also a more systemic form of competition. A further, albeit more complex, possibility is that our conscious selves might suffer from characteristic uncertainty about our true values, and gather information about them from choices we make (the Jamesian: "How do I know what I like until I see what I pick?"; Tversky & Thaler, 1990) – one could, for instance, imagine that the model-based system attempts to infer the Platonic utility at least partly by observing Pavlovian preferences.

We should finally note a more radical challenge to the concept of Platonic utility that arises from nascent work in the reinforcement learning field under the rubric of intrinsic motivation (Barto, Singh & Chentanez, 2004; Oudeyer, Kaplan & Hafner, 2007; Singh, Barto & Chentanez, 2005; Singh, Lewis & Barto, 2009). One idea is that the "true" evolutionarily appropriate metric for behavior is the extremely sparse one of propagating ones genes. What we think of as a Platonic utility over immediate rewards such as food or water, would merely be a surrogate that helps overcome the otherwise insurmountable credit assignment path associated with procreation. In these terms, even the Platonic utility is the same sort of heuristic expedient as the Pavlovian controller itself, with evolutionary optimality molding approximate economic rationality to its own ends. It is a sober thought that understanding values may be less important as a way of unearthing the foundations of choice than we might have expected.

Acknowledgments

I am very grateful to my collaborators on the work described here, notably Nathaniel Daw, Ray Dolan, Emrah Düzel, Marc Guitart-Masip, Quentin Huys, Yael Niv and Ben Seymour. I particularly thank Benedetto De Martino and Ben Seymour for helpful comments on a previous version of the manuscript. This work was funded by the Gatsby Charitable Foundation.

References

Ainslie, G. (2001). *Breakdown of will*. Cambridge, UK: Cambridge University Press.

Alexander, G. E., & Crutcher, M. D. (1990). Functional architecture of basal ganglia circuits: Neural substrates of parallel processing. *Trends in Neurosciences, 13*(7), 266–271.

Balleine, B. W. (2005). Neural bases of food-seeking: Affect, arousal and reward in corticostriatolimbic circuits. *Physiology & Behavior, 86*(5), 717–730.

Balleine, B. W., & O'Doherty, J. P. (2010). Human and rodent homologies in action control: Corti-costriatal determinants of goal-directed and habitual action. *Neuropsychopharmacology, 35*(1), 48–69.

Barto, A. (1995). Adaptive critics and the basal ganglia. In J. Houk, J. Davis & D. Beiser (Eds.), *Models of information processing in the Basal Ganglia* (pp. 215–232). Cambridge MA: MIT Press.

Barto, A., Sutton, R., & Anderson, C. (1983). Neuron-like adaptive elements that can solve difficult learning control problems. *IEEE Transactions on Systems, Man, and Cybernetics, 13*(5), 834–846.

Barto, A., Singh, S., & Chentanez, N. (2004). Intrinsically motivated learning of hierarchical collections of skills. In *Proceedings of international conference of developmental learning*, San Diego, CA.

Berridge, K. C. (2004). Motivation concepts in behavioral neuroscience. *Physiology & Behavior, 81*, 179–209.

Berridge, K. C. (2007). The debate over dopamine's role in reward: The case for incentive salience. *Psychopharmacology (Berl), 191*(3), 391–431.

Berridge, K. C., Robinson, T. E., & Aldridge, J. W. (2009). Dissecting components of reward: "Liking", "wanting", and learning. *Current Opinion in Pharmacology, 9*(1), 65–73.

Blanchard, D. C., & Blanchard, R. J. (1988). Ethoexperimental approaches to the biology of emotion. *Annual Review of Psychology, 39*, 43–68.

Bogacz, R., Brown, E., Moehlis, J., Holmes, P., & Cohen, J. D. (2006). The physics of optimal decision making: A formal analysis of models of performance in two-alternative forced-choice tasks. *Psychological Review, 113*(4), 700–765.

Bolles, R. C. (1970). Species-specific defense reactions and avoidance learning. *Psychological Review, 77*, 32–48.

Boureau, Y.-L., & Dayan, P. (2011). Opponency revisited: Competition and cooperation between dopamine and serotonin. *Neuropsychopharmacology, 36*, 74–97.

Bushong, B., King, L., Camerer, C., & Rangel, A. (2010). Pavlovian processes in consumer choice: The physical presence of a good increases willingness-to-pay. *American Economic Review, 100*, 1–18.

Cardinal, R. N., Parkinson, J. A., Hall, J., & Everitt, B. J. (2002). Emotion and motivation: The role of the amygdala, ventral striatum, and prefrontal cortex. *Neuroscience and Biobehavioral Reviews, 26*(3), 321–352.

Correia, S. P. C., Dickinson, A., & Clayton, N. S. (2007). Western scrub-jays anticipate future needs independently of their current motivational state. *Current Biology, 17*(10), 856–861.

Daw, N. D., Gershman, S. J., Seymour, B., Dayan, P., & Dolan, R. J. (2011). Model-based influences on humans' choices and striatal prediction errors. *Neuron, 69*(6), 1204–1215.

Daw, N. D., Niv, Y., & Dayan, P. (2005). Uncertainty-based competition between prefrontal and dorsolateral striatal systems for behavioral control. *Nature Neuroscience, 8*(12), 1704–1711.

Dayan, P., Niv, Y., Seymour, B., & Daw, N. D. (2006). The misbehavior of value and the discipline of the will. *Neural Networks, 19*(8), 1153–1160.

Dayan, P. (2008). The role of value systems in decision-making. In C. Engel & W. Singer (Eds.), *Better than conscious: Decision making, the human mind, and implications for institutions* (pp. 51–70). Ernst Strüngmann Forum, Cambridge, MA: MIT Press.

Dayan, P., & Daw, N. D. (2008). Decision theory, reinforcement learning, and the brain. *Cognitive, Affective & Behavioral Neuroscience, 8*(4), 429–453.

Dayan, P., & Huys, Q. J. M. (2008). Serotonin, inhibition, and negative mood. *PLoS Computational Biology*, *4*(2), e4.

Dayan, P., & Seymour, B. (2008). Values and actions in aversion. In P. Glimcher, C. Camerer, R. Poldrack & E. Fehr (Eds.), *Neuroeconomics: Decision making and the brain* (pp. 175–191). New York, NY: Academic Press.

Dayan, P., & Huys, Q. J. M. (2009). Serotonin in affective control. *Annual Review of Neuroscience*, *32*, 95–126.

De Martino, B., Kumaran, D., Seymour, B., & Dolan, R. J. (2006). Frames, biases, and rational decision-making in the human brain. *Science*, *313*(5787), 684–687.

Dearden, R., Friedman, N., & Russell, S. (1998). Bayesian Q-learning. In *Proceedings of the fiteenth National/tenth Conference on Artificial intelligence/Innovative Applications of Artificial Intelligence Table of Contents*, pp. 761–768. Menlo Park, CA: American Association for Artificial Intelligence.

Dickinson, A. (1980). *Contemporary animal learning theory*. Cambridge, UK: Cambridge University Press.

Dickinson, A., & Balleine, B. (2002). The role of learning in the operation of motivational systems. In R. Gallistel (Ed.), *Stevens' handbook of experimental psychology* (Vol. 3, pp. 497–534). New York, NY: Wiley.

Elster, J. (1979). *Ulysses and the sirens: Studies in rationality and irrationality*. Cambridge, UK: Cambridge University Press.

Faure, A., Reynolds, S. M., Richard, J. M., & Berridge, K. C. (2008). Mesolimbic dopamine in desire and dread: Enabling motivation to be generated by localized glutamate disruptions in nucleus accumbens. *Journal of Neuroscience*, *28*(28), 7184–7192.

Foster, D. J., & Wilson, M. A. (2006). Reverse replay of behavioural sequences in hippocampal place cells during the awake state. *Nature*, *440*(7084), 680–683.

Frank, M. J., Seeberger, L. C., & O'Reilly, R. C. (2004). By carrot or by stick: Cognitive reinforcement learning in parkinsonism. *Science*, *306*(5703), 1940–1943.

Frank, M. J., & Claus, E. D. (2006). Anatomy of a decision: Striato-orbitofrontal interactions in reinforcement learning, decision making, and reversal. *Psychological Review*, *113*(2), 300–326.

Fudenberg, D., & Tirole, J. (1994). *Game theory*. Cambridge, MA: MIT Press.

Gläscher, J., Daw, N., Dayan, P., & O'Doherty, J. P. (2010). States versus rewards: Dissociable neural prediction error signals underlying model-based and model-free reinforcement learning. *Neuron*, *66*(4), 585–595.

Gupta, A. S., van der Meer, M. A. A., Touretzky, D. S., & Redish, A. D. (2010). Hippocampal replay is not a simple function of experience. *Neuron*, *65*(5), 695–705.

Hazy, T. E., Frank, M. J., & O'Reilly, R. C. (2007). Towards an executive without a homunculus: Computational models of the prefrontal cortex/basal ganglia system. *Philosophical Transactions of the Royal Society of London. Series B, Biological Sciences*, *362*(1485), 1601–1613.

Hershberger, W. A. (1986). An approach through the looking-glass. *Animal Learning & Behavior*, *14*, 443–451.

Hertwig, R., Barron, G., Weber, E. U., & Erev, I. (2004). Decisions from experience and the effect of rare events in risky choice. *Psychological Science*, *15*(8), 534–539.

Hertwig, R., & Erev, I. (2009). The description-experience gap in risky choice. *Trends in Cognitive Sciences*, *13*, 517–523.

Holland, P. C. (2004). Relations between Pavlovian-instrumental transfer and reinforcer devaluation. *Journal of Experimental Psychology. Animal Behavior Processes*, *30*(2), 104–117.

Huys, Q. J. M., & Dayan, P. (2009). A Bayesian formulation of behavioral control. *Cognition*, *113*, 314–328.

Kahneman, D. (2003). A perspective on judgment and choice: Mapping bounded rationality. *American Psychologist*, *58*(9), 697–720.

Kahneman, D., & Frederick, S. (2007). Frames and brains: Elicitation and control of response tendencies. *Trends in Cognitive Sciences*, *11*, 45–46.

Killcross, S., & Coutureau, E. (2003). Coordination of actions and habits in the medial prefrontal cortex of rats. *Cerebral Cortex*, *13*(4), 400–408.

Lengyel, M., & Dayan, P. (2007). Hippocampal contributions to control: A normative perspective. *Computational and Systems Neuroscience.*

Mackintosh, N. J. (1983). *Conditioning and associative learning.* Oxford, UK: Oxford University Press.

Montague, P. R., Dayan, P., & Sejnowski, T. J. (1996). A framework for mesencephalic dopamine systems based on predictive hebbian learning. *Journal of Neuroscience*, *16*(5), 1936–1947.

Morris, G., Nevet, A., Arkadir, D., Vaadia, E., & Bergman, H. (2006). Midbrain dopamine neurons encode decisions for future action. *Nature Neuroscience*, *9*(8), 1057–1063.

Morrison, S. E., & Salzman, C. D. (2010). Re-valuing the amygdala. *Current Opinion in Neurobiology*, *20*(2), 221–230.

Niv, Y. (2009). Reinforcement learning in the brain. *Journal of Mathematical Psychology*, *53*(3), 139–154.

Niv, Y., Daw, N. D., Joel, D., & Dayan, P. (2007). Tonic dopamine: Opportunity costs and the control of response vigor. *Psychopharmacology (Berl)*, *191*(3), 507–520.

O'Doherty, J. P. (2004). Reward representations and reward-related learning in the human brain: Insights from neuroimaging. *Current Opinion in Neurobiology*, *14*(6), 769–776.

O'Doherty, J. P. (2007). Lights, camembert, action! The role of human orbitofrontal cortex in encoding stimuli, rewards, and choices. *Annals of the New York Academy of Sciences, USA*, *1121*, 254–272.

Oudeyer, P., Kaplan, F., & Hafner, V. (2007). Intrinsic motivation systems for autonomous mental development. *IEEE Transactions on Evolutionary Computation*, *11*(2), 265–286.

Padoa-Schioppa, C. (2007). Orbitofrontal cortex and the computation of economic value. *Annals of the New York Academy of Sciences, USA*, *1121*, 232–253.

Peciña, S. (2008). Opioid reward "liking" and "wanting" in the nucleus accumbens. *Physiology & Behavior*, *94*(5), 675–680.

Peciña, S., & Berridge, K. C. (2005). Hedonic hot spot in nucleus accumbens shell: Where do mu-opioids cause increased hedonic impact of sweetness? *Journal of Neuroscience*, *25*(50), 11777–11786.

Puterman, M. L. (2005). *Markov decision processes: Discrete stochastic dynamic programming (Wiley series in probability and statistics).* Wiley-Interscience.

Rescorla, R. A., & Wagner, A. R. (1972). A theory of Pavlovian conditioning: Variations in the effectiveness of reinforcement and non-reinforcement. In A. H. Black & W. F. Prokasy (Eds.), *Classical conditioning II: Current theory and research* (pp. 64–99). New York, NY: Appleton-Century-Crofts.

Reynolds, S. M., & Berridge, K. C. (2001). Fear and feeding in the nucleus accumbens shell: Rostrocaudal segregation of GABA-elicited defensive behavior versus eating behavior. *Journal of Neuroscience*, *21*(9), 3261–3270.

Reynolds, S. M., & Berridge, K. C. (2002). Positive and negative motivation in nucleus accum-bens shell: Bivalent rostrocaudal gradients for GABA-elicited eating, taste "liking"/"disliking" reactions, place preference/avoidance, and fear. *Journal of Neuroscience*, *22*(16), 7308–7320.

Roesch, M. R., Calu, D. J., & Schoenbaum, G. (2007). Dopamine neurons encode the better option in rats deciding between differently delayed or sized rewards. *Nature Neuroscience*, *10*(12), 1615–1624.

Rummery, G., & Niranjan, M. (1994). *On-line Q-learning using connectionist systems.* Technical Report CUED/F-INFENG-TR 166. Cambrdige, UK: Cambridge University.

Samejima, K., Ueda, Y., Doya, K., & Kimura, M. (2005). Representation of action-specific reward values in the striatum. *Science*, *310*(5752), 1337–1340.

Samuel, A. (1959). Some studies in machine learning using the game of checkers. *IBM Journal of Research and Development, 3*, 210–229.

Schultz, W. (2002). Getting formal with dopamine and reward. *Neuron, 36*(2), 241–263.

Sheffield, F. (1965). Relation between classical conditioning and instrumental learning. In W. Prokasy (Ed.), *Classical conditioning* (pp. 302–322). New York, NY: Appleton-Century-Crofts.

Singh, S., Barto, A., & Chentanez, N. (2005). Intrinsically motivated reinforcement learning. In *Advances in Neural Information Processing Systems* (pp. 1281–1288). Cambridge, MA: MIT Press.

Singh, S., Lewis, R., & Barto, A. (2009). Where do rewards come from? In *Proceedings of the thirty-first Annual Conference of the Cognitive Science Society* (pp. 2601–2606). Amsterdam, The Netherlands.

Smith, Y., & Villalba, R. (2008). Striatal and extrastriatal dopamine in the basal ganglia: An overview of its anatomical organization in normal and parkinsonian brains. *Movement Disorders, 23*(Suppl 3), S534–S547.

Suddendorf, T., & Corballis, M. C. (1997). Mental time travel and the evolution of the human mind. *Genetic, Social, and General Psychology Monographs, 123*(2), 133–167.

Suri, R. E., & Schultz, W. (1999). A neural network model with dopamine-like reinforcement signal that learns a spatial delayed response task. *Neuroscience, 91*(3), 871–890.

Sutton, R. (1988). Learning to predict by the methods of temporal differences. *Machine Learning, 3*(1), 9–44.

Sutton, R. (1990). Integrated architectures for learning, planning, and reacting based on approximating dynamic programming. *Proceedings of the seventh international conference on machine learning, 216*: 224.

Sutton, R. S., & Barto, A. G. (1998). *Reinforcement learning: An introduction.* Adaptive Computation and Machine Learning. Cambridge, MA: The MIT Press.

Todorov, E. (2009). Efficient computation of optimal actions. *Proceedings of the National Academy of Sciences of the United States of America, 106*(28), 11478–11483.

Tolman, E. (1949). There is more than one kind of learning. *Psychological Review, 56*, 144–155.

Tricomi, E., Balleine, B. W., & O'Doherty, J. P. (2009). A specific role for posterior dorsolateral striatum in human habit learning. *The European Journal of Neuroscience, 29*(11), 2225–2232.

Tversky, A., & Thaler, R. H. (1990). Anomalies: Preference reversals. *Journal of Economic Perspective, 4*, 201–211.

Usher, M., & McClelland, J. L. (2001). The time course of perceptual choice: The leaky, competing accumulator model. *Psychological Review, 108*(3), 550–592.

Valentin, V. V., Dickinson, A., & O'Doherty, J. P. (2007). Determining the neural substrates of goal-directed learning in the human brain. *Journal of Neuroscience, 27*(15), 4019–4026.

Wallis, J. D., & Kennerley, S. W. (2010). Heterogeneous reward signals in prefrontal cortex. *Current Opinion in Neurobiology, 20*(2), 191–198.

Watkins, C. (1989). *Learning from delayed rewards.* PhD thesis, Cambridge, UK: University of Cambridge.

Weber, E. U., Shafir, S., & Blais, A.-R. (2004). Predicting risk sensitivity in humans and lower animals: Risk as variance or coefficient of variation. *Psychological Review, 111*(2), 430–445.

Widrow, B., & Hoff, M. (1960). Adaptive switching circuits. In *Western Electric Show and Convention Record* (Vol. 4, pp. 96–104). New York, NY.

Williams, D. R., & Williams, H. (1969). Auto-maintenance in the pigeon: Sustained pecking despite contingent non-reinforcement. *Journal of the Experimental Analysis of Behavior, 12*(4), 511–520.

Wittmann, B. C., Daw, N. D., Seymour, B., & Dolan, R. J. (2008). Striatal activity underlies novelty-based choice in humans. *Neuron, 58*(6), 967–973.

Predicting Emotional Reactions: Mechanisms, Bias and Choice

Tali Sharot

Welcome Trust Centre for Neuroimaging, University College London, London

INTRODUCTION

The fundamental question we address here is how humans, and other animals, make decisions. Traditional theories of choice behavior view individuals as rational decision-makers. According to these models, people make decisions based on an imperative that prescribes a maximization of reward (von Neumann & Morgenstern, 1944), such as money and nourishment, and a minimization of loss and pain (Fishburn, 1970). The well-known problem with this account is that it ignores the fact that humans often display behaviors that seem irrational, and exercise choice patterns that do not necessarily lead to an increase in gains or diminish losses (Kahneman & Tversky, 2000). For example, De Martino describes how people often reject an opportunity to play the lottery with a positive expected value, such as a lottery with an equal likelihood of winning $150 and losing $100 (De Martino, Camerer & Adolphs, 2010). People also make choices that seem to enhance, rather than reduce, pain. For example, when given a choice between receiving an electric shock immediately or a slightly smaller one after a delay, people often choose the immediate, more painful, shock (Bern, Chappelow, Cekic et al., 2006). These decisions do not align well with the normative prescription that an agent should always choose an option that will lead to an increased likelihood of reward and a decreased likelihood of pain. Why then do we reject opportunities to earn money and choose to suffer more rather than wait and suffer less?

One answer is that these behaviors emerge out of the operation of an emotional system. On this basis, if we understand how emotions impact on human judgment and choice, these seemingly irrational behaviors can be explained. For example, the reason people select a larger shock immediately rather than a smaller one later is that by doing so they minimize the unpleasant emotion of dread. By receiving the pain now we can avoid time spent in fear anticipating the shock, and so we choose to get it over and done with as soon as possible (Loewenstein, 1992). We reject a lottery with a positive expected value because we anticipate that if we lose $100 we will feel regretful. True, if we are lucky and win a larger sum we will be elated, but an anticipated aversive emotional reaction to loosing carries greater weight in our decision-making than the prospect of a positive emotional reaction to winning.

These examples illustrate a fundamental principle that guides everyday human behavior. Often, it is not present emotions felt in response to experienced outcomes that drive behavior; rather it is *anticipated* emotions to *future* outcomes that carry a greater impact on our actions (Loewenstein, 1987; for a review, see Baumeister, Vohs, De Wall & Zhang, 2007). I am not implying that experienced emotions do not shape actions. They do. If we are afraid we run away, if we are angry we retaliate, if we are happy we jump with joy. However, we spend a large amount of time and effort anticipating the things that will make us happy, angry or sad. It is these forecasts that considerably influence choices.

A key question is how we generate affective predictions. How do we foresee what will bring us joy and what will bring sorrow? This problem is relatively easy if what we are trying to predict is an event we have already experienced in the past. However, the problem becomes complex when the reaction we are trying to predict is in response to an event/stimuli/environment we had never experienced before and/or one that is highly volatile.

In this chapter, I begin by presenting a model for how humans can estimate future emotional reactions to complex novel scenarios, and provide behavioral, neuroimaging and pharmacological evidence in support. I then ask how accurate these predictions are, and show that humans exhibit systematic biases in their predictions that impacts on how accurately they assess what is likely to occur in the future. I conclude by offering a mechanistic explanation for why these biases exist and speculate on their likely adaptive function.

HOW WE PREDICT EMOTIONAL REACTIONS

Imagine you are the lucky winner of a photography contest held by Condé Nast Traveller magazine. Your prize is an all expenses paid vacation. You can choose between a week in Paris and a week in Venice. This is a tough decision. You have never been to either city. How do you decide which destination you prefer?

Perhaps the best way to determine preferences is to sample the options. To solve the "vacation dilemma" you may want to spend a day in Paris and a day in Venice before selecting one. In some instances it is indeed possible to try out the alternatives before making a decision; most ice cream parlors, for example, will let you taste different flavors prior to making a purchase. However, in many instances, sampling is not an option. Landlords will not allow sleepovers before you sign the lease, romantic partners will not wait around while you try out other options, and no magic carpet will waft you between Paris and Venice for a sneak preview.

How then, in the absence of prior experience, do we determine which alternative we are likely to prefer? One tool our brain uses to solve this problem is imagination. Indeed, there are compelling reasons to believe that the adaptive function of imagination is to allow mental simulation of a range of scenarios in order to provide a prediction of the pleasure or pain they are likely to entail. Below I describe how the brain creates images of possible future events and how it estimates reactions to those events.

How the Brain Creates Images of Possible Future Events

Brain imaging studies show that the neural substrates that allow us to construct images of the future overlap those engaged when we recollect the past. These regions include structures in medial temporal and prefrontal cortices (Addis, Wong & Schacter, 2007; Okuda, Fujii, Ohtake et al., 2003; Sharot, Riccardi, Raio & Phelps, 2007a; Szpunar, Watson & McDermott, 2007). Such findings are not surprising insofar as imaginative reconstruction of the future and recollection of the past are long believed to draw on similar information processes (Tulving, 2002). To imagine your upcoming trip to Venice, for example, you need a system that can take bits and pieces of past memories (your last vacation to a foreign country, images of canals, your partner smiling in the sunshine, picture postcards and movies of Venice) and flexibly bind them to create something new (you and your loved one on a boat ride in Venice next month), an event that has not yet occurred. It is suggested that retrieving and binding information relies on the medial temporal cortex, particularly the hippocampus (Slotnick, 2010).

Other evidence that the neural systems important for re-experiencing the past are critical for pre-experiencing the future comes from studies of amnesic patients, children, and non-human animals. Amnesic patients, with damage to the hippocampus and/or frontal regions, have impairments not only in recollection but also in imagination (Hassabis, Kumaran, Vann & Marguire, 2007; Tulving, 1985). These patients are unable to create detailed images of possible future occurrences or past events, providing compelling evidence that recollection and imagination share similar mechanisms. It is also known that children develop an ability to recollect episodes from their past around the same time they acquire an ability to simulate future events (Atance & O'Neill, 2004), indicating that common brain maturation processes are needed for both functions.

Although non-human animals are unlikely to create complex, detailed, simulation of future events, they too exhibit forms of "preplay." The mouse hippocampus, for example, shows activity sequences that are predictive of subsequent activity in environments never visited before (Dragoi & Tonegawa, 2011). These data have been interpreted as evidence that the mouse simulates the path it is about to take, prior to embarking on it. Such predictive simulation can be fundamental for action selection. In humans, predictive simulation, together with episodic memory, allows imagination of the complex scenarios people are likely to encounter.

Results such as the above have led to an argument that neural systems responsible for recollecting episodes from our past might not have been developed for that purpose at all. Instead, the core function of a system, that many had heretofore believed evolved for memory, may in fact serve a primary function of simulating the future (Schacter, Addis & Buckner, 2007). Mental simulation in this framework is in essence a learning mechanism that enables us to estimate likely outcomes associated with a yet to be experienced event. Such flexibility goes beyond what can be learned from past behavior alone, as it allows us to prepare for novel events and make decisions involving unfamiliar stimuli.

A Neural Signature that Predicts, and is Altered by, Choice

Predicting Choice

Who among us has not sought sanctuary from a tedious lecture by a mental simulation of our next trip abroad. These vivid images invariably produce emotional reactions, allowing us to "prefeel" the likely pleasures and pains future events are likely to engender (Gilbert, 2006; Gilbert & Wilson, 2007, 2009). These reactions are proxies for what it would be like to embark on the different options in real life, and thus provide a guide as to what to avoid and what to embrace, enabling us to choose the option associated with a greater hedonic value (Figure 3.1).

If our model is correct, we should be able to identify a neural signal that tracks estimates of future emotional reaction during imagination, a neural signal that will predict choice. To characterize this signal my colleagues and I conducted an fMRI study where subjects imagined vacationing in different destinations and provided ratings of how happy they would be if they actually vacationed in those locations (Sharot, De Martino & Dolan, 2009a). As expected, imagining the future engaged regions of medial temporal lobe and prefrontal cortex. However, the novel finding was that during imagination, activity in the bilateral caudate nucleus correlated with a subject's explicit expectations of future

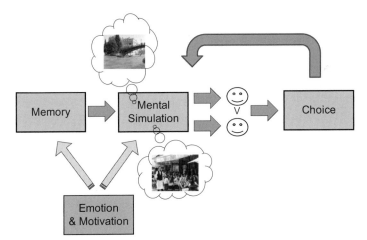

FIGURE 3.1 **Memories of past experiences are used to create mental simulations of novel events.** Both memories and simulations are biased, as they are susceptible to modulation by emotion and motivation. During simulation, emotional reactions are generated, which are used as a basis for estimating the value of future states. These are then compared to determine choice. In a feedback loop, choice alters simulations and estimated reactions, enhancing commitment to the chosen option.

pleasure. This signal specifically tracked estimation of hedonic reaction rather than other features of the stimuli, such as vividness, familiarity with location, emotional arousal or previous visits to these destinations.

This finding is consistent with the documented role of the caudate nucleus in reward processing, reward expectation and reward learning (Delgado, 2007). For example, increased activity in the caudate is associated with anticipation of both primary (Gottfried, O'Doherty & Dolan, 2003) and monetary (Knutson, Adams, Fong & Hommer, 2001) rewards. According to our model this signal should predict the choices subjects will make at a later time. Indeed, we observed greater activity in the caudate nucleus when participants imagined vacationing in destinations they later selected (relative to those they later rejected). This difference was observed before subjects knew they would be required to make a later decision, and predicted choice even when selecting between destinations that were previously rated equally. In other words, the caudate nucleus coded hedonic preference with a greater sensitivity than the rating scale used in the study. If a subject rated both Thailand and Greece as a "5" on a scale from 1 (will be unhappy to vacation in this destination) to 6 (will be extremely happy to vacation in this destination), but later selected "Thailand" when forced to make a decision between the two, caudate activity was found to be greater when the subject initially imagined vacationing in Thailand than when imagining Greece. This suggests that decisions are predetermined by differences in expected hedonic outcome as reflected in caudate activity during imagination.

Altered by Choice

In Chapter 6, Johansson and colleagues discuss how choices not only reflect expectancies of hedonic outcome, but also alter such expectations (see also Ariely & Norton, 2008). A classic observation here is that, after making a difficult choice between two equally valued options, such as between two vacation destinations, we tend to value the selected alternative more strongly and the discarded one less so (Brehm, 1956; Egan, Santos & Bloom, 2007; for a review, see Harmon-Jones & Mills, 1999). This phenomenon is most commonly explained in terms of a need to reduce cognitive dissonance. According to cognitive dissonance theory, having to make a choice between two equally desirable alternatives engenders psychological discomfort. This is because the decision conflicts with the desirable aspects of the rejected alternative, and with the undesirable aspects of the selected alternative (Festinger, 1957). Such psychological tension is reduced by re-evaluating the options post-choice.

Although it is possible that post-choice re-evaluation is driven by the need to reduce cognitive dissonance (Festinger, 1957), other processes are likely to play a role. According to self perception theory people infer their preferences by observing their own choices and update their explicit ratings accordingly (Bem, 1967). It is also feasible that re-evaluation occurs because envisioning competing possibilities provides a new context in which the stimuli are assessed (Sharot et al., 2009a). Specifically, making a choice highlights the unique aspects of the two alternatives (Houston, Sherman & Baker, 1991; Tversky, 1972). Thus, the act of choosing may change our preferences by providing new weightings to aspects of the stimuli that may not have been considered beforehand.

In any case, if making a decision modifies predictions of future hedonic outcomes, we can hypothesize that following choice caudate activity tracking these expectancies will also be altered. Indeed, after making a decision activity in the caudate nucleus increases when subjects imagine vacationing in destinations they had selected, and decreases when they imagine vacationing in destinations they rejected (Sharot et al., 2009a). These changes parallel those evident in subjects' behavioral ratings. Thus, after choosing Paris over Venice, for example, participants rate Paris higher than they did before the decision, and rate Venice as worse.

The conclusion from these findings is that the physiological representation of a stimulus' expected hedonic value is altered by a commitment to it. This idea has since been supported by other studies. Izuma, Matsumoto, Murayama et al. (2010) reported a post-decision decrease in caudate nucleus activity when subjects thought about food items they had rejected a few moments ago, a decrease accompanied by an explicit devaluation of the item. It is interesting to speculate that in the absence of a rapid update of value that concurs with choice, we are likely to second-guess our decisions and actions, and this may promote negative affect and interfere with our daily function.

Choice-induced revaluation is not constrained to positive stimuli. In a study where subjects were required to select between aversive events (such as medical conditions) a modulation in caudate nucleus activity, contingent on choice, was observed when subjects imagined the negative events (Sharot, Shiner & Dolan, 2010a). Relative to pre-choice levels, caudate activity was enhanced when subjects imagined having a medical condition they had selected (hypothetically, of course), and decreased while imagining a medical condition they rejected.

One mechanistic suggestion is that a value signal in the caudate nucleus is modulated by anterior cingulate cortex (ACC) to align with the decision made. Izuma et al. (2010) showed that ACC activity conveyed the degree of discrepancy between a subject's past preference and his/her choice. In particular, ACC activity was greater when a subject rejected a highly preferred item, and when selecting a low preferred item. This is consistent with a view that the ACC signals levels of cognitive dissonance (see also van Veen, Krug, Schooler & Carter, 2009). Furthermore, we have reported that choice-induced changes in ACC activity while imaging negative events mirrored those observed in the caudate nucleus. Importantly, the functional connectivity between these two regions increased after subjects completed the decision-making task relative to before (Sharot et al., 2010a). Although these results do not imply directionality, they accord with the notion that ACC modulates a value signal in the caudate after a decision is made. The extent of the modulation may be contingent on the degree of cognitive dissonance that is generated by the act of making a choice.

The Role of Dopamine

The caudate nucleus is one of the major targets for the brain's dopaminergic system (Wise & Rompre, 1989). Dopamine is a principal neuromodulator for reward learning and reward-seeking behaviour (Bayer & Glimcher, 2005; Belin & Everitt, 2008; Schultz, 2001). In particular, dopamine is suggested to signal predictions of rewards, and errors in such prediction – providing a learning signal when predictions do not align with outcomes (Montague, Hymann & Cohen, 2004; Schultz, 2001, 2007). Evidence in support of this comes from single cell recording of midbrain dopaminergic neurons in non-human primates (Schultz, 2001). In humans, drugs enhancing dopaminergic function (e.g., dihydroxy-L-phenylalanine; L-DOPA) augment a striatal signal that expresses reward prediction errors, thereby increasing the likelihood of choosing stimuli associated with greater monetary gains (Pessiglione, Seymour, Flandin et al., 2006).

If dopamine modulates reward prediction, then its enhancement during imagination of future events should impact on subjective estimations of future pleasure to be derived from those events. To test this hypothesis we administered L-DOPA, which increases dopaminergic function, while subjects imagine vacationing in distinct sets of destinations (Sharot, Shiner, Brown et al., 2009b). Subjects' expectations of pleasure from vacationing in these destinations were recorded both before and after the pharmacological manipulation. We showed that subjects rated destinations higher after imagining them under L-DOPA. The effect was long lasting in that an enhancement in estimations of pleasure was observed 24 hours after the administration of L-DOPA, by which time the drug had been fully metabolized and eliminated. Importantly, changes in estimated pleasure were not observed when subjects imagined the destinations under placebo, but were specific to instances when events were imagined under L-DOPA.

The precise mechanisms through which dopamine alters predictions of future hedonic reaction during imagination remain unclear.

The possibility that dopaminergic function increases the experience of pleasure during simulation of a future event by directly altering its hedonic impact seems unlikely, as a large body of evidence contradicts the notion that dopamine acts as a pleasure neurotransmitter (see Berridge, 2007 and Berridge, Robinson & Aldridge, 2009 for review). A more likely explanation is that by affecting learning mechanisms, L-DOPA strengthens the association between the stimuli (e.g., "Venice") and the brief hedonic reaction generated during imagination of the pleasurable event. Dopamine may also strengthen, via hippocampus-dependent loops, a memory trace of pleasant associations made during the imagination stage, or alternatively enhance the incentive salience attributed to imagined stimuli, consistent with suggestions that dopamine increases "wanting" (Berridge, 2007; Berridge et al., 2009). Alternatively, increasing dopamine function may directly affect the signal that mediates predictive information during an act of imagination. If this latter mechanism is correct then one would expect a signal in the caudate that tracks expected pleasure during imagination (Sharot et al., 2009a) to be increased after the administration of L-DOPA. This prediction has yet to be tested.

In summary, to estimate emotional reactions to complex novel future events, humans are endowed with an ability to mentally simulate those events (Gilbert, 2006). Creating such detailed images of likely future occurrences relies on similar neural systems to those that are engaged during recollection, including the hippocampal complex and prefrontal cortex (Addis et al., 2007; Tulving, 1985). Once these images are formed they generate emotional reactions, which can be used as proxies for future reactions (Gilbert & Wilson, 2007). These estimations are tracked within caudate nucleus, as well as other parts of the striatum, and predict the choices we are likely to make at a later time (Sharot et al., 2009b). Enhancing dopaminergic function during simulation alters estimations of future hedonic reactions, suggesting a role for dopamine in predicting hedonic outcome (Sharot et al., 2009b). A feedback loop, once a decision had been taken, alters people's estimation of future reaction and its associated neural signal (Sharot et al., 2009a; Sharot, Velasquez & Dolan, 2010a).

BIASES IN PREDICTING EMOTIONAL REACTIONS

The evolution of predictive simulation through which the brain forecasts the emotional consequences of possible future events attests to the clever hand of nature, but the question remains as to the reliability of these estimates. The accuracy of our predictions relies predominantly on two factors. Firstly, the validity of our expected reaction to an event depends on our ability to reliably simulate that event (Gilbert & Wilson,

2007, 2009). If our simulation is different from what transpires in reality, the emotional reaction to the simulation will be different from our experienced emotional reaction to the reality of the event itself. For example, we might imagine ourselves cruising the canals of Venice in brilliant sunshine, a simulation that will engender a positive emotion. We thus predict that a vacation in Venice will make us happy. However, if we arrive in Venice in fog and rain, this is likely to make us feel unpleasant emotion. Thus, our emotional reaction will be different from what we anticipated simply because we did not foresee the fog. Secondly, even if we had perfectly simulated the event, our predicted reaction to it will be accurate only if our reaction to the simulation aligns with our reaction to the experienced reality.

Below I consider each of these elements in turn. I will first address the question of how accurate human simulations are, and then consider potential obstacles that prevent us from predicting our reactions to them, even when events are accurately simulated.

Why Mental Simulations are Biased

In 1989, a year before Saddam Hussein invaded Kuwait, project "Operation Internal Look" – a computer program that can simulate war in the Gulf area was developed by the US government (Kelly, 1995). To effectively simulate scenarios, which may unfold during armed conflict in the region, a large amount of data had to be compiled. The exact terrain of the area, and its climate, was fed into the program along with detailed data on number of troops, types of weaponry, aircraft and tanks available to each side. Countless other details were gathered from different sources. By the date of invasion (3 August 1990), the program was ready to run an infinite number of "what-if" scenarios. Alternative strategies tried out on this powerful computer network spit out likely number of casualties, monetary costs and length of battle. By running numerous simulations, with different iterations, US commanders learned that the critical factor for winning the war was airpower, leading to a conclusion that conflict would be brief and lead to minimum casualties if air superiority was established. The predicted outcome generated by Operation Internal Look gave the US commanders the confidence needed to launch a heavy air campaign (Kelly, 1995). Although in reality differences emerged between what was forecast and what transpired, by and large the war unfolded as predicted.

Operation Internal Look was successful at simulating the Gulf War for a number of reasons. Firstly, the program had access to vast amounts of data. Secondly, it had the capacity to consider all these variables simultaneously and account for their interactions. Finally, the program had the ability to run thousands of different scenarios in a relatively short amount of time. This enabled the program to generate statistics of the likely outcomes that would result from various actions the commanders may take.

As the future is inherently uncertain, neither a computer program nor a biological organism can predict it with precision. Yet, if the human brain were to simulate future scenarios in a manner similar to Operation Internal Look, people's estimations of possible outcomes would probably be more precise. The problem is that when simulating the future people consider only a selective set of data, give unrepresentative weights to certain variables and run a restricted number of different scenarios. Importantly, the scenarios we do come up with do not necessarily resemble the most frequently experienced variation of the event. Take, for example, the response of a subject in a study we conducted (Sharot et al., 2007a) who, when asked to imagine herself getting a haircut a few years from now, said: *"I projected that I was getting my hair cut to donate to Locks of Love. It had taken me years to grow it out and my friends were all there to help celebrate. We went to my favorite hair place in Brooklyn and then went to lunch at our favorite restaurant."* This is hardly your typical haircut scenario.

Now imagine we were to build a computer program that simulates future haircuts. We will feed it with information regarding all past haircuts, haircuts experienced by others, the latest developments in the shampoo and hairstyling industry and so on. We then let it run to simulate all possible scenarios of future haircuts. Indeed, in one scenario the simulator may come up with what is described above. In which case, we anticipate a future haircut to be extremely exciting. However, the most frequent scenario the program is likely to generate will be a variation of this one: *"go to hairdresser, get hair shampooed, read magazine and have coffee while hair dresser cuts away, get hair dried, check yourself in the mirror, pay and leave."* In which case, we will not predict an intense emotional reaction in response.

As humans do not create thousands of mental simulations to generate statistics of possible outcomes, significant inaccuracy in prediction is to be expected. What is striking about our mis-predictions, however, is that they are not random. To the contrary, people mis-predict future outcomes in a predictable fashion. As I describe below, this is because the scenarios humans choose to simulate are inherently biased. Our images are unrepresentative because: (a) they are based on biased memories, and (b) they are susceptible to strong motivational modulation.

Biased Memories

When predicting future outcomes, people rely on memories of past outcomes to similar events. Although in theory this can be an effective strategy, our memories are not an accurate representation of the past.

In particular, we remember highly emotional events more vividly than neutral events (Neisser & Harsch, 1992; Phelps & Sharot, 2008; Sharot, Verfaellie & Yonelinas, 2007b; Sharot, Martorella, Delgado & Phelps, 2007c; Sharot & Yonelinas, 2008; Talarico & Rubin, 2003). This enhancement has been shown to be mediated by the amygdala (Richardson, Strange & Dolan, 2004; Sharot, Delgado & Phelps, 2004). The amygdala is a small almond shaped structure in the anterior medial temporal lobe that is important in processing emotion and modulating perception, evaluation and memory (for a review, see Phelps, 2006). It is adjacent to the hippocampus, which places it in a strategic location to alter the consolidation or storage of memories for arousing events (McGaugh, 1992; LeDoux, 1996), so that they are more likely to be retained over time and easily retrieved (Cahill, Babinsky, Markowitsch & McGaugh, 1995; Sharot et al., 2007c; for a review, see Phelps, 2006). Having extremely vivid memories of past emotional experiences, and only weak memories of past everyday events, means we maintain a biased perception of our past. We tend to recall the past as a concentrated timeline of emotionally exciting events. This perception of the past can bias our perception of the future, because we ultimately rely on our memories to construct simulations.

For example, when asked to imagine a subway ride we retrieve episodes of past subway rides. However, it has been shown that the past rides we recall are not the most common ones (Morewedge, Gilbert & Wilson, 2005). The uneventful rides have been forgotten, but unique, emotionally arousing, subway rides are retrieved and incorporated when we imagine future rides. As a result future rides are predicted to be more eventful and exciting than the average subway ride (Morewedge et al., 2005).

Motivational Modulation

The human brain is disposed to adopt the most rewarding (or least aversive) perspective on situations. In the present, our desires and preferences bias visual perception in a manner that best fits our goals (Balcetis & Dunning, 2007). For example, thirsty participants are more likely to perceive transparency, a characteristic of water, in an ambiguous visual stimulus in comparison with hydrated participants (Changizi & Hall, 2001). When recalling the past, we tend to remember positive self-relevant events with more details than negative self-relevant events (D'Argembeau, Xue, Lu et al., 2008). As the future is less constrained than either the past or present, the way we imagine it is strongly influenced by the overwhelming motivation to realize positive states and avoid negative ones.

Consistent with this notion, we find that people mentally approach positive future events and distance themselves from negative ones (Sharot et al., 2007a). Specifically, humans tend to imagine positive future events in more detail and closer in time than negative events. They also feel that when imaging positive events they are truly "pre-experiencing" the event, but less so when imaging negative events.

These differences are mediated by enhanced activation in the amygdala and ACC, and enhanced functional connectivity between the two regions, when subjects imagine positive future events relative to negative events (Sharot et al., 2007a). Greater amygdala activation may be related to the heightened arousal reported by subjects when imagining positive events relative to negative events and/or to greater recollection/vividness of emotional details. The involvement of the amygdala in imaging future events seems specific to emotional events, and has not been observed when imagining neutral events (Okuda et al., 2003; Addis et al., 2007; Szpunar et al., 2007). It has been suggested that activity in the amygdala, and other areas involved in emotion and motivation, is modulated by the ACC (Cunningham, Raye & Johnson, 2005). During imagination of future events, enhanced ACC activation may reflect a self-regulatory focus that underlies a bias in attention and vigilance towards positive future events and away from negative ones.

The inclination to imagine positive events in more detail, and closer in time, has been directly related to a subject's level of trait optimism (Sharot et al., 2007a). Optimism is the tendency to overestimate the likelihood of positive future outcomes and underestimate the likelihood of negative outcomes (Weinstein, 1980, for review see Sharot, 2011). It has been suggested that people determine the likelihood of an event based on how easy it is to picture mentally (Kahneman & Tversky, 1982). Thus, a bias in how people simulate positive and negative future events may result in a tendency to assign unrealistically optimistic probabilities to future outcomes. The reverse causation, however, may also be true.

Not only are mental simulations of positive events qualitatively different from negative events, people also tend to reconstruct neutral, and even aversive events, in a positive light during simulation. When people consider future adverse situations they often imagine themselves getting out of the muddle. Consider, for example, the response of a subject from our study (Sharot et al., 2007a) who was asked to describe what it would be like to lose her apartment keys: *"Getting locked out is always very annoying, but I always have a spare key somewhere and/or with someone (a roommate). Even though I have never owned an apartment I assume the landlord has a key, so I envisioned myself going downstairs and asking for a spare."* Clearly, imagining how things may go wrong (being locked out of your apartment) can aid in identifying actions (giving a spare key to your neighbor) that will help in avoiding adverse situations.

To conclude, emotion and motivation modulate mental simulation, such that future occurrences are imagined as: (a) more exciting, less mundane, than average occurrences, and (b) positive events are imagined with a greater sense of "pre-experiencing" than negative events. These two biases alter expected outcome in the following manner; people over-predict the intensity of future emotional reactions (this is known as the "impact bias" – Wilson & Gilbert, 2003), because predictions are based on imagined events, and those are more unique and arousing than real events: people overestimating the likelihood of positive events (this is known as the "optimism bias" – Weinstein, 1980; Sharot, 2011), because positive events are imagined with greater detail, enhanced sense of "pre-experiencing," and closer in time than negative events (Sharot et al., 2007a).

For positive events, the two biases go hand in hand; we predict desirable outcomes as more likely than they really are (optimism bias), and overestimate the emotional benefit to be derived from them (impact bias). On the surface, for negative events, the two contradict. An important distinction between the two phenomena, however, allows them to co-exist. According to the optimism bias we will underestimate the likelihood of negative events. Thus, we will assume it unlikely that we will become ill, unemployed, or divorced (Weinstein, 1980). However, in the unlikely event that we were to become severely ill or get a divorce, we imagine feeling worse than we actually end up feeling (Albrecht & Devlieger, 1999).

Why Reactions to Simulations are Different from Reactions to Real Outcomes

Even if imagination was to faithfully represent the event it was simulating, our predicted emotional reaction to it would not necessarily be an accurate estimate of our real reaction. Several mechanisms are critical in producing this discrepancy. They include reduced emotional adaptation (Gilbert & Wilson, 2007, 2009) and reduced positive re-evaluation (Sharot, 2010a; Sharot, Velasquez & Dolan, 2010b), during simulation relative to real experiences, as well as a mismatch between one's motivational state during simulation and real experience (Gilbert & Wilson, 2007, 2009).

Adaptation and Re-evaluation

In accord with the impact bias, when people predict how they would feel if they were to suffer from different medical conditions, they estimate feeling worse than patients who actually suffer from those conditions report feeling (Albrecht & Devlieger, 1999; Ubel, Loewenstein, Schwarz & Smith, 2005a). One reason for this discrepancy is that in reality we adapt to novel situations, whether good or bad (Brickman, Coates & Janoff-Bulman, 1978). Our emotional reactions are dampened with time as we adjust to our loss or fortune (Diener & Biswas-Diener, 2002). This dampening, however, is not experienced during brief simulations, and therefore not predicted. Thus, when we imagine becoming paraplegic, for example, we experience an intense negative reaction (Gilbert, Pine, Wilson et al., 1998). If we were to become paraplegic we may indeed experience this reaction at first, but over time we adapt to our disability and our well-being returns to previous levels (Brickman et al., 1978).

Adaptation to aversive events is partially due to a positive re-evaluation of the event. Once a negative situation becomes reality people tend to engage in re-evaluation of the event in an attempt to seek the silver lining (Sharot et al., 2010a). For example, after being fired one may rationalize that unemployment offers an opportunity to learn new skills. Although some positive re-evaluation may take place during mental simulation, it is not as profound as in real life. Thus, estimated negative reactions to future aversive events are worse than experienced reactions. Consistent with this account, we have shown that aversive events experienced in the past are predicted to generate less of a negative reaction if experienced again in the future than never experienced events (Sharot et al., 2010a). This interaction, between past experience and a decrease in expectations of aversive outcome, is expressed in the striatum (both the caudate and putamen) and the ACC (Sharot et al., 2010a). It is speculated that the ACC modulates activity in the striatum – such that negative expectations of aversive outcomes are decreased once they are encountered.

An overvaluation of predicted emotional reaction is adaptive insofar as it motivates us to shun hardship and avoid unnecessary danger. However, once adversity becomes reality, an overly negative valuation may no longer be in our interest, and thus we perceive such circumstance as less negative than before (Ubel et al., 2005a; Ubel, Loewenstein & Jepson, 2005b).

Mismatch of Motivational States

Our emotional reaction to a simulated event will be dissimilar from our reaction to a real event if our motivational state is different in both points in time. For example, if we think about tonight's dinner when we are hungry, a positive reaction is likely to ensue and thus we will anticipate enjoyment. If, however, we just had a full breakfast we will not find imagining dinner as enjoyable and may mistakenly under-predict our positive reaction to the meal (Gilbert et al., 2002). Thus, whether we are satiated, cold, tired, energetic, or cheerful, each will impact on our expected reaction to a future outcome.

It is important to note, however, that although a mismatch of motivational states can cause mis-predictions, humans are able to take into account future motivational states when making decisions. We go grocery shopping, even if we had recently eaten a large meal, because we can anticipate being hungry in a few hours and are aware that the fridge is empty (although we might purchase less than we eventually need). This ability may be absent in our less sophisticated evolutionary ancestors. When given food, non-human primates eat until they are full and trash the rest. They do so even if they are fed only once a day and will certainly be hungry in a few hours. When they are allowed to choose different amount of foods (for example 2, 4, 8, 10, or 20 dates) they do not always choose the largest amount, but will often pick the amount they can eat there and then. This suggests that non-human primates cannot anticipate a future motivational state that is different from their present state (McKenzie, Cherman, Bird et al., 2005).

CONCLUSION

Although not always accurate, humans are able to generate estimates of emotional reactions to future occurrences that drive their decisions. Predicting our reactions to complex, novel, events involves a "tête-à-tête" between the frontal cortex and subcortical regions that process emotion and value. These subcortical structures, particularly the striatum, convey the expected pleasure and pain from a simulated event. This signal reflects a later choice when an individual is faced with selecting between different options. The frontal regions are involved not only in creating simulations (together with medial temporal lobe structures), but also in modulating the signal that conveys expectations of emotional reaction. These alterations are driven by motivations and goals, such as the need to increase commitment to a chosen action and to adopt the most rewarding perspective of reality.

Our predictions of future emotional reactions are thus biased partially because emotion and motivation modulate the processes involved in their generation. Emotion enhances the vividness of memories, which are the ingredients of future projections, resulting in simulations that are more exciting than real life and consequently in over-estimation of our reactions. Motivation biases imagination such that simulations of positive events are especially compelling, which is partially why humans tend to be unrealistically optimistic. A moderate optimistic illusion can motivate behaviour in the present towards a future goal, and has been related to mental and physical health (Taylor & Brown, 1988). In fact, realistic expectations of future life events have been associated with mild depression (Strunk, Lopez & DeRubeis, 2006), suggesting that biased expectations are a good thing and can serve to keep our minds healthy and at ease.

Acknowledgments

Thanks to Sara Bengtsson, Benedetto De Martino, Ray Dolan and Steve Fleming for helpful comments on a previous version of the manuscript. This work was funded by the British Academy and the Wellcome Trust.

References

Addis, D. R., Wong, A. T., & Schacter, D. L. (2007). Remembering the past and imagining the future: Common and distinct neural substrates during event construction and elaboration. *Neuropsychologia, 45,* 1363–1377.

Albrecht, G. L., & Devlieger, P. J. (1999). The disability paradox: High quality of life against all odds. *Social Science & Medicine, 48*(8), 977–988.

Ariel, D., & Norton, M. I. (2008). How actions create – not just reveal – preferences. *Trends in Cognitive Sciences, 12,* 13–16.

Atance, C. M., & O'Neill, D. K. (2004). Acting and planning on the basis on a false belief: Its effects on 3-year-old children's reasoning about their own false beliefs. *Developmental Psychology, 6,* 953–964.

Balcetis, E., & Dunning, D. (2007). Cognitive dissonance and the perception of natural environments. *Psychological Science, 10,* 917–921.

Baumeister, R. F., Vohs, K. D., DeWall, C. N., & Zhang, L. (2007). How emotion shapes behavior: Feedback, anticipation, and reflection, rather than direct causation. *Personality and Social Psychology Review, 11*(2), 167–203.

Bayer, H. M., & Glimcher, P. W. (2005). Midbrain dopamine neurons encode a quantitative reward prediction error signal. *Neuron, 47,* 129–141.

Belin, D., & Everitt, B. J. (2008). Cocaine seeking habits depend upon dopamine-dependent serial connectivity linking the ventral with the dorsal striatum. *Neuron, 57,* 432–441.

Bem, D. J. (1967). Self-perception: An alternative interpretation of cognitive dissonance phenomena. *Psychological Review, 74,* 183–200.

Berns, G. S., Chappelow, J., Cekic, M., Zink, C. F., Pagnoni, G., & Martin-Skurski, M. E. (2006). Neurobiological substrates of dread. *Science, 312,* 704–706.

Berridge, K. C. (2007). The debate over dopamine's role in reward: The case for incentive salience. *Psychopharmacology, 191,* 391–431.

Berridge, K. C., Robinson, T. E., & Aldridge, J. W. (2009). Dissecting components of reward: "liking", "wanting" and learning. *Current Opinion in Pharmacology, 1,* 65–73.

Brehm, J. W. (1956). Post-decision changes in the desirability of choice alternatives. *Journal of Abnormal Psychology, 52,* 384–389.

Brickman, P., Coates, D., & Janoff-Bulman, R. (1978). Lottery winners and accident victims: Is happiness relative? *Journal of Personality and Social Psychology, 36,* 917–927.

Cahill, L., Babinsky, R., Markowitsch, H. J., & McGaugh, J. L. (1995). The amygdala and emotional memory. *Nature, 377,* 295–296.

Changizi, M. A., & Hall, W. G. (2001). Thirst modulates a perception. *Perception, 30,* 1489–1497.

Cunningham, W. A., Raye, C. L., & Johnson, M. K. (2005). Neural correlates of evaluation associated with promotion and prevention regulatory focus. *Cognitive, Affective & Behavioral Neuroscience, 5,* 202–211.

D'Argembeau, A., Xue, G., Lu, Z. L., Van der Linden, M., & Bechara, A. (2008). Neural correlates of envisioning emotional events in the near and far future. *NeuroImage, 40*(1), 398–407.

Delgado, M. R. (2007). Reward-related responses in the human striatum. *Annals of the New York Academy of Sciences, 1104,* 70–88.

De Martino, B., Camerer, C. F., & Adolphs, R. (2010). Amygdala damage eliminates monetary loss aversion. *Proceedings of the National Academy of Sciences of the United States of America, 107*(8), 3788–3792.

Diener, E., & Biswas-Diener, R. (2002). Will money increase subjective well-being? *Social Indicators Research, 57*(2), 119–169.

Dragoi, G., & Tonegawa, S. (2011). Preplay of future place cell sequences by hippocampal cellular assemblies. *Nature, 469*, 397–401.

Egan, L. C., Santos, L. R., & Bloom, P. (2007). The origins of cognitive dissonance: Evidence from children and monkeys. *Psychological Science, 11*, 978–983.

Festinger, L. (1957). *A theory of cognitive dissonance*. Stanford, CA: Stanford University Press.

Fishburn, P. C. (1970). *Utility theory for decision-making*. New York: Wiley.

Gilbert, D. (2006). *Stumbling on happiness*. New York: Knopf.

Gilbert, D. T., Driver-Linn, E., & Wilson, T. D. (2002). The trouble with Vronsky: Impact bias in the forecasting of future affective states. In L. F. Barrett & P. Salovey (Eds.), *The wisdom in feeling: Psychological processes in emotional intelligence* (pp. 114–143). New York: Guilford.

Gilbert, D. T., Pine, E. C., Wilson, T. D., Blumberg, S. J., & Wheatley, T. P. (1998). Immune neglect: A source of durability bias in affective forecasting. *Journal of Personality and Social Psychology, 75*, 617–638.

Gilbert, D. T., & Wilson, T. D. (2007). Prospection: Experiencing the future. *Science, 317*, 1351–1354.

Gilbert, D. T., & Wilson, T. D. (2009). Why the brain talks to itself: Sources of error in emotional prediction. *Philosophical Transactions of the Royal Society B, 364*, 1335–1341.

Gottfried, J. A., O'Doherty, J. P., & Dolan, R. J. (2003). Encoding predictive reward value in human amygdala and orbitofrontal cortex. *Science, 301*, 1104–1107.

Harmon-Jones, E., & Mills, J. (1999). *Cognitive dissonance progress on a pivotal theory in social psychology*. Washington, DC: Braun-Brumfield, Inc.

Hassabis, D., Kumaran, D., Vann, S. D., & Marguire, E. A. (2007). Patients with hippocampal amnesia cannot imagine new experiences. *Proceedings of the National Academy of Sciences USA, 104*, 1726–1731.

Houston, D. A., Sherman, S. J., & Baker, S. M. (1991). Feature matching, unique features, and the dynamics of the choice process: Predecision conflict and postdecision satisfaction. *Journal of Experimental Social Psychology, 27*, 411–414.

Izuma, K., Matsumoto, M., Murayama, K., Samejima, K., Sadato, N., & Matsumoto, K. (2010). Neural correlates of cognitive dissonance and choice-induced preference change. *Proceedings of the National Academy of Sciences of the United States of America, 107*(51), 22014–22019.

Kahneman, D., & Tversky, A. (1982). On the study of statistical intuitions. *Cognition, 11*(2), 123–141.

Kahneman, D., & Tversky, A. (2000). *Choices, values, and frames*. New York: Cambridge University Press.

Kelly, K. (1995). *Out of control: The new biology of machines, social systems, and the economic world*. Basic Books.

Knutson, B., Adams, C. M., Fong, G. W., & Hommer, D. (2001). Anticipation of increasing monetary reward selectively recruits nucleus accumbens. *The Journal of Neuroscience, 21*, RC159.

LeDoux, J. (1996). *The emotional brain*. New York: Simon and Schuster.

Loewenstein, G. (1987). Anticipation and the valuation of delayed consumption. *Economic Journal, 97*, 666–684.

Loewenstein, G. (1992). *Choice over time* (1st ed.). Russell Sage Foundation Publications.

Mantague, P. R., Hyman, S. E., & Cohen, J. D. (2004). Computational roles for dopamine in behavioural control. *Nature, 431*, 760–767.

McGaugh, J. L. (1992). Affect, neuromodulatory systems and memory storage. In S. Christianson (Ed.), *The handbook of emotion and memory: Research and theory* (pp. 269–288). Hillsdale, NJ: Erlbaum.

McKenzie, T. L. B., Cherman, T., Bird, L. R., Naqshbandi, M., & Roberts, W. A. (2005). Can squirrel monkeys (*Saimiri sciureus*) plan for the future? Studies of temporal myopia in food choice. *Learning & Behavior*: A Psychonomic Society Publication 32, no. 4 (November 2004): 377–390.

Morewedge, C. K., Gilbert, D. T., & Wilson, T. D. (2005). The least likely of times: How remembering the past biases forecasts of the future. *Psychological Science, 16*(8), 626–630.

Neisser, U., & Harsch, N. (1992). Phantom flashbulbs: False recollections of hearing the news about Challenger. In E. Winograd & U. Neisser (Eds.), *Affect and accuracy in recall: Studies of "flashbulb" memories* (pp. 9–31). New York: Cambridge University Press.

Okuda, J., Fujii, T., Ohtake, H., Tsukiura, T., Tanji, K., Suzuki, K., et al. (2003). Thinking of the past and past: The roles of the frontal pole and the medial temporal lobe. *NeuroImage, 19,* 1369–1380.

Pessiglione, M., Seymour, B., Flandin, G., Dolan, R. J., & Frith, C. D. (2006). Dopamine-dependent prediction errors underpin reward-seeking behaviour in humans. *Nature, 442,* 1042–1045.

Phelps, E. A. (2006). Emotion and cognition: Insights from studies of the human amygdala. *Annual Review of Psychology, 57,* 27–53.

Phelps, E. A., & Sharot, T. (2008). How (and why) emotion enhances the subjective sense of recollection. *Current Directions in Psychological Science, 17*(2), 147–152.

Richardson, M. P., Strange, B. A., & Dolan, R. J. (2004). Encoding of emotional memories depends on amygdala and hippocampus and their interactions. *Nature Neuroscience, 7*(3), 278–285. Epub 2004 Feb 1.

Schacter, D. L., Addis, D. R., & Buckner, R. L. (2007). Remembering the past to imagine the future: The prospective brain. *Nature Reviews Neuroscience, 9,* 657–661. Review.

Schultz, W. (2001). Reward signaling by dopamine neurons. *The Neuroscientist, 7,* 293–302.

Schultz, W. (2007). Behavioral dopamine signals. *Trends in Neurosciences, 30,* 203–210.

Sharot, T., Delgado, M. R., & Phelps, E. A. (2004). How emotion enhances the feeling of remembering. *Nature Neuroscience, 7*(12), 1376–1380.

Sharot, T., Riccardi, M. A., Raio, C. M., & Phelps, E. A. (2007). Neural mechanisms mediating optimism bias. *Nature, 450*(7166), 102–105.

Sharot, T., Verfaellie, M., & Yonelinas, A. P. (2007). How emotion strengthens the recollective experience: A time-dependent hippocampal process. *PLoS One, 2*(10), e1068.

Sharot, T., Martorella, E. A., Delgado, M. R., & Phelps, E. A. (2007). How personal experience modulates the neural circuitry of memories of September 11. *Proceedings of the National Academy of Sciences of the United States of America, 1*(104), 389–394.

Sharot, T., & Yonelinas, A. P. (2008). Differential time-dependent effects of emotion on the recollective experience and memory for contextual information. *Cognition, 106*(1), 538–547.

Sharot, T., De Martino, B., & Dolan, R. (2009a). How choice reveals and shapes expected hedonic reaction. *The Journal of Neuroscience, 29*(12), 3760–3765.

Sharot, T., Shiner, T., Brown, A. C., Fan, J., & Dolan, R. J. (2009b). Dopamine enhances expectation of pleasure in humans. *Current Biology, 24*(19), 2077–2080.

Sharot, T., Shiner, T., & Dolan, R. J. (2010a). Experience and choice shape expected aversive outcome. *Journal of Neuroscience, 30*(27), 9209–9215.

Sharot, T., Velasquez, C. M., & Dolan, R. J. (2010b). Do decisions shape preference? Evidence from blind choice. *Psychological Science, 9,* 1231–1235.

Sharot, T. (2011). The Optimum Bias. Pantheon, New York, NY.

Slotnick, S. D. (2010). Does the hippocampus mediate objective binding or subjective remembering? *NeuroImage, 49*(2), 1769–1776.

Strunk, D. R., Lopez, H., & DeRubeis, R. J. (2006). Depressive symptoms are associated with unrealistic negative predictions of future life events. *Behaviour Research and Therapy, 44,* 861–882.

Szpunar, K. K., Watson, J. M., & McDermott, K. B. (2007). Neural substrates of envisioning the future. *Proceedings of the National Academy of Sciences of the United States of America, 104*(2), 642–647.

Talarico, J. M., & Rubin, D. C. (2003). Confidence, not consistency, characterizes flashbulb memories. *Psychological Science, 14,* 455–461.

Taylor, S. E., & Brown, J. D. (1988). Illusion and well-being: A social psychological perspective on mental health. *Psychological Bulletin, 103,* 193–210.

Tulving, E. (1985). Memory and consciousness. *Canadian Psychology, 26,* 1–12.

Tulving, E. (2002). Episodic memory: From mind to brain. *Annual Review of Psychology, 53,* 1–25. doi:10.1146/annurev.psych.53.100901.135114.

Tulving, E. (2005). Episodic memory and autonoesis: Uniquely human? In H. S. Terrace & J. Metcalfe (Eds.), *The missing link in cognition* (pp. 3–56). New York: Oxford University Press.

Tversky, A. (1972). Elimination by aspects: A theory of choice. *Psychological Review, 79,* 281–299.

Ubel, P. A., Loewenstein, G., Schwarz, N., & Smith, D. (2005). Misimagining the unimaginable: The disability paradox and health care decision making. *Health Psychology, 4,* S57–62.

Ubel, P. A., Loewenstein, G., & Jepson, C. (2005). Disability and sunshine: Can hedonic predictions be improved by drawing attention to focusing illusions or emotional adaptation?. *Journal of Experimental Psychology: Applied, 11,* 111–123.

van Veen, V., Krug, M. K., Schooler, J. W., & Carter, C. S. (2009). Neural activity predicts attitude change in cognitive dissonance. *Nature neuroscience, 12*(11), 1469–1474.

Von Neumann, J., & Morgenstern, O. (1944). *Theory of games and economic behavior.* Princeton: Princeton University Press.

Weinstein, N. D. (1980). Unrealistic optimism about future life events. *Journal of Personality and Social Psychology, 39,* 806–820.

Wilson, T. D., & Gilbert, D. T. (2003). Affective forecasting. In M. Zanna (Ed.), *Advances in experimental social psychology* (Vol. 35, pp. 345–411). New York: Elsevier.

Wise, R. A., & Rompre, P. P. (1989). Brain dopamine and reward. *Annual Review of Psychology, 40,* 191–225.

PART II

CONTEXTUAL
FACTORS

The Evolution of Our Preferences: Insights from Non-human Primates

Venkat Lakshminarayanan and Laurie R. Santos

Department of Psychology, Yale University, New Haven, USA

INTRODUCTION

Some of the most idealistic views of human psychology are – sadly – often the most empirically mistaken. One of the most striking cases of this involves the view that humans are rational decision-makers par excellence. Under this optimistic account, humans are rational beings, creatures that develops rational preferences and makes decisions that perfectly maximize such preferences. Unfortunately, recent empirical

Neuroscience of Preference and Choice
DOI: 10.1016/B978-0-12-381431-9.00004-8

work suggests that humans often fail to live up to this optimistic view. Both in the lab and in the real world, humans appear to show a number of systematic biases that cause us to violate rational assumptions. We, for example, let context shape our preferences (e.g., Ariely & Norton, 2008), react differently to payoffs depending on how they're worded (e.g., Tversky & Kahneman, 1981), and change our future preferences to fit with our past behavior (e.g., Brehm, 1956).

In the past few decades, researchers have learned much about the many ways in which context can bias our decisions (see reviews in Ariely & Norton, 2008; Camerer, 1998; Kahneman, Slovic & Tversky, 1982). Unfortunately, although much is known about the nature of human decision-making biases, less work has examined the question of where these biases come from in the first place. How do adult consumers come to be motivated by sales and promotions? How do we end up overvaluing objects that we own, or altering our preferences to fit with our decisions? Do we learn these biased strategies? Or are these strategies experience-independent, emerging as a basic aspect of human decision-making? Additionally, are such biases unique to our own species, or are these strategies a more ancient part of our psychological make-up? In short, what are the origins of our biased valuation strategies?

The question of where biased decision-making comes from is important for several reasons. First, a better understanding of how human biases originate will allow researchers a deeper insight into the nature of how valuation processes work more generally. As with any human cognitive process, our decision-making strategies are shaped by the processes by which they developed. Second, understanding how biased decision-making emerges can provide researchers with hints about how to teach people to overcome these strategies. If decision-making researchers knew that some experiences were likely to promote choice biases, they could intervene in ways that might promote better decision-making. If researchers instead learned that our biases were experience-independent, and an inherent part of our evolved psychology, such findings would advise different types of policy interventions, such as those that try to circumvent the biases people typically exhibit (see Thaler & Sunstein, 2008 for this type of approach). For these reasons, developing a more informed science of the origins of human decision-making is critical both for the science and policy of human decision-making.

Thankfully, the last few years have seen the integration of work examining the origins of valuation biases, both from a developmental and evolutionary perspective. Here, we review some of this recent work, focusing on studies from our own lab examining how contexts affect decisions in both young children and capuchin monkeys.

We begin by examining the origins of choice-induced preference change and argue that children and primates share common mechanisms for re-evaluating a choice's value after a decision. We then go on to review recent work examining the evolutionary origins of framing effects on choice, discussing our recent work demonstrating that capuchin monkeys exhibit framing biases that are virtually identical to those seen in adult humans. We then conclude by exploring what these findings mean for the science of human decision-making and the implications this work has for policy.

HOW CHOICE CHANGES PREFERENCES IN ADULT HUMANS

Perhaps the earliest account of biased evaluation strategies was penned back in 600 BC. Around this time Aesop, an enslaved storyteller, wrote his famous fable about "The Fox and the Grapes." In the fable, a fox sees a delicious looking bunch of grapes that he soon realizes is inaccessible. Eventually, after unsuccessful attempts to get the grapes, the fox decides not to extend any more effort. Soon after this decision, his evaluation of the grapes abruptly changes – the fox remarks that the grapes are probably not ripe and questions why he would need to eat "sour grapes" at all.

In the current century, our biased evaluations have been subject to scientific documentation. Since the early 1950s, researchers have identified numerous methodological situations in which choices, actions, and decisions compel participants to change what were otherwise strong prior evaluations (see review in Ariely & Norton, 2008; Harmon-Jones & Mills, 1999). To take a few examples, participants who are induced to espouse a preference that conflicts with their current attitudes will change their attitudes to fit with the preference they were forced to espouse (e.g., Bem & McConnell, 1970; Elliott & Devine, 1994; Linder, Cooper & Jones, 1967; Steele & Liu, 1983). In addition, making participants work harder on a disliked task leads them to evaluate that task as more enjoyable (e.g., Aronson & Mills, 1959).

One of the most striking examples of decisions affecting our evaluations, however, is a phenomenon researchers refer to as a choice-based preference change (Brehm, 1956; Lieberman, Ochsner, Gilbert & Schacter, 2001; Lyubomirsky & Ross, 1999; Steele, 1988). In one famous account of this phenomenon, Brehm (1956) presented participants with a set of household items and asked them to rate each item based on how much they liked it. After performing these ratings, participants were given a choice between two of the items. The two items picked for this choice phase were not random. Instead, Brehm forced participants to make what might be considered a difficult choice, choosing between two objects that they had already rated equally. After making this enforced choice, participants were asked to re-rate all items. Brehm found that participants changed their ratings post-choice in several ways. He first found that the item participants had chosen increased in value relative to the other items. More surprising, however, was what happened to the non-chosen item – it substantially decreased in value. Rejecting an object thus appeared to change people's future preferences.

The phenomenon of choice-induced preference change has now been documented across a number of real-world and experimental situations (see review in Egan, Bloom & Santos, 2010). Indeed, in just the last few years, researchers have begun learning more about the robust nature of this phenomenon. Sharot, Velasquez and Dolan (2010), for example, have observed that choice can affect preferences even in cases that involve extremely subtle or "blind" choices. For example, they presented participants with a cover story which led them to think that an arbitrary choice between two strings of symbols represented a true "subliminal" choice between different vacation options. Participants later decreased their valuation of vacation options they thought they had chosen against. In this way, choice-based preference change can occur even in cases where participants only think they made an intentional choice. Additionally, Lieberman and his colleagues (2001) observed that choice-based preference changes occur even under situations of cognitive load and bad memory. Taken together, this work suggests that choice-induced preferences can be elicited even without rich working memory resources, and even in the absence of a true choice being made.

Although psychologists have learned much about the nuanced nature of choice-based preference changes, little work to date had addressed the origins of these cognitive processes. In spite of over fifty years of work on this phenomenon, we know little about how these biases first start to emerge. Do choice-based preference reversals require experience before they develop? Or are these strategies a more fundamental part of the way we establish preferences and make decisions?

THE ORIGINS OF CHOICE-BASED PREFERENCE REVERSALS

To get at the origins of choice-based preference reversals, Egan, Santos and Bloom (2007) decided to explore whether similar phenomena could be observed in two populations that lacked the kinds of experience that adult humans have with choices and their consequences. In their first set of studies, they explored whether four-year-old children would switch their preferences after making a choice between two equally-preferred options. Adapting the original design by Brehm (1956), they presented children with novel stickers and examined whether the children's preferences for these stickers changed after they had made a decision. Each child was first asked to rate the stickers and then, after he or she had given ratings, to choose one of two possible stickers to take home. However, the two stickers presented at this stage were ones to which the child had just given equal ratings. In this way, the researchers were able to set up a choice situation between two equally preferred stickers. After the child had made a choice between the two stickers, he or she was then asked to make one additional decision – they had to choose between the sticker that they had just rejected in the first choice, and a novel sticker that they had originally rated as equal in value to the rejected sticker. If children, like adults, derogate options that they are forced to choose against, then they should prefer the novel sticker in this second choice. Egan and colleagues observed exactly this pattern of performance – children rejected the sticker they had chosen against in the first phase, suggesting that their choice against this sticker in the initial decision resulted in a change of preference. Importantly, this effect seemed to hold only in cases where children made their own decision. In a control condition, where the experimenter made the initial decision between the two stickers, children showed no subsequent change in preference for the rejected sticker. In this way, a child's commitment to a decision seems to affect future preferences in much the same way as occurs in adults.

Egan and colleagues (2007) went on to explore whether similar choice-based preference changes occur in a non-human population: capuchin monkeys. Using a similar design to the one performed with children, they gave capuchins a choice between novel food objects: differently colored M&M candies. They first allowed the monkeys to make a choice between two colored M&Ms, and then gave the monkeys a subsequent choice between the color M&M that was initially rejected and a novel colored, but identical tasting, M&M. Just like children, capuchins preferred the novel M&M color in the subsequent choice, suggesting that the act of rejecting an option in the first decision caused them to dislike this rejected option subsequently. In addition, like children, monkeys only showed this derogation effect in cases where they themselves made the decision. In a control condition, in which monkeys were forced to choose one color by the experimenter, they showed no change in preference to the color they did not receive. In this way, both monkeys and children appear to share a characteristic human choice bias: they let their decisions affect their future preferences.

Having established that children and monkeys experience choice-induced preference changes, Egan and colleagues (2010) went on to explore the boundary conditions of this bias. For example, do children and monkeys require specific kinds of choices to experience preference reversals, or is the phenomenon of choice-induced preference changes just as robust in these populations as it is in human adults? To get at this issue, they explored whether children and monkeys would exhibit preference change even in a case in which the features they were choosing were less obvious. In one study, they gave children a "blind" choice that mimicked the one that Sharot et al. (2010) used with human adults. Four-years-olds were given a choice between different toys that were hidden under occluders. Thus children had to make this first choice blind to the options' features. After making this choice, the children were then given a second choice between the toy they had blindly rejected and another hidden toy. The authors observed that children continued to avoid the toy they previously rejected, even though they had no new information about the features of this rejected toy. These results demonstrate that choice-induced preferences are robust in children, just as they are in adults, operating even where initial choices were not based on any true preference (see Chen & Risen, 2009 for discussion of this issue).

Egan and colleagues (2010) explored whether similar effects hold for capuchins' choices. Monkeys were given an option to forage for different foods that were hidden in a bin of wood-shavings. Previously, the monkeys had learned that they could forage for one piece of food before they had to leave the testing area. Using this feature of the foraging task, the experimenters were able to engineer a situation where the monkey was led to think there were two different kinds of food (differently colored Skittles candy) hidden in the bin, when in reality there was only one kind of food. This set-up created a situation in which the monkeys believed they could "choose" between two colored Skittles, but in fact the one they found was the only one available. The question of interest was whether this simulated choice would be enough to drive a choice-based preference change. To examine this possibility, the authors then presented the monkeys with a subsequent choice between the Skittle that they thought they had previously "chosen" against, and a novel colored Skittle. Just as with children, capuchins continued avoiding the option they thought they had chosen against. In this way, both children and capuchins' choice-based preference changes appear quite robust – they operate even when the features of the choice options are unknown and in cases where the choices themselves are not actually real decisions.

Taken together, the results of children and monkeys in these choice studies suggest several important features about the origins of choice-based preference changes. First, the results suggest that preference changes can occur outside the context of adult human decision-making, and thus that this phenomenon may be a more fundamental aspect of human choice than we had previously thought. Choice-based preference biases appear to operate in populations that lack rich experience with decisions, as well as in populations who lack access to the kinds of complex decisions that adult humans typically make. Egan and colleagues have thus argued that choice-induced preference changes may be an evolutionarily-older aspect of decision-making than researchers previously thought. In addition, the results from children and capuchins suggest that choice-based preference changes are incredibly robust – in all populations, these biases occur in cases where a decision-maker thought they made a decision and were blind to their choice options. Finally, children's and capuchins' performance on these choice tasks suggests that it is unlikely that experience and training can be used to teach adult human decision-makers to overcome these biases, a point we return to in the final section of our chapter.

HOW FRAMING AFFECTS CHOICE IN ADULT HUMANS

A second domain where researchers have examined the origins of choice biases involves framing effects. Although classical economists may want to believe otherwise, real human decision-makers often make different choices depending on how a problem is described or framed. Indeed, subtle differences in how choice outcomes are worded can cause people to make sets of choices that violate standard economic assumptions such as preference transitivity and invariance (Ariely & Norton, 2008; Camerer, 1998; Kahneman et al., 1982; Kahneman & Tversky, 1979).

Consider, for example, a classic demonstration by Kahneman and Tversky (1979) where participants were presented with the following choices.

You have been given $1000. You are now asked to choose between:

(A) a 50% chance to receive another $1000 and 50% chance to receive nothing, or

(B) receiving another $500 with certainty.

When presented with the above scenarios, 84% of participants choose the less risky option B. One might interpret this result as evidence that participants would rather have a certain payoff of $1500 than a risky payoff of either $2000 or $1000. However, such a conclusion would not fit with participants' performance when they are presented with a different framing of similar final outcomes.

You have been given $2000. You are now asked to choose between:

(A) a 50% chance to lose $1000 and 50% chance to lose nothing, or

(B) losing $500 with certainty.

Here, participants again had a choice between a certain payoff of $1500 and a risky payoff of either $2000 or $1000, but now they showed exactly the opposite preference as they did in the original scenario. Here, more than half of the participants choose the risky gamble, avoiding the safe bet. In this, and a number of other studies (see Kahneman & Tversky, 1979; Kahneman et al., 1982; Tversky & Kahneman, 1981, 1986), people become more risk-seeking when their final payoffs are viewed as losses, than they do when their final payoffs are seen as gains. This bias – known as the *reflection effect* – demonstrates several features of how framing effects work. First, people tend to evaluate their choices not only in terms of the absolute value of their payoffs, but also based on how their payoffs compare to some arbitrary point or context, a value often referred to as a *reference point*. In this way, people view their choices as either gains or losses relative to their reference point. Second, people behave differently when they perceive payoffs as gains than they do when they perceive payoffs as losses. Specifically, they appear to work harder to avoid losses than they do to seek out correspondingly sized gains, a bias known as *loss aversion*. Loss aversion is evident in the above scenario insofar as people tend to take on more risk in an attempt to avoid experiencing any loss relative to their reference point.

Reference dependence and loss aversion lead to a number of inconsistencies both in the laboratory and in the real world. One phenomenon, thought to be due to these biases, is termed the *endowment effect*, a bias wherein ownership causes decision-makers to attach extra value to possessions (Kahneman, Knetsch & Thaler, 1990, 1991; Thaler, 1980). Consider two prizes of equal value – for example, a mug and a box of pens. When a participant is asked to evaluate these two prizes separately, he or she would likely consider them equivalent in value, and when asked to assign a cash value to each of these items, might rate them as being worth about the same – say, $5.50 each. Presumably, then, when they received one of these items as a gift, they would be just as likely to keep the item received as they would be to trade it for the equally valued alternative. In practice, however, when asked to give up the mug in exchange for the pen set (or vice versa), decision-makers only choose to make the swap about 10% of the time. When asked to estimate the cash value of the item in their possession, they tend to overvalue the owned object. Researchers have explained this tendency to overvalue possessions in terms of loss aversion. In order to trade an owned object, a decision-maker must give up (i.e., lose) that item in order to acquire another one. Since losses impact well-being more than gains, decision-makers demand more in order to give up an object than they're willing to pay to obtain one.

Similar framing effects appear to influence real-world decision-making when the stakes are higher than mugs and pens. One example is a problem economists have referred to as the "equity premium puzzle." When stocks are underperforming, investors tend to hold on to them longer than they should – they incur a chance of a large loss in order to have a chance to make back past losses. In contrast, investors tend to sell winning stocks early because this guarantees them a sure gain (Odean, 1998). The same biases between losing and gaining assets leads to asymmetries in the housing market, where homeowners are reluctant to sell their homes when doing so would be seen as a loss relative to purchase price (Genesove & Mayer, 2001).

Although there is substantial work exploring how framing effects wreak havoc on human decision-making, until recently little work had explored where these biases come from in the first place. One possibility is that framing effects reflect learned strategies, ones that are developed via experience with certain specific features of our markets. Alternatively, framing effects may reflect something more fundamental. Indeed, it is possible that framing effects may emerge even without human-specific experiences with markets and gambles. One way to distinguish between these two alternatives is to examine whether similar biases are present in individuals who lack experience with human-like markets. To this end we, and our colleagues, have begun exploring whether loss aversion and reference dependence are also evident in capuchin monkeys (Chen, Lakshminarayanan & Santos, 2006; Lakshminarayanan, Chen & Santos, 2008; Lakshminarayanan, Santos & Chen, 2011).

Unfortunately, establishing framing effects in a population that does not naturally engage in economic transactions comes with formidable challenges. In the next section, we detail how we introduced our monkeys to a token-based market where they could buy and sell food rewards. We then detail how we used this market to establish whether capuchins share with humans irrational economic preferences such as loss-aversion, a reflection effect, and the endowment effect.

FRAMING EFFECTS IN CAPUCHIN MONKEYS

Our first task in exploring the nature of capuchin framing effects was figuring out a method that could be used with non-verbal subjects. To this end, we decided to train our monkeys on a novel fiat currency – a set of metal disc tokens – that they could trade with human experimenters for different kinds of food. Although establishing a fiat currency in monkeys may seem like a lofty goal, numerous previous studies had successfully trained primates to trade tokens with human experimenters (e.g., Addessi, Crescimbene & Visalberghi, 2007; Brosnan & de Waal, 2003; Westergaard, Liv, Chavanne & Suomi, 1998). After a short training period in which the monkeys learned how to use their tokens, we were able to place our capuchin subjects in an experimental market where they could buy food at different prices. In each experiment, monkeys entered a testing area where they were given a wallet of tokens. The tokens could be exchanged with one of two experimenters who each offered a different food reward. In this way, we allowed monkeys to reveal their preferences between the two options by spending more of their tokens on whichever option they preferred.

The initial experiments involving this experimental set-up were intended to establish that monkeys understood the market (Chen et al., 2006). We first explored whether the monkeys would buy more food from an experimenter who offered a greater reward at the same price. To do this, we gave the monkeys a choice between two experimenters who differed in terms of their average payoff. The first experimenter gave the monkeys an average payoff of one and a half apple chunks, while the second experimenter provided an average payoff of only one apple chunk. Like smart shoppers, our monkeys selectively traded with the first experimenter, suggesting that they both understood the market set-up and tried to shop in ways that gave them a better deal overall.

We then turned to the question of whether the same market could be modified to reveal the monkeys' preferences between two options that are equivalent but treated differently due to the nature of the frame (Chen et al., 2006). We first gave the monkeys a choice between two options that were, in terms of their overall payoffs, equivalent. The only difference between these two options was that one was framed as an opportunity for a gain while the other was framed as a possible loss. This was accomplished by having each experimenter begin the act of trading by holding a small dish that contained different amounts of food. We reasoned that the number of chunks displayed on the dish would serve as a reference point against which the monkeys might frame their decision. We could then vary whether what the experimenter offered seemed like a gain or loss relative to that initial reference point by having them add or subtract pieces of food from the number they initially displayed. In the first study, we gave monkeys a choice between an experimenter who framed his offer as a gain – he started with one chunk of apple but then half the time added an extra chunk – and an experimenter who framed his offer as a loss – he started by offering two chunks of apple and half the time removed one. If monkeys in this study simply computed over-all expected value, then they would have no reason to form a preference between these two options. Instead, consistent with human performance, our monkeys showed a preference to trade with the experimenter who provided a possible gain and avoided the experimenter who provided a potential loss. Just as in human studies, our monkeys appeared to both frame their final payoff relative to an arbitrary reference point and to avoid options that were framed as losses relative to that reference point.

To explore whether this reference dependence and loss aversion also affected monkeys' risk preferences (e.g., Kahneman & Tversky, 1979; Tversky & Kahneman, 1981), our next experiment explored whether monkeys also exhibited a reflection effect when risky and safe options were framed differently (Lakshminarayanan et al., 2011). Specifically, we gave monkeys a choice between an experimenter who provided his reward consistently on every trial – he was a safe bet – and another experimenter – the risky experimenter – who provided a variable reward that was, on average, equivalent to that of the safe experimenter. In the first condition, both the safe and risky experimenters initially displayed three chunks of apple. The risky experimenter would subtract either nothing or two chunks from this initial display, thereby representing a risky shot at either three or one apple chunk. The safe experimenter, in contrast, always removed one chunk of apple from his initial display. The safe experimenter therefore represented a certain loss of a single apple chunk. When presented with this loss framing, monkeys were risk-seeking; they selectively chose to trade with the risky experimenter over the safe experimenter. We observed a different pattern of performance, however, when the same final payoffs were framed as gains instead of losses. Here, we presented monkeys with a choice between a risky and safe experimenter who both started out with one chunk of apple but added more. The risky experimenter would then either add nothing or add two chunks of apple, for an average final payoff of two chunks. The safe experimenter initially displayed a single apple chunk, but always added one chunk to this initial display. In contrast to how they per-formed with loss framing, monkeys were risk-averse when dealing with gains. Just like humans, capuchins exhibited a reflection effect: they were risk-seeking when their prospects were framed as losses and risk-averse when the same final outcomes were framed as gains.

We next explored whether framing led monkeys to show an endowment effect (Lakshminarayanan et al., 2008). Do monkeys also over-value objects they own and refuse to trade them for an equivalent good? To investigate this, we switched the monkeys' market by endowing them not with tokens but instead with one of two foods: either cereal or fruit pieces. We chose these two types of food as stimuli because the monkeys value them about equally. We could therefore use these equivalent goods to explore how monkeys' valuation changes when they own one of the foods. If monkeys valued owned foods just as much as the ones they didn't yet own, then we might expect them to keep about half the food and trade about half away for an equivalent food. Our monkeys, however, showed a very different pattern of performance – they overwhelmingly preferred to keep the food they owned. This pattern of results hinted that monkeys might experience something like an endowment effect, overvaluing objects that they own (see Brosnan, Jones, Lambeth et al., 2007 for a similar effect in chimpanzees). Nevertheless, a few key alternative explanations for this result demanded follow-up studies. One alternative was that monkeys chose to retain their food because they didn't understand how to trade foods like they did tokens. To rule this out, we endowed capuchins with the same foods as in the initial study but offered them the chance to trade these foods for a much higher valued alternative. When offered a much higher valued food, capuchins were willing to trade the food they owned, suggesting that monkeys understood that food could be traded like their tokens. A second alternative, however, was that monkeys refused to trade their endowed food not because of an endowment effect but because of the additional physical effort required for trading. To address this alternative, we estimated the physical cost of the transaction by calculating the minimum possible compensation capuchins required to trade a token, and then added this extra compensation to the equivalent food reward we had offered in our original study. Even when compensating for the cost of the transaction, our subjects were still willing to retain their endowed foods rather than trade them away. In this way, our capuchins' endowment effect persisted despite a transaction-cost compensation. Finally, we tested whether the observed endowment effect was due to temporal discounting; specifically, would monkeys still exhibit an endowment effect even when it took them more time to eat the endowed food than it would to trade it away? If the observed endowment effect was, in reality, because the food in their endowment took less time to eat than the food available for trade, then subjects should be willing to trade away their endowment if doing so results in a shorter time to obtain their payoff. Even when subjects were offered the opportunity to trade for a faster-to-eat food, they nevertheless refused to trade their endowed food. Taken together, our results suggest that capuchins exhibit an endowment effect even in cases in which we controlled for discounting effects and transaction costs.

In summary, capuchins' choice behavior in situations of framing mirrors that of humans in remarkable ways – they exhibit reference dependence and loss aversion, and such biases also lead them to over-value objects they own and exhibit reflection effects. As with similarities between capuchins and humans in choice-based preference change studies, these framing results suggest a number of conclusions about framing effects in human decision-making – such effects are likely to be due to fundamental cognitive mechanisms, ones that are likely to be experience-independent, evolutionary ancient, and possibly harder to overcome than we'd like to think.

HOW STUDIES OF THE ORIGIN OF CHOICE BIASES INFORM ADULT HUMAN DECISION-MAKING

While we have recognized for centuries that our decision-making is reliably error-prone, only recently have we started to understand why humans exhibit these biased decision-making strategies. The previous sections have shown that adult humans share their preference biases with both children and monkeys, and thus are not unique in how their preferences are influenced by irrelevant factors, such as contextual information. Here, we will conclude by discussing how this comparative and developmental work can help explain why we might be so prone to these context effects.

Comparisons across age and species often shed light on mechanisms that drive cognitive processing. In the case of cognitive errors, similarities across development and phylogeny demonstrate that common intuitive mechanistic explanations for why adults have these biases are not the most plausible. For example, it is often assumed that biases arise because adult decision-makers have grown up in a world of complex choices and economic options. This assumption cannot account for the fact that young children possess similar biases, despite far less experience with markets and choices. In addition, the fact that capuchins also show context-based preferences indicates that experiences unique to adult humans – having economic markets and wide ranges of choice options – are not required for the development of these biases. A second account of choice biases – one that has come up particularly when trying to account for framing effects – has argued that these tendencies are due to verbal task confounds. Under this account, verbal aspects of gambles presented in surveys induce subjects to attend to irrelevant aspects of the problem in ways that lead to framing effects. This interpretation, however, would predict that non-verbal populations would not exhibit identical framing effects. Comparative results showing framing in capuchins demonstrate that such effects cannot result from linguistic aspects of the task. Finally, some have hypothesized that choice biases result from rather sophisticated cognitive mechanisms. For example, some psychologists propose that a sophisticated self-construct is required for biases like choice-induced preference changes (e.g., Gawronski, Bodenhausen & Becker, 2007; Steele & Liu, 1983) and the endowment effect (e.g., Beggan, 1992; Gawronski et al., 2007). However, we have shown that capuchin monkeys – who very likely do not share a rich sense of self – also denigrate unchosen alternatives and over-value objects they own. These results make it unlikely that high-level cognitive explanations, involving a rich sense of self, can fully account for these biases. In this way, research on the origins of choice biases has narrowed the hypothesis space for how and why such biases occur in adult decision-makers.

In addition to constraining mechanistic interpretations of choice biases, the current work on monkeys' choice errors begins to allow researchers new ways to examine biases at the neural level. To date, neuroscientists interested in the nature of choice biases have had to study these phenomena in humans using functional neuroimaging techniques. Although these non-invasive techniques have provided important insights into the nature of choice biases (e.g., De Martino, Kumaran, Seymour & Dolan, 2006; Lee, 2006; Sharot, Martino & Dolan, 2009; Tom, Fox, Trepel & Poldrack, 2007), the nature of human neuroimaging techniques precludes studying choice biases at the level of individual neurons and circuits (but see Chapter 5 for a discussion of similar issues by De Martino). As such, researchers interested in understanding the nature of choice errors at the level of single neurons have faced a methodological challenge; indeed, such researchers could greatly benefit from developing primate model of these biases, one that would allow researchers to examine neural activity using more refined techniques (e.g., single-cell recordings). Having learned that monkeys' choice biases at the behavioral level are similar to those of adult humans, neuroscientists are now poised to develop just such a neurophysiological model of choice biases. In this way, learning more about the evolutionary origins of choice biases has potentially provided a new methodological window into which researchers can begin studying the neural bases of these biases.

A final insight that stems from studying the origins of choice biases concerns ways to overcome these biases. Because humans and capuchins are disposed to make choices in strikingly similar ways, it is likely that these two species' common ancestor – a species that existed approximately 35 million years ago – exhibited many of the same choice patterns. In this way, our choice strategies may be evolutionarily quite old.

This insight provides hints as to why such biases might be so pervasive in modern economies. Evolutionarily old strategies are often the trickiest ones to overcome – just consider our evolutionarily-selected predilections for sweets and aversion to snakes and spiders. The finding that choice biases may be similarly old suggests that policymakers would be well-served to incorporate an evolutionary perspective into their efforts to modify the behavior of human decision-makers through laws and incentives. The early emergence of choice biases hints that such biases might persist in the face of extensive economic exposure or market disciplining. Our origins approach therefore recommends that the optimal role of policy may be to work around these systematic errors rather than attempt to change them. Approaches like "nudging" consumers to avoid their choice biases (e.g., Thaler & Sunstein, 2008) or using framing effects to a consumer's advantage (e.g., Thaler & Bernartzi, 2004) may therefore be the only way to make the best of our rationally-inadequate evolutionary endowment.

Acknowledgments

This chapter is based upon work supported by the National Science Foundation under Grant No. 0624190 and a McDonnell Foundation Scholar Award.

References

Addessi, E., Crescimbene, L., & Visalberghi, E. (2007). Do capuchin monkeys (Cebus *apella*) use tokens as symbols? *Proceedings of the Royal Society of London, Series B, 274*, 2579–2585.

Ariely, D., & Norton, M. I. (2008). How actions create – not just reveal – preferences. *Trends in Cognitive Sciences, 12*(1), 13–16.

Aronson, E., & Mills, J. (1959). The effect of severity of initiation on liking for a group. *Journal of Abnormal and Social Psychology, 59*, 177–181.

Beggan, J. K. (1992). On the social nature of nonsocial perception: The mere ownership effect. *Journal of Personality and Social Psychology, 62*, 229–237.

Bem, D. J., & McConnell, H. K. (1970). Testing the self-perception explanation of dissonance phenomena: On the salience of premanipulation attitudes. *Journal of Personality and Social Psychology, 14*(1), 23–31.

Brehm, J. W. (1956). Postdecision changes in the desirability of alternatives. *Journal of Abnormal and Social Psychology, 52*, 384–389.

Brosnan, S. F., & de Waal, F. B. M. (2003). Monkeys reject unequal pay. *Nature, 425*, 297–299.

Brosnan, S. F., Jones, O. D., Lambeth, S. P., Mareno, M. C., Richardson, A. S., & Schapiro, S. J. (2007). Endowment effects in chimpanzees. *Current Biology, 17*(19), 1704–1707.

Camerer, C. F. (1998). Bounded rationality in individual decision making. *Journal of Personality and Social Psychology, 1*(2), 163–183.

Chen, M. K., & Risen, J. L. (2009). Is choice a reliable predictor of choice? A comment on Sagarin and Skowronski. *Journal of Experimental Social Psychology, 45*, 425–427.

Chen, M. K., Lakshminarayanan, V. R., & Santos, L. R. (2006). The evolution of our preferences: Evidence from capuchin monkey trading behavior. *The Journal of Political Economy, 114*(3), 517–537.

De Martino, B., Kumaran, D., Seymour, B., & Dolan, R. J. (2006). Frames, biases, and rational decision-making in the human brain. *Science, 313,* 684–687.

Egan, L. C., Santos, L. R., & Bloom, P. (2007). The origins of cognitive dissonance: Evidence from children and monkeys. *Psychological Science, 18,* 978–983.

Egan, L. C., Bloom, P., & Santos, L. R. (2010). Choice-induced preferences in the absence of choice: Evidence from a blind two choice paradigm with young children and capuchin monkeys. *Journal of Experimental Social Psychology, 46,* 204–207.

Elliot, A. J., & Devine, P. G. (1994). On the motivational nature of cognitive dissonance: Dissonance as psychological discomfort. *Journal of Personality and Social Psychology, 67,* 382–394.

Gawronski, B., Bodenhausen, G. V., & Becker, A. P. (2007). I like it, because I like myself: Associative self-anchoring and post-decisional change of implicit evaluations. *Journal of Experimental Social Psychology, 43*(2), 221–232.

Genesove, D., & Mayer, C. (2001). Loss aversion and seller behavior: Evidence from the housing market. *The Quarterly Journal of Economics, 116,* 1233–1260.

Harmon-Jones, E., & Mills, J. (1999). *Cognitive dissonance: Progress on a pivotal theory in social psychology.* Washington, DC: American Psychological Association.

Kahneman, D., Slovic, P., & Tversky, A. (Eds.) (1982). *Judgment under uncertainty: Heuristics and biases.* Cambridge, UK: Cambridge University Press.

Kahneman, D., Knetsch, J. L., & Thaler, R. H. (1990). Experimental tests of the endowment effect and the Coase theorem. *The Journal of Political Economy, 98,* 1325–1348.

Kahneman, D., Knetsch, J. L., & Thaler, R. H. (1991). Anomalies: The endowment effect, loss aversion, and status quo bias. *The Journal of Economic Perspectives, 5,* 193–206.

Kahneman, D., & Tversky, A. (1979). Prospect theory: An analysis of decision under risk. *Econometrica, 47,* 263–292.

Lakshminarayanan, V., Chen, M. K., & Santos, L. R. (2008). Endowment effect in capuchin monkeys (Cebus apella). *Philosophical Transactions of the Royal Society, Series B, 363,* 3837–3844.

Lakshminarayanan, V., Santos, L. R., & Chen, M. K. (2011). The evolution of decision-making under risk: Framing effects in monkey risk preferences. *Journal of Experimental Social Psychology, 47*(3), 689–693.

Lee, D. (2006). Neural basis of quasi-rational decision making. *Current Opinion in Neurobiology, 16,* 191–198.

Lieberman, M. D., Ochsner, K. N., Gilbert, D. T., & Schacter, D. L. (2001). Do amnesics exhibit cognitive dissonance reduction? The role of explicit memory and attention in attitude change. *Psychological Science, 12,* 135–140.

Linder, E. D., Cooper, J., & Jones, E. E. (1967). Decision freedom as a determinant of the role of incentive magnitude in attitude change. *Journal of Personality and Social Psychology, 6,* 245–254.

Lyubomirsky, S., & Ross, L. (1999). Changes in attractiveness of elected, rejected, and precluded alternatives: A comparison of happy and unhappy individuals. *Journal of Personality and Social Psychology, 76,* 988–1007.

Odean, T. (1998). Are investors reluctant to realize their losses? *Journal of Finance, 5,* 1775–1798.

Sharot, T., Velasquez, C. M., & Dolan, R. J. (2010). Do decisions shape preference? Evidence from blind choice. *Psychological Science, 21*(9), 1231–1235.

Sharot, T., De Martino, B., & Dolan, R. J. (2009). How choice reveals and shapes expected hedonic reaction. *The Journal of Neuroscience, 29*(12), 3760–3765.

Steele, C. M. (1988). The psychology of self-affirmation: Sustaining the integrity of the self. In L. Berkowitz (Ed.), *Advances in experimental social psychology* (pp. 261–301). San Diego, CA: Academic Press, Inc.

Steele, C. M., & Liu, T. J. (1983). Dissonance processes as self-affirmation. *Journal of Personality and Social Psychology, 45,* 5–19.

Thaler, R. H. (1980). Toward a positive theory of consumer choice. *Journal of Economic Behavior & Organization, 1,* 39–60.

Thaler, R. H., & Bernartzi, S. (2004). Save more tomorrow: Using behavioral economics in increase employee savings. *The Journal of Political Economy, 112*(1), S164–S187.

Thaler, R. H., & Sunstein, C. R. (2008). *Nudge: Improving decisions about health, wealth, and happiness.* New Haven, CT: Yale University Press.

Tom, S., Fox, C. R., Trepel, C., & Poldrack, R. A. (2007). The neural basis of loss aversion in decision making under risk. *Science, 315,* 515–518.

Tversky, A., & Kahneman, D. (1981). The framing of decisions and the psychology of choice. *Science, 211,* 453–458.

Tversky, A., & Kahneman, D. (1986). Rational choice and the framing of decisions. *Journal of Business, 59,* 251–278.

Westergaard, G. C., Liv, C., Chavanne, T. J., & Suomi, S. J. (1998). Token mediated tool-use by a tufted capuchin monkey (*Cebus apella*). *Animal Cognition, 1,* 101–106.

The Effect of Context on Choice and Value

Benedetto De Martino

University College of London, IBN Department of Psychology, London
and California Institute of Technology, Pasadena

Neuroscience of Preference and Choice
DOI: 10.1016/B978-0-12-381431-9.00005-X

93

INTRODUCTION

It is Monday morning: you have just started a diet, and are at the supermarket shopping for healthy food to see you through the week. En route to the checkout you spy a mouthwatering chocolate bar – you are aware that this doesn't actually belong to the category of "healthy food" but then the package tells you that it is eighty percent fat free! You buy the bar and enjoy its sweet flavor without guilt. But would you have behaved differently if the package was reminding you that the very same chocolate bar is actually twenty percent full fat?

The everyday situation described here is one of countless examples of how decision-makers are vulnerable to the way in which information is presented to them. In this chapter I will explain how such behavior is a direct consequence of how choices are constructed and values computed by our brains. The chapter is divided in four parts. The first part presents a brief historical overview of relevant research in economics and psychology. The second part reviews the current research in neuroscience which addresses the neurobiology of the contextual modulation of choice: in particular, I will look at the neural computations underlying choice behavior in situations where the manipulation of contextual information is either manipulated exogenously by the experimenter (i.e. the "framing effect") or arises endogenously (i.e. the "status quo bias"). The third part will discuss how the underlying value signals are modulated by contextual information, focusing in particular on how reference-dependent and reference-independent values are represented in the orbitofrontal cortex. The last part of the chapter will dissect the neural circuitry underpinning loss aversion.

BRIEF HISTORICAL OVERVIEW

The susceptibility of humans to the way in which information is presented has been intensively studied by psychologists. In 1981, Amos Tversky and Daniel Kahneman showed that contextual information can bias decision-making in a highly predictable fashion, and they called this behavioral tendency the *Framing effect* (Tversky & Kahneman, 1981). The framing effect has consequences which extend beyond advertising – consequences which impinge on the welfare of our society. For instance, framing can affect the quality of the medical treatment we receive, as shown by a study conducted by McNeil, Pauker, Sox and Tversky (1982). In this study experienced physicians were presented with a hypothetical case of a patient affected by lung cancer, and were asked to choose between radiotherapy and surgery as a mode of treatment. In this scenario, radiotherapy was the safest option but also the one with higher risk of cancer recurrence following treatment. Meanwhile, surgery gave the patient a higher chance of complete remission but was the more dangerous procedure. Exactly the same statistics were given to all doctors. However, in one group the patient's risk of the surgery was presented in terms of survival rate, while to the other group were presented in terms of mortality rate. The disquieting result was that doctors assigned to the "survival rate" group were approximately twice as likely to recommend surgery as the ones assigned to the "mortality rate" group. The question of why such important decisions are so strongly influenced by the simply rewording of a sentence is the subject of this chapter.

For over a century, theories of economic choice have often found it difficult to confront the irksome evidence that humans are often inconsistent in their decision-making. For example, to avoid logical circularity in their axiomatic systematization of expected utility theory (EUT), Johnny von Neuman and Oskar Morgenstren (1944) explicitly prescribed that irrelevant information should not affect the outcome of a decision. This prescription is at the very core of EUT (Luce, 1959; Ray, 1973) and is summarized by the *axiom of substitution (a.k.a. independence from irrelevant alternatives)*:

$$\text{if } X > Y \quad pX + (1 - p)Z > pY + (1 - p)Z \quad \text{for all Z and } p \in (0,1) \quad (1)$$

This axiom states that a decision-maker who strongly prefers option X to option Y should still prefer a bundle of option X (multiplied for any probability p) and option Z (multiplied for complementary probability $1 - p$) to the bundle of option Y (multiplied for the same probability p) and option Z (multiplied for $1 - p$). In other words, if you prefer a bottle of red wine over a bottle of white, you should still prefer the bottle of red sold together with a corkscrew, to the bottle of white sold together with the same corkscrew. This axiom seems to capture a general principle of logic, but shortly after it was formulated its behavioral untenability was demonstrated by the elegant work of the French economist Maurice Allais (1953), who designed a simple choice problem that was subsequently referred to as the *Allais paradox*. This paradox is a strict violation of the axiom of independence. Its paradoxical nature is evident when we compare the response to two different experiments in which subjects are required to choose between different bundles of monetary gambles. Table 5.1 shows the pay-off matrix for one formulation of this problem.

The paradox arises from the experimental evidence that many people will choose option 1 in experiment A and option 2 in the experiment B. This is in sharp contrast to the prescription of independence.

The violation of the independence axiom is not the only discrepancy between the prescription of EUT and the empirical data but it is probably the best known, and the one with most experimental support. The evidence

TABLE 5.1   Allais paradox payoff matrix

Experiment A		Experiment B	
Option 1	Option 2	Option 1	Option 2
89% chance of winning $1 million	89% chance of winning $1 million	89% chance of winning nothing	89% chance of winning nothing
11% chance of winning $1 million	10% chance of winning $2.5 million	11% chance of winning $1 million	10% chance of winning $2.5 million
	1% chance of winning nothing		1% chance of winning nothing

that independence is achieved only with difficulty has brought some scholars to formulate alternative versions of EUT such as Rank Dependent Expected Utility in which this axiom has been relaxed (Quiggin, 1993). However, the most systematic attempt to reconcile these and other anomalies of expected utility in a coherent theoretical framework was carried out at the end of the 1970s by Kahneman and Tversky (2000) and Tversky and Kahneman (1974). Most of the recent work in the field of neuroeconomics, which I will present later in this chapter, owes a great debt to this vast and systematic research agenda that produced a successful model of choice behavior under risk termed Prospect Theory (PT). In particular, PT introduced a handful of novel concepts not originally present in classic EUT, including a new functional form of utility that incorporated the asymmetry between the way humans perceive losses and gains.

In the standard formulation of EUT (Figure 5.1A), the critical parameter is "change in total wealth" (x-axis) and the difference between gains and losses is simply disregarded. Instead in PT this asymmetry is captured by a kink in the utility function (Figure 5.1B) which generates a steeper negative utility associated with losses when compared with the positive utility associated with gains.

The main consequence of this feature of the utility function introduced in PT is that the subjective negative utility assigned to the loss for a given amount is larger than the positive utility assigned to a gain of the same magnitude.

This feature of PT explains the pattern of choice behavior known as loss aversion. Loss aversion accounts for the fact that many people would refuse to play a coin-flip gamble unless the potential win is at least twice as large as the possible loss, comfortably giving up on other gambles with a smaller (but still positive) expected payoff. PT also introduced the critical notion that utility is not calculated in an absolute reference system, but the outcomes of a choice are evaluated either as a gain or as a loss relative to a reference point. While the precise nature of this reference point

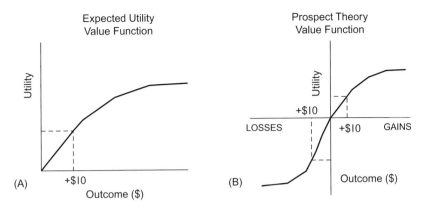

FIGURE 5.1 Panel A depicts the value function at the core of classic expected utility theory (EUT). Panel B depicts the value function introduced by Prospect Theory (PT) in which the dis-utility for losses is usually larger in magnitude than the utility for gains.

is still unclear (Koszegi & Rabin, 2006), it is known to be rather unstable, as the framing manipulation starkly demonstrates. A final feature of the value function introduced by PT is that curvature of the function is concave in the gain domain (as it was for expected utility value function) but convex for potential losses. This characteristic form of the value function accommodates another key aspect of human choice: risk-averse behavior for potential gains (i.e. the preference for a small but safe gain over a large but risky gamble) but risk-seeking behavior for potential losses (i.e. the preference for a risky gamble over a small but sure loss).

Since the first formulation of these ideas some four decades ago, concepts such as loss aversion and reference dependency have had a significant impact on the fields of Psychology and Economics. But it is only in the last few years that neuroscientists have started to look at the neural computations underlying these phenomena. I will now move on to review some of the most recent findings in the burgeoning field of *Neuroeconomics*, arguing that some of the behavioral deviations from classic EUT can provide an opportunity for neuroscientists to understand the process behind the construction of preferences. In turn, an understanding of the neural processes underpinning these behavioral "anomalies" may provide decision theorists with a series of empirical constraints for the development of a more comprehensive theory of human decision processes.

EFFECT OF CONTEXT ON CHOICES

"Context" is a loaded and and often ambiguous word. According to the Oxford English Dictionary *"Context is circumstances that form the setting for an event, statement, or idea."* In the domain of decision-making we often use the term context to refer to all the extra – and sometime redundant – information that is not necessary for the decision-maker to make a fully informed choice. Here's an example: imagine you are writing a simple computer algorithm that selects from two or more gambles the one with the highest payoff. The only information needed in the algorithm is the magnitude of a potential reward and the probability of getting that reward. The task that the algorithm needs to perform is simply to multiply these two pieces of information, calculate what is called the expected value (EV) of the gamble and choose the gamble with higher EV. It is irrelevant to the outcome of the algorithm whether the probability is entered as a number between 0 and 1 or stated as a percentage or, similarly, whether the monetary reward is expressed in dollars or euros. Changing this information will not affect the outcome of the algorithm. However, for humans it is often unavoidable to process this additional information and incorporate it into the choice. It is not always the case that processing such "redundant" information constitutes a disadvantage: for example, the use of contextual cues can be extremely beneficial during verbal communication to clarify the meaning of a sentence (Duranti & Goodwin, 1992) or for cuing memory capacities (Tulving, 1973). However, as we see later, in some circumstances this contextual information may bias the process of choice, generating in some cases suboptimal decisions. This problem is exacerbated by the fact that many choices that are relevant for economics use abstract concepts like probability and secondary reinforcers like money, requiring a level of abstraction. This issue is also not exclusive to the decision-making domain – similar effects have been reported in other areas of neuroscience. A classic example comes from visual perception. When asked to judge the luminosity of a visual stimuli, the contextual surrounding can have a strong bias in the perception of the target stimulus (see Figure 5.2). This phenomenon is due to visual adaptation and has to do with the intrinsic constraints of the visual system (Clifford, Webster, Stanley et al., 2007; Laughlin, 1989).

The Framing Effect: the Role of Amygdala in Mediating Appetitive-Aversive Responses

The Framing Effect is probably the best known example of the biasing effect of context on choice (Tversky & Kahneman, 1981). This effect has been shown to play a key role in many different fields such as stock market forecasting (Haigh & List, 2005; Tovar, 2009), rate of organ donation (Johnson & Goldstein, 2003) and even international conflicts (Mercer, 2005). In 2006 we designed an fMRI study that aimed to elucidate the neural computations associated with the framing effect (De Martino, Kumaran, Seymour & Dolan, 2006). Volunteers that took part in the study

FIGURE 5.2 Optical illusion: while the square B is perceived to be of a lighter color than square A, both squares are exactly the same shade of gray. This perceptual illusion is a consequence of the fact that the visual system estimates color and brightness of the square not in isolation but in relation to the other surrounding squares. This property of the visual system guarantees that a familiar object will appear the same color regardless of the amount of light reflecting from it.

were asked to make a decision between a safe option (e.g. get £20 for sure) or a gamble matched in expected value (e.g. 40% chance of getting £50 versus a 60% chance of getting nothing). The critical manipulation consisted in the rewording of the safe option in the following way: Participants were told at the beginning of each trial that they had been given a certain amount of money to play with in that trial (e.g. "You receive £50"); in half of the trials the safe option was either presented in a "Gain" frame by using the word *Keep* (e.g. Keep £20); in the other half it was presented in a "Loss" frame using the word *Lose* (e.g. Lose £30) (Figure 5.3).

This simple manipulation elicited a robust framing effect with participants consistently preferring the safe option when it was presented in a Gain frame (57.1% of the trials); as opposed to when the identical option was presented in a Loss frame (38.4% of the trials). This pattern of choice is exactly the one predicted by the value function introduced in prospect theory (see Figure 5.1B): risk-averse for gains (the convex shape of the function), and risk-seeking for losses (the concave shape of the function). At the neural level, the bilateral amygdala (almond-shaped nuclei located deep within the medial temporal lobe – see Figure 5.4) showed a pattern of activity that mirrored the asymmetric pattern of decisions elicited by the framing manipulation. In other words the amygdala showed an increased activity when subjects chose the safe option in the context of the gain frame, but showed a reverse pattern of activity in the context of the loss frame (by increasing activation for the gamble).

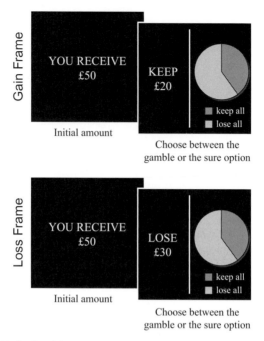

FIGURE 5.3   Task. Participants were shown a message indicating the amount of money received to play in that trial (e.g., "you receive £50"). Subjects then had to choose between a "sure" or a "gamble" option presented in the context of two different "frames." The "sure" option was formulated as either the amount of money retained from the initial starting amount (e.g., keep £20 of a total of £50–gain frame) or as the amount of money lost from the initial amount (e.g., lose £30 of a total of £50–loss frame). The "gamble" option was identical in both frames and represented as a pie chart depicting the probability of winning or losing.

Much work on the amygdala has revealed that this region is responsible for fear conditioning (LeDoux, 1996). A wealth of studies across many different species (rodents, monkeys and humans) has now expanded its role to the domain of learning and value computation (Baxter & Murray, 2002; Dolan, 2002; Everitt, Parkinson, Olmstead et al., 1999; LeDoux, 2000; Morrison & Salzman, 2010; Phelps & LeDoux, 2005; Salzman & Fusi, 2010). Specifically, this region has been shown to compute both positive and negative value signals during Pavlovian conditioning (Paton, Belova, Morrison & Salzman, 2006) and to integrate convergent information about rewarding and punishing stimuli (Morrison & Salzman, 2010). There is an emerging general consensus that this region plays a key role in the representation of the motivational value of a conditioned stimulus (CS) associated with an appetitive or aversive unconditioned stimulus (US). Such information can be used to guide behavior during instrumental tasks (Tye & Janak, 2007). Furthermore amygdala enables animals

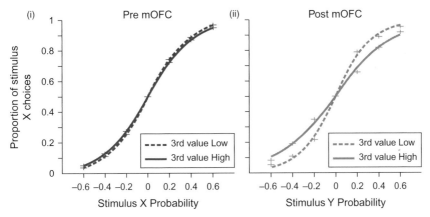

FIGURE 5.4 Proportion of trials on which monkeys chose options as a function of the value difference with respect to one other option in the context of a high (solid line) or low (dashed line) value third option before (i) and after (ii) mOFC lesion. (Adapted from Noonan et al., 2010.)

to update the value signal as a consequence of a different contingency between CS and US as shown in devaluation tasks (Baxter & Murray, 2002). According to this view, it is possible that in the context of the framing task described above the amygdala may be updating the value of the same option presented in two different emotional contexts (Morrison & Salzman, 2010). A plausible hypothesis is that amygdala response, triggered by the framing manipulation, emits a Pavlovian approach-avoidance signal that affects the subjects' instrumental action (i.e. biasing the choice). It is possible that such appetitive-aversive signals emitted by the amygdala at the time of the decision interact with the computation of value of the available options, generating the characteristic framing bias. It is also possible that this Pavlovian signal acts directly at the time of the instrumental behavior (choice) without necessarily changing the actual value of the option.

Whichever of the above possibilities holds true, the view that the amygdala generates the framing effect through an appetitive-aversive response is in keeping with a theoretical proposition that different controllers guide the decision process and that the most simple and automatic of these controllers (the Pavlovian controller) can bias the agent's response by interfering with more sophisticated computations carried by other controllers (Dayan, 2009; Dayan, Niv, Seymour & Daw, 2006) (see also Chapter 2 written by Dayan).

A powerful example of how this automatic Pavlovian response can interfere with instrumental learning is provided by behavioral experiments in pigeons. In these studies the experimenters trained a group of pigeons using an experimental schedule called *automaintenance* (or autoshaping) (Brown & Jenkins, 1968; Eldridge & Pear, 1987): a light was repeatedly presented to the animals paired with a reward (food or water). Due to classic Pavlovian conditioning response the animals increased pecking every time the light was turned on. Once established, this behavior tended to be maintained even when the light ceased to be predictive of the reward. In a variation of this original setup, called *negative automaintenance* (Williams & Williams, 1969), a "malevolent" experimenter actively denied the reward every time the pigeon pecked in response to the flashing light. In this situation the animal continued to peck in response to the light, thereby losing out on valuable rewards.

The effect of a Pavlovian signal on instrumental behavior has also been shown in setups where it does not necessarily trigger maladaptive behavior (as in the case of negative automaintenance) but rather exerts a motivational influence over the instrumental performance. This effect, which has mostly been studied in the field of animal learning, is called Pavlovian-Instrumental-Transfer (PIT) (Colwill & Rescorla, 1988; Estes, 1943; Rescorla, 1994; Rescorla & Solomon, 1967; Rescorla & Wagner, 1972). Here an animal has been previously conditioned on a CS+ using a classical Pavlovian conditioning paradigm. In the PIT *test phase* the CS+ is presented in concomitance with an instrumental task. In this way the CS+ exerts a motivational influence over instrumental performance energizing the rate of the instrumental response. At the neurobiological level PIT requires an intact neural circuit connecting the nuclei of basolateral amygdala with the nucleus accumbens (Balleine & Killcross, 2006; Cardinal et al., 2002; Corbit & Balleine, 2005; Mogenson et al., 1980). In this circuitry the amygdala is supposed to maintain the information about the motivational significance of the CS+ at the time of the instrumental task (Tye & Janak, 2007), influencing the magnitude instrumental response and consequently generating the PIT effect. A similar result is seen in human subjects where an amygdala signal is positively correlated with the magnitude of Pavlovian-instrumental transfer (PIT) (Talmi et al., 2008).

Another automatic approach-avoidance influence on an instrumental action has been shown in humans using a manipulation of semantic information. Here subjects were instructed to either pull or push away a lever in response to the presentation of a word on the screen. The words presented had either positive valence (e.g. puppy) or negative valence (e.g. cancer). Even if the valence of the words was irrelevant for the task, the experimenters found that for congruent trials (i.e. negative word/push or positive word/pull) subjects were faster than for incongruent trials (Chen & Bargh, 1999; Kahneman & Frederick, 2007).

It is very tempting to see a close analogy between the examples described above (automaintenance, PIT, pulling-push response) and the framing effect, where a similar signal is elicited in the amygdala in response to the emotional valence of the word "keep" or "lose," conditioned stimuli interact on a decision (instrumental behavior). A recent study has directly tested the role of these Pavlovian influences in generating the framing effect (Guitart-Masip, Talmi & Dolan, 2010). In this study, the authors modified the framing task presented earlier (Figure 5.3); by substituting the safe option (presented in either "keep" or "lose" frame) with an abstract fractal image, which in a previous task phase had been paired with either gains or losses using a classic Pavlovian conditioning paradigm. What the authors found was that the conditioned (positive or negative) stimuli were able to elicit an approach-avoidance response and consequently generated a framing effect. They also replicated a similar pattern of amygdala response as that shown in the previous study (De Martino et al., 2006), thus confirming the key role of this region in the framing bias.

The Framing Effect: Inter-Subjects Variability and Neural Modulation

The discussion so far has explored the computations and neural mechanisms underpinning the framing effect. However, this still leaves unanswered the questions of why people show different degrees of susceptibility to the framing bias. In a recent study led by Jon Roiser (Roiser, De Martino, Tan and colleagues 2009) we examined how variability in the genetic background modulates the susceptibility to the framing effect. In this study we focused on a specific polymorphism (a change in the sequence of DNA associated with a large portion of the population) of the serotonin transporter gene (SERT) that is known to affect amygdala reactivity in response to emotional stimuli (Hariri, Mattay, Tessitore et al., 2002; Lesch, Bengel, Heils et al., 1996). People carrying a short(s) allelic form of this gene have increased levels of serotonin (due to a less efficient clearance of serotonin from the synaptic cleft) which is associated with higher amygdala reactivity to emotional stimuli compared with people carrying the long (l) allelic variant of the same gene. The participants to the study were screened for these two polymorphic forms of the SERT gene and separated into two groups, one comprising those subjects who were homozygous for short allele (ss) of the SERT gene and the other those subjects who were homozygous for the long allele (ll). Subjects in the ss group showed an increased susceptibility to the framing effect compared with the ll group. Critically, this increased susceptibility was associated with increased activity in the amygdala response to the framing manipulation.

But which neural circuits are involved in the modulation of the amygdala's biasing influence? A between-subject analysis of the original fMRI framing study started to address this question (De Martino et al., 2006). The analysis yielded a significant correlation between increased resistance to the framing effect (measured at the individual level) and activity in the medial orbitofrontal (mOFC) cortex, a brain region which has been shown to play a central role in value-based decision-making (Rangel, Camerer & Montague, 2008; Rangel & Hare, 2010). In other words, activity in mOFC was a good predictor of the degree to which decisions were uncontaminated by the contextual irrelevant information embedded in the framing manipulation (i.e. more rational behavior in economics terms). The role played by the mOFC in enabling individuals to make more consistent and less context-dependent decisions has recently been confirmed by a lesion study in macaques (Noonan, Walton, Behrens et al., 2010). Here the animals were presented with three visual stimuli, each one associated with a variable reward probability. A logistic regression showed that before the lesion of the mOFC the monkeys' choices were dependent solely on the difference in value between the best and second best options, and were entirely uninfluenced by the worst option (the irrelevant 3rd option) (Figure 5.4). However, after lesioning the mOFC, the monkeys' choices become systematically influenced by the value of the irrelevant option.

Exactly how the mOFC exerts a control on behavior, enabling a less context-dependent pattern of choice, has still not been fully understood. However, some recent data has hinted a likely neural circuitry. Anatomically, this region of the orbitofrontal cortex is part of a network – the medial network – which is reciprocally interconnected to a number of limbic structures. The connection between the mOFC and amygdala is one of the most prominent in this network. A number of other connections have been identified between the mOFC and the nuclei of the basal ganglia; in particular the shell and the core of the nucleus accumbens (Amaral, Price, Pitkanen & Carmichael, 1992; Ferry, Öngür, An & Price, 2000; Ongur & Price, 2000) (see Figure 5.5).

At the functional level these connections have been shown as critical in computing the value of a reward and in processing this information to control goal-directed behavior (Schoenbaum et al., 2006; Schoenbaum et al., 2009). In particular, the functional role of the connections between the amygdala (specifically the BLA nucleus) and the orbitofrontal cortex in goal-directed action has been shown using reinforcer devaluation tasks (Baxter, Parker, Lindner et al., 2000; Gallagher, McMahan & Schoenbaum, 1999; Hatfield, Han, Conley et al., 1996; Pickens, Saddoris, Setlow et al., 2003). In such tasks animals are trained to associate a CS+ stimulus (e.g. light or a sound) with an appetitive reward (e.g. food), before the food is devaluated by pairing it with an aversive outcome (e.g. illness). Animals with lesions in the OFC and BLA are impaired in responding to reinforcer devaluation (Pickens et al., 2003). Another line of research has shown that in both animals and humans the same OFC-amygdala circuit mediates fear extinction (Phelps, Delgado, Nearing &

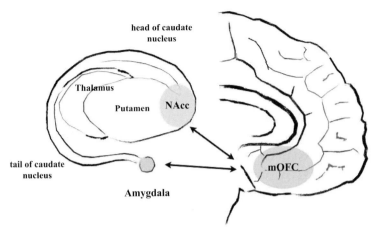

FIGURE 5.5 Schematic diagram of the connection between mOFC, Amygdala and Ventral Striatum. The increased connectivity between mOFC and Amygdala has been shown to be crucial in the modulation of the Framing effect (Roiser et al., 2009).

LeDoux, 2004; Quirk & Mueller, 2007; Sotres-Bayon, Cain & LeDoux, 2006) and that the mOFC modulation of the amygdala response is at the core of emotional regulation strategies (Delgado, Nearing, LeDoux & Phelps, 2008; Hartley & Phelps, 2009; Phelps & LeDoux, 2005).

In keeping with the above function of mOFC in modulating the amygdala response, Roiser and colleagues (2009) found that subjects carrying the long allele of the SERT gene (i.e. those who showed reduced susceptibility to the framing manipulation associated with a reduced amygdala response), had an increased functional connectivity between the mOFC and amygdala in trials where they successfully overcame the framing bias. These data suggest that control exerted by mOFC on the amygdala is a candidate mechanism through which the amygdala Pavlovian approach-avoidance influence might be reduced and a more consistent "rational" behavior be achieved.

It is important to acknowledge, however, that the ability to remain uninfluenced by contextual information might not be optimal in all circumstances. For instance the ability to incorporate a broad range of contextual cues into the decision process may on some occasions endow humans with a clear evolutionary edge. Such advantages are particularly evident in social environments in which subtle contextual cues carry critical information. This tension was evident in a recent behavioral study involving subjects affected by autism, a condition characterized by marked behavioral inflexibility and a severe social impairment. Autistic individuals showed a decreased susceptibility to the framing effect associated with a lack of emotional arousal in response to the loss frame (De Martino, Harrison, Knafo et al., 2008). This results in an unusual enhancement in logical consistency that is paradoxically more in line with the normative prescriptions of rationality at the core of the classical economics theory (Luce, 1959; Ray, 1973; von Neumann & Morgestern, 1944). However, the same inability to incorporate emotion contextual information is also likely to be the root of many of the social impairments brought by this condition.

The *Status Quo Bias*

In the examples described so far contextual information has been exogenously manipulated by the experimenters. However, a framing effect can also occur endogenously. In this case, a default frame arises naturally from the current disposition of the decision-maker at the time of the decision – the *status quo*. This type of endogenous framing effect is called the *Status Quo bias* and has been shown to be extremely perva- sive. A striking real-life example of this effect was provided by Johnson and Goldstein who discovered that the rate of organ donation regis- tered in a given country was highly dependent on the default option on the donation agreement (Johnson & Goldstein, 2003). Thus, in coun- tries where the default option is to *not donate* (i.e. people are not donors unless they decide to opt-in) the rate of donation for transplants is sig- nificantly smaller than those countries where being a donor is the default (i.e. people are required to opt-out if they do not want to be a donor). The neurobiological mechanisms mediating this effect have been investigated only recently, by two fMRI studies. In the first study Fleming, Thomas and Dolan (2010) devised a perceptual "tennis line-judgment" task in which subjects were required to decide whether or not a ball landed within two lines. The critical manipulation was changing the default: in one condition subjects were required to hold down a button if the ball landed inside the two lines (IN) but to switch and press another button if the ball was outside (OUT). In the second condition this default was reversed (i.e. keep holding if OUT, switch if IN). In the difficult cases, when the position of the ball was more ambiguous, subjects showed a strong default bias agreeing more often with whichever was the current default. At the neural level an increase in BOLD signal was detected in the bilateral subthalamic nuclei (STN) when the *status quo* was rejected in the face of heightened decision difficulty. Furthermore, functional con- nectivity analysis (DCM analysis) showed that in the context of difficult choice, connectivity between the right inferior frontal cortex (rIFC) and the STN increased in trials in which the default was rejected, suggest- ing a model of action control where information from the rIFC drives the STN to override the default bias. This finding is in keeping with a number of computational models which propose that this region is a hub in decision-making and action control (Frank, 2006; Gurney, Prescott &

Redgrave, 2001a, 2001b), and empirical findings that show a key role for the STN in switching between action alternatives (Isoda & Hikosaka, 2008).

A second study extended the investigation of the default bias beyond the perceptual domain, investigating this bias in the context of value-based judgment (Yu, Mobbs, Seymour & Calder, 2010). In this experiment, participants were presented with two cards that when flipped could result in either a win or a loss. One of the two cards was then assigned to the subject by default. Participants were required either to stay with the default card or switch to the other card. After the decision was made the cards were "flipped" and the outcomes presented to the subject. As predicted by the default bias, subjects preferred the default option more frequently (57.2% stay vs 42.8% switch). Also, they reported being more frustrated if they lost after switching cards, but also more satisfied if they won after switching. Analysis of the imaging data at the decision time revealed that increased activity in the insular cortex was associated with switching trials, while activity in the ventral striatum correlated with not switching (i.e. staying with the default card). The authors also showed that both these activations in the insula and striatum overlapped with activity at the outcome level in response to losses (insula) and gains (striatum). According to these data the authors proposed that the neural mechanism underpinning the default bias included an anticipatory somatic signal in the insula to potential losses (a proxy of loss aversion) and a "reward" signal encoded by the striatum associated with the choice of not switching. The discrepancies between the two studies presented above might be ascribed to the focus on perceptual difficulty in the former case, and emotionally-laden processes in the latter. Consistent with this interpretation, when explicit feedback is added to the perceptual task used by Fleming et al. (2010) modulation of the default bias is associated with error-related insula activity (Nicolle, Fleming, Bach et al., 2011). Thus insula and striatal activity may modulate the strength of the default bias, with action control circuitry (including STN) being important in overcoming this bias. Further studies are required to compare and contrast the role of emotional and visuomotor circuitry in the default bias.

EFFECT OF CONTEXT ON VALUES

It is well established that context influences choices. But are the actual *values* behind choices also modulated by contextual information? More precisely, does the brain compute values in a reference-dependent or reference-independent fashion? There is a general consensus in decision theory that choices should be a direct consequence of the underlying values. In simple terms, a decision-maker assigns values to different options and then chooses the one with the highest value. According to this view, our choices should simply reflect the values assigned to different options. The theory of *Revealed Preference* (Houthakker, 1950; Samuelson, 1948) introduced by the American economist Paul Samuelson rests tacitly upon this assumption. It may well be that a similar contextual modulation seen at the choice level acts also at the value level; however, there are several caveats before assuming this is true. First, at the pure experimental level we need direct evidence of the modulation of these value signals. Second, there is much new evidence that the relation between choice and value is not always so straightforward (see chapter by Dayan). It would, for example, be equally plausible to assume that the contextual modulation of choice elicited by the framing manipulation arises through intervening heuristics that act at the time of choice but yet leave value computations unchanged. To address this question we have to look at studies that directly measure these value signals.

The Role of Ventromedial Prefrontal Cortex in Representing Contextual Information

Plassmann and colleagues started to address this question by looking at hedonic value signals generated by wine tasting (Plassmann, O'Doherty, Shiv & Rangel, 2008). Participants in the study were asked to test samples of Cabernet Sauvignon from many bottles that varied widely in retail price ($5 to $90), while being scanned in an fMRI experiment. In fact, unknown to the subjects, only three types of wine were being tasted, with the same wine being presented sometimes as an expensive wine and sometimes as a cheap wine. Thus, the actual stimulus remained unchanged allowing the authors to investigate how contextual information (i.e. retail price) interacted with the hedonic value assigned by the subject to the wine's degustation. At the behavioral level subjects significantly preferred the high priced compared to the low priced wine (although of course in reality these wines were exactly the same). At the neural level, activity in ventromedial prefrontal cortex (VPFC), a region of the prefrontal cortex known to encode goal-values, correlated with an increase in the perceived pleasantness of the wine. Crucially, this same region was more active for the high price compared to the low priced wine. This data shows that the VMPFC, in encoding the hedonic value elicited by the wine testing, incorporates the information about prices that somehow seems irrelevant for the judgment about the taste of the wine. However, it is important to remember that in the face of uncertainty it is highly recommendable to use all available sources of information to make a choice. For example, in the case of the wine-tasting experiment discussed earlier, the information about the quality of the wine which is sent to the brain from the taste-buds is imprecise and difficult to decode (in particular for those of us who are not trained sommeliers). In a case like this it would not be unreasonable for a decision-maker to incorporate information about prices in her Bayesian estimate of the wine's value.

A similar contextual modulation of a value signal encoded in the VMPFC appears to account for a phenomenon known as "money illusion" (Weber, Rangel, Wibral & Falk, 2009). This is another example of a decision bias which causes people to be influenced by the "nominal value" of money rather than by its actual purchasing power (i.e. the "real" value). Money illusion is thought to be one of the causes of *price stickiness*, whereby nominal prices are slow to change even when inflation has caused the underlying real prices to rise. The authors of this study showed that the value encoded in VMPFC was highly responsive to an increase in nominal value even when the actual purchasing value remained unchanged.

Reference-Dependent Values in Orbitofrontal Cortex

Research in primates has also addressed the role of the orbitofrontal cortex in reference-dependent value computation. Tremblay and Schultz measured the firing rates of OFC neurons in response to fractal images which were predicative of a liquid or food reward: they showed that the response of many neurons in the OFC discriminated between different rewards, showing a clear reference-dependent pattern of activity (Tremblay & Schultz, 1999). More specifically these neurons coded for *relative* and not *absolute* preference. For example in an animal with an order of preference between rewards such as A > B > C, the same neuron showed low activity for B when it was presented in the same block as A (the most preferred reward), but high activity for the same identical B reward when it was presented together with C (the least preferred reward). These results have been challenged by Padoa-Schioppa and Assad, who showed that neurons in the same region of the OFC exhibit a pattern of activity that is invariant to the change in menu (i.e. reference-independent values) (Padoa-Schioppa & Assad, 2008). Three different types of rewards (A, B, C in decreasing order of preference) were used in the study. The monkey was required to choose between different quantities of two of these rewards presented at the same time on the screen. Changing the respective proportions of the different rewards allowed the authors to construct a psychometric curve and find the idiosyncratic indifference points between the different types of reward (i.e. 1A = 1.3B; 1B = 3C; 1A = 4C). Electrophysiological recordings from OFC showed that if a monkey chose between juices A, B and C offered pairwise, the activity of neurons encoding the value of juice B was independent of whether juice B was offered as an alternative of juice A or as an alternative of juice C.

The inconsistencies between these two studies have puzzled the neuroeconomics community, but a subsequent paper by Padoa-Schioppa has gone some way to reconciling these discrepancies (Padoa-Schioppa, 2009). This paper put forward the idea that neurons in the OFC adapt to the range of available rewards across blocks of trials; however, at trial-by-trial level these same neurons respond only to the relevant item (the chosen item) but their activity does not depend on the value of the other item (the unchosen item) presented at the same time. A re-analysis of the electrophysiological data supported this model. The author argued that the differences between his previous study and that of Tremblay and Schultz arose from the timescale of the range adaptation phenomenon. While in Tremblay and Schultz's study (1999) the value range remained constant for a large number of trials (highlighting a slower range adaptation phenomenon), in the study of Padoa-Schioppa and Assad (2008) the menu invariance was elicited by randomly interleaving trials with different juice pairs (emphasizing transitivity – a property of a short timescale). It is striking that the same population of neurons show such complementary capacities (i.e. reference-dependent *and* reference-independent value computations) that make them computationally efficient, while retaining a degree of preference transitivity.

The Endowment Effect: A Case of Reference-Dependent Value Computation

A paradigmatic case of reference-dependent value computation is the "endowment effect" – that is, the behavioral tendency to value an item that one already owns substantially more than an identical item that is available for purchase (Kahneman, Knetsch & Thaler, 1990, 1991; Thaler, 1980). Such apparently inconsistent behavior can be explained by assuming that prices are computed as a shift from a changing reference-point that is set by the position of the subject in the transaction, i.e. buyer or seller. To investigate its neurobiology, we devised a *within-subjects* version of the original endowment task suitable for fMRI (De Martino, Kumaran, Holt & Dolan, 2009). Subjects were endowed with a bunch of lottery tickets that they had selected from a choice of two bunches of tickets that were identical in expected value but different in color. During the scanning phase they had the opportunity to buy tickets belonging to the other bunch (i.e. the one they hadn't chosen) and sell their own tickets. In the transaction they were required to state a minimum selling price for each one of their tickets and a maximum buying price for the tickets belonging to the bunch they had not chosen. An incentive compatible mechanism, a BDM auction (Becker, Degroot & Marschak, 1964) was implemented specifically to isolate the value signal associated with the discrepancy between selling and buying prices. In keeping with previous studies, subjects assigned higher prices to tickets they owned compared to the ones they were required to buy. A parametric analysis showed that activity in the medial part of the OFC correlated with an increase in willingness to pay for the tickets (WTP – buying prices). Such a signal has been isolated in a number of other studies (Chib, Rangel, Shimojo & O'Doherty, 2009; Plassmann, O'Doherty & Rangel, 2007). In a more lateral sector of OFC activity correlated with subjects' willingness to accept a payment for tickets they already owned (WTA – selling prices). Interestingly, Hare, O'Doherty, Camerer and colleagues (2008) found that while the mOFC encoded the value of the items (WTP), a more lateral area of the OFC encoded the net value of the purchases, and thus reflected the costs of the stimuli. An interpretation of our findings that dovetails with that of Hare and colleagues is that during WTA elicitation subjects may be engaged in a computation of loss, reflected in a cost-like signal in lateral OFC (Hare et al., 2008). Indeed, this interpretation of our data is in agreement with a proposal that the WTP–WTA discrepancy arises because the transactions are perceived as a potential loss from the perspective of the seller (Kahneman et al., 1991). By contrast the dorsal striatum showed an increase for both WTP and WTA value signals. Critically, in the OFC (as opposed to the dorsal striatum) these two values signals (WTP and WTA – which should be indistinguishable in a *reference-independent* theoretical framework) were encoded by different subregions (medial and lateral OFC). This shows that information about the role (seller or buyer) of the subject in the transaction is incorporated into the computation of the object values. Furthermore the ventral portion of the striatum encoded a fully reference-dependent signal that mirrored the subject's behavior. In fact, this signal correlated with both an increase in value in the selling domain and a decrease in value in the buying domain. In other words, this area encoded the WTP–WTA price discrepancy as a function of the subject's role in the transaction, showing a striking trial-by-trial psychometric–neurometric match between neural signal and behavior. Notably, activity in this same region predicted the degree of susceptibility to the endowment manipulation at the individual level – supporting a key functional role of this computation in eliciting the observed behavior.

Taken together, these different studies provide support for a hypothesis that contextual information is incorporated directly into coding of value. However, there still remains a lack of knowledge about the specific role of the different regions and neural populations involved in the computations of these values. How precisely, from a mechanistic perspective, are reference-dependent and reference-independent values generated, and how are these integrated in the brain to influence decision-making?

THE NEUROBIOLOGY OF LOSS AVERSION

Loss aversion is proposed as a likely cause of most of the effects such as framing effect, the *status quo bias* and the endowment effect. The question then is what neurobiological circuit underpins loss aversion? Tom, Fox, Trepel and Poldrack (2007) addressed this issue using fMRI in a study where subjects were presented with a series of mixed gambles that offered a 50/50 chance to either gain or lose a given amount of money. Potential gains (ranging between $10 to $40, with $2 increments) and potential losses (ranging between –$5 to $20, with increments of $1) were manipulated independently and subjects were required to either accept or reject each proposed gamble. This setup enabled the authors to estimate the individual behavioral loss aversion (λ) computed as the ratio of the (absolute) loss response to the gain response, which yielded a median $\lambda = 1.93$, a degree of loss aversion consistent with many previous studies. The values of the potential gains and losses were entered into a regression analysis to identify brain areas showing a parametric change in the response to increasing magnitude of losses or gains. Regions encoding for potential reward responded to the increase in potential gains and this included the striatum, OFC, and dopaminergic midbrain regions. However, potential losses were also coded (deactivation in BOLD signal) by the same network. The observed pattern was suggestive of a signal that computed the net value of the gamble, by averaging the value of potential gains and potential losses. However, the most significant finding was that although a net value signal was monotonic, it was not linear: rather, it reflected the asymmetry between gains and losses shown behaviorally and predicted by prospect theory. In other words, for the majority of participants the decrease in activity in this network was steeper than the increase. Using the parameter estimates from the BOLD signal, the authors constructed a parameter that they called "neural loss aversion" that reflected the ratio between two slopes that correlated with individual behavioral loss aversion (see Figure 5.6).

A key question raised by this study concerns how such asymmetry is generated when the gamble net value encoded by the VMPFC and striatum is computed. A plausible candidate mechanism would be that the potential loss triggers an aversive "breaking" signal that is communicated to VMPFC and/or the striatum. This signal would overweight the loss component of the gamble, thereby affecting the computation of the net value to produce the kink observed in the neural signal and reflected in the value function. Consistent with a number of previous studies on framing and loss processing (some of them presented earlier), the amygdala is a likely candidate to emit such a signal. Indeed, in support of an hypothesis that an emotional response is at the root of the gain-loss asymmetry there is experimental evidence that implementation

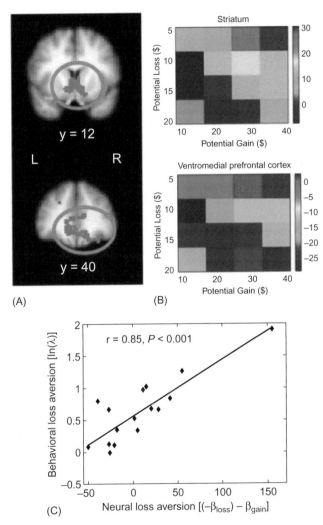

FIGURE 5.6 (A) Map showing regions with conjointly significant positive gain response and negative loss response. (B) Heatmaps were created by averaging parameter estimates versus baseline within each cluster in the conjunction map for each of the 16 cells (of 16 gambles each) in the gain/loss matrix; color coding reflects strength of neural response for each condition, such that dark red represents the strongest activation and dark blue represents the strongest deactivation. (C) Scatterplot of correspondence between neural loss aversion and behavioral loss aversion in ventral striatum. Beta-loss and beta-gain are the unstandardized regression coefficients for the loss and gain variables, respectively. (Adapted from Tom et al., 2007.)

of standard cognitive appraisal strategies (aimed to down-regulate the emotional response) produces a reduction in autonomic skin conductance response, as well as a marked reduction in loss aversion (Sokol-Hessner, Hsu, Curley et al., 2009).

Quite surprisingly, however, Tom and colleagues did not find any signal in the amygdala. One explanation for this null finding is that the range of potential losses used might not have been large enough to evoke detectable BOLD signal in the amygdala. Another reason might be the peculiar nature of the BOLD signal that is known to reflect local field potential rather than output activity (Logothetis, 2002; Logothetis, Pauls, Augath et al., 2001). If, as proposed here, the amygdala's contribution to loss aversion is mainly through its inputs to other brain structures like the VMPFC or the striatum, BOLD activity may well fail to reflect this neural signal.

To directly test a hypothesis that the amygdala is part of a computational network that produces loss aversion, we studied individuals with focal damage to the amygdala (De Martino, Camerer & Adolphs, 2010). Two subjects affected by a rare genetic syndrome that produces lesions of the amygdala nuclei were tested on a modified version of the task used by Tom and colleagues (2007). Each subject was compared with a control group closely matched in IQ, age, gender, education and monetary income. Both amygdala-lesioned participants showed a dramatic absence of loss aversion compared with their own control group and in fact one of the lesion subjects showed a mild loss *seeking* behavior. However, both individuals retained a monotonic sensitivity to reward magnitude (i.e., they preferred larger gains and smaller losses) and showed a marked dislike for increases in variance and risk. Given the well-known role of the amygdala in inhibiting instrumental behavior in response to potential threats (Adolphs, Tranel, Damasio & Damasio, 1995; LeDoux, 2000) it is plausible that the same mechanism is elicited in response to the prospect of a potential monetary loss. Such a "withdrawal" signal elicited by the amygdala would affect the value computation (in the VMPFC or striatum) and generate loss aversion. While this hypothetical model accommodates most of the experimental results currently available it will require direct testing by further studies.

CONCLUSIONS

In this chapter I reviewed some recent work that has investigated the computational and neural mechanisms that contribute to a contextual modulation of choice. In particular, this work has addressed the role that contextual information plays in value computation and the neural circuitry involved in the asymmetric computations of gains and losses.

What emerges is the idea that context is a fundamental aspect of the decision process, which needs to be explored further to have a more realistic understanding of how humans make choices. In the same way that unraveling the genetic code made sense of the great variability in the natural world, the mechanistic understanding of the way preferences are constructed has the potential to resolve in a coherent framework many of the anomalies and apparent inconsistencies of human choice.

Acknowledgments

I would like to thank Peter Dayan and Jessica Hughes for their insightful comments on the manuscript.

References

Adolphs, R., Tranel, D., Damasio, H., & Damasio, A. R. (1995). Fear and the human amygdala. *The Journal of Neuroscience, 15*(9), 5879–5891.

Allais, M. (1953). La psychologie de l'home rationnel devant le risque: Critique des postulats et axiomes de l'école Américaine. *Econometrica, 21*(4), 503–546.

Amaral, P., Price, J. L., Pitkanen, A., & Carmichael, S. T. (1992). *The amygdala: Neurobiological aspects of emotion, memory and mental dysfunction.* New York, NY: Wiley-Liss, Inc.

Balleine, B., & Killcross, S. (2006). Parallel incentive processing: An integrated view of amygdala function. *Trends in Neurosciences, 29*(5), 272–279.

Baxter, M., Parker, A., Lindner, C., Izquierdo, A., & Murray, E. (2000). Control of response selection by reinforcer value requires interaction of amygdala and orbital prefrontal cortex. *The Journal of Neuroscience, 20*(11), 4311–4319.

Baxter, M. G., & Murray, E. A. (2002). The amygdala and reward. *Nature Reviews Neuroscience, 3*(7), 563–573.

Becker, G. M., Degroot, M. H., & Marschak, J. (1964). Measuring utility by a single-response sequential method. *Behavioral Science, 9*(3), 226–232.

Brown, P., & Jenkins, H. (1968). Auto-shaping of the pigeon's key-peck. *Journal of the Experimental Analysis of Behavior, 11*(1), 1–8.

Cardinal, R., Parkinson, J., Hall, J., & Everitt, B. (2002). Emotion and motivation: The role of the amygdala, ventral striatum, and prefrontal cortex. *Neuroscience and Biobehavioral Reviews, 26*(3), 321–352.

Chen, M., & Bargh, J. (1999). Consequences of automatic evaluation: Immediate behavioral predispositions to approach or avoid the stimulus. *Personality and Social Psychology Bulletin, 25*(2), 215–224.

Chib, V., Rangel, A., Shimojo, S., & O'Doherty, J. (2009). Evidence for a common representation of decision values for dissimilar goods in human ventromedial prefrontal cortex. *The Journal of Neuroscience, 29*(39), 12315–12320.

Clifford, C., Webster, M., Stanley, G., Stocker, A., Kohn, A., Sharpee, T., et al. (2007). Visual adaptation: Neural, psychological and computational aspects. *Vision Research, 47*(25), 3125–3131.

Colwill, R., & Rescorla, R. (1988). Associations between the discriminative stimulus and the reinforcer in instrumental learning. *Journal of Experimental Psychology Animal Behavior Processes, 14*(2), 155–164.

Corbit, L. H., & Balleine, B. W. (2005). Double dissociation of basolateral and central amygdala lesions on the general and outcome-specific forms of Pavlovian-instrumental transfer. *The Journal of Neuroscience, 25*(4), 962–970.

Dayan, P. (2009). Goal-directed control and its antipodes. *Neural Networks, 22*(3), 213–219.

Dayan, P., Niv, Y., Seymour, B., & Daw, N. (2006). The misbehavior of value and the discipline of the will. *Neural Networks, 19*(8), 1153–1160.

De Martino, B., Kumaran, D., Seymour, B., & Dolan, R. (2006). Frames, biases, and rational decision-making in the human brain. *Science, 313*(5787), 684–687.

De Martino, B., Harrison, N., Knafo, S., Bird, G., & Dolan, R. (2008). Explaining enhanced logical consistency during decision making in autism. *The Journal of Neuroscience, 28*(42), 10746–10750.

De Martino, B., Kumaran, D., Holt, B., & Dolan, R. (2009). The neurobiology of reference-dependent value computation. *The Journal of Neuroscience, 29*(12), 3833–3842.

De Martino, B., Camerer, C., & Adolphs, R. (2010). Amygdala damage eliminates monetary loss aversion. *Proceedings of the National Academy of Sciences of the United States of America, 107*(8), 3788–3792.

Delgado, M., Nearing, K., LeDoux, J., & Phelps, E. (2008). Neural circuitry underlying the regulation of conditioned fear and its relation to extinction. *Neuron, 59*(5), 829–838.

Dolan, R. J. (2002). Emotion, cognition, and behavior. *Science, 298*(5596), 1191–1194.

Duranti, A., & Goodwin, C. (1992). Rethinking context: An introduction. In C. Goodwin & A. Duranti (Eds.), *Rethinking context: Language as an interactive phenomenon* (pp. 1–42). New York, NY: Cambridge University Press.

Eldridge, G., & Pear, J. (1987). Topographical variations in behavior during autoshaping, automaintenance, and omission training. *Journal of the Experimental Analysis of Behavior, 47*(3), 319–333.

Estes, W. (1943). Discriminative conditioning. I. A discriminative property of conditioned anticipation. *Journal of Experimental Psychology, 32*(2), 150–155.

Everitt, B. J., Parkinson, J. A., Olmstead, M. C., Arroyo, M., Robledo, P., & Robbins, T. W. (1999). Associative processes in addiction and reward. The role of amygdala-ventral striatal subsystems. *Annals of the New York Academy of Sciences United States of America, 877*, 412–438.

Ferry, A., Öngür, D., An, X., & Price, J. (2000). Prefrontal cortical projections to the striatum in macaque monkeys: Evidence for an organization related to prefrontal networks. *The Journal of Comparative Neurology, 425*(3), 447–470.

Fleming, S., Thomas, C., & Dolan, R. (2010). Overcoming *status quo bias* in the human brain. *Proceedings of the National Academy of Sciences of the United States of America, 107*(13), 6005–6009.

Frank, M. (2006). Hold your horses: A dynamic computational role for the subthalamic nucleus in decision making. *Neural Networks, 19*(8), 1120–1136.

Gallagher, M., McMahan, R., & Schoenbaum, G. (1999). Orbitofrontal cortex and representation of incentive value in associative learning. *The Journal of Neuroscience, 19*(15), 6610–6614.

Guitart-Masip, M., Talmi, D., & Dolan, R. J. (2010). Conditioned associations and economic decision biases. *Neuroimage, 53*(1), 206–214.

Gurney, K., Prescott, T., & Redgrave, P. (2001). A computational model of action selection in the basal ganglia. I. A new functional anatomy. *Biological Cybernetics, 84*(6), 401–410.

Gurney, K., Prescott, T., & Redgrave, P. (2001). A computational model of action selection in the basal ganglia. II. Analysis and simulation of behaviour. *Biological Cybernetics, 84*(6), 411–423.

Haigh, M., & List, J. (2005). Do professional traders exhibit myopic loss aversion? An experimental analysis. *Journal of Finance, 60*(1), 523–534.

Hare, T. A., O'Doherty, J., Camerer, C. F., Schultz, W., & Rangel, A. (2008). Dissociating the role of the orbitofrontal cortex and the striatum in the computation of goal values and prediction errors. *The Journal of Neuroscience, 28*(22), 5623–5630.

Hariri, A., Mattay, V., Tessitore, A., Kolachana, B., Fera, F., Goldman, D., et al. (2002). Serotonin transporter genetic variation and the response of the human amygdala. *Science, 297*(5580), 400–403.

Hartley, C., & Phelps, E. (2009). Changing fear: The neurocircuitry of emotion regulation. *Neuropsychopharmacology, 35*(1), 136–146.

Hatfield, T., Han, J., Conley, M., Gallagher, M., & Holland, P. (1996). Neurotoxic lesions of basolateral, but not central, amygdala interfere with Pavlovian second-order conditioning and reinforcer devaluation effects. *The Journal of Neuroscience, 16*(16), 5256–5265.

Houthakker, H. (1950). Revealed preference and the utility function. *Economica, 17*(66), 159–174.

Isoda, M., & Hikosaka, O. (2008). Role for subthalamic nucleus neurons in switching from automatic to controlled eye movement. *The Journal of Neuroscience, 28*(28), 7209–7218.

Johnson, E., & Goldstein, D. (2003). Do defaults save lives? *Science, 302*, 1338–1339.

Kahneman, D., Knetsch, J. L., & Thaler, R. H. (1990). Experimental tests of the endowment effect and the Coase theorem. *The Journal of Political Economy, 98*(6), 1325–1348.

Kahneman, D., Knetsch, J. L., & Thaler, R. H. (1991). Anomalies – the endowment effect, loss aversion, and *status-quo bias. The Journal of Economic Perspectives, 5*(1), 193–206.

Kahneman, D., & Tversky, A. (2000). *Choices, values, and frames.* New York, NY: Cambridge University Press.

Kahneman, D., & Frederick, S. (2007). Frames and brains: Elicitation and control of response tendencies. *Trends in Cognitive Sciences, 11*(2), 45–46.

Koszegi, B., & Rabin, M. (2006). A model of reference-dependent preferences. *The Quarterly Journal of Economics, 121*(4), 1133–1165.

Laughlin, S. (1989). The role of sensory adaptation in the retina. *The Journal of Experimental Biology, 146*, 39–62.

LeDoux, J. E. (1996). *The emotional brain.* New York, NY: Simon & Schuster.

LeDoux, J. E. (2000). Emotion circuits in the brain. *Annual Review of Neuroscience, 23*, 155–184.

Lesch, K., Bengel, D., Heils, A., Sabol, S., Greenberg, B., Petri, S., et al. (1996). Association of anxiety-related traits with a polymorphism in the serotonin transporter gene regulatory region. *Science, 274*(5292), 1527.

Logothetis, N. K. (2002). The neural basis of the blood-oxygen-level-dependent functional magnetic resonance imaging signal. *Philosophical Transactions of the Royal Society of London Series B: Biological Sciences, 357*(1424), 1003–1037.

Logothetis, N. K., Pauls, J., Augath, M., Trinath, T., & Oeltermann, A. (2001). Neurophysiological investigation of the basis of the fMRI signal. *Nature, 412*, 150–157.

Luce, R. (1959). *Individual choice behavior: A theoretical analysis* (115). New York, NY: Wiley. 191–243

McNeil, B. J., Pauker, S. G., Sox, H. C., Jr., & Tversky, A. (1982). On the elicitation of preferences for alternative therapies. *The New England Journal of Medicine, 306*(21), 1259–1262.

Mercer, J. (2005). Prospect theory and political science. *Political Science, 8*(1), 1–21.

Mogenson, G., Jones, D., & Yim, C. (1980). From motivation to action: Functional interface between the limbic system and the motor system. *Progress in Neurobiology, 14*(2–3), 69–97.

Morrison, S. E., & Salzman, C. D. (2010). Re-valuing the amygdala. *Current Opinion in Neurobiology, 20*(2), 221–230.

Nicolle, A., Fleming, S., Bach, D., Driver, J., & Dolan, R. (2011). A regret-induced *status quo bias. The Journal of Neuroscience, 31*(9), 3320.

Noonan, M., Walton, M., Behrens, T., Sallet, J., Buckley, M., & Rushworth, M. (2010). Separate value comparison and learning mechanisms in macaque medial and lateral orbitofrontal cortex. *Proceedings of the National Academy of Sciences of the United States of America, 107*(47), 20547–20552.

Ongur, D., & Price, J. (2000). The organization of networks within the orbital and medial prefrontal cortex of rats, monkeys and humans. *Cerebral Cortex, 10*(3), 206.

Padoa-Schioppa, C. (2009). Range-adapting representation of economic value in the orbitofrontal cortex. *The Journal of Neuroscience, 29*(44), 14004–14014.

Padoa-Schioppa, C., & Assad, J. A. (2008). The representation of economic value in the orbitofrontal cortex is invariant for changes of menu. *Nature Neuroscience, 11*(1), 95–102.

Paton, J. J., Belova, M. A., Morrison, S. E., & Salzman, C. D. (2006). The primate amygdala represents the positive and negative value of visual stimuli during learning. *Nature, 439*(7078), 865–870.

Phelps, E. A., Delgado, M. R., Nearing, K. I., & LeDoux, J. E. (2004). Extinction learning in humans: Role of the amygdala and vmPFC. *Neuron, 43*(6), 897–905.

Phelps, E. A., & LeDoux, J. E. (2005). Contributions of the amygdala to emotion processing: From animal models to human behavior. *Neuron, 48*(2), 175–187.

Pickens, C. L., Saddoris, M. P., Setlow, B., Gallagher, M., Holland, P. C., & Schoenbaum, G. (2003). Different roles for orbitofrontal cortex and basolateral amygdala in a reinforcer devaluation task. *The Journal of Neuroscience, 23*(35), 11078–11084.

Plassmann, H., O'Doherty, J., & Rangel, A. (2007). Orbitofrontal cortex encodes willingness to pay in everyday economic transactions. *The Journal of Neuroscience, 27*(37), 9984–9988.

Plassmann, H., O'Doherty, J., Shiv, B., & Rangel, A. (2008). Marketing actions can modulate neural representations of experienced pleasantness. *Proceedings of the National Academy of Sciences of the United States of America, 105*(3), 1050–1054.

Quiggin, J. (1993). *Generalized expected utility theory: The rank-dependent model.* Boston, MA: Kluwer Academic Pub.

Quirk, G., & Mueller, D. (2007). Neural mechanisms of extinction learning and retrieval. *Neuropsychopharmacology, 33*(1), 56–72.

Rangel, A., Camerer, C., & Montague, P. R. (2008). A framework for studying the neurobiology of value-based decision making. *Nature Reviews Neuroscience, 9*(7), 545–556.

Rangel, A., & Hare, T. (2010). Neural computations associated with goal-directed choice. *Current Opinion in Neurobiology, 20*(2), 262–270.

Ray, P. (1973). Independence of irrelevant alternatives. *Econometrica: Journal of the Econometric Society, 41*(5), 987–991.

Rescorla, R. (1994). Transfer of instrumental control mediated by a devalued outcome. *Animal Learning & Behavior, 22*(1), 27–33.

Rescorla, R., & Solomon, R. (1967). Two-process learning theory: Relationships between Pavlovian conditioning and instrumental learning. *Psychological Review, 74*(3), 151–182.

Rescorla, R. A., & Wagner, A. R. (1972). A theory of Pavlovian conditioning: Variations in the effectiveness of reinforcement and nonreinforcement. In A. H. Black & W. F. Prokasy (Eds.), *Classical conditioning II: Current research and theory* (pp. 64–99). New York: Appleton Century Crofts.

Roiser, J., De Martino, B., Tan, G., Kumaran, D., Seymour, B., Wood, N., et al. (2009). A genetically mediated bias in decision making driven by failure of amygdala control. *The Journal of Neuroscience, 29*(18), 5985–5991.

Salzman, C. D., & Fusi, S. (2010). Emotion, cognition, and mental state representation in amygdala and prefrontal cortex. *Annual Review of Neuroscience, 33*(1), 173–202.

Samuelson, P. (1948). Consumption theory in terms of revealed preference. *Economica, 15*(60), 243–253.

Schoenbaum, G., Roesch, M., & Stalnaker, T. (2006). Orbitofrontal cortex, decision-making and drug addiction. *Trends in Neurosciences, 29*(2), 116–124.

Schoenbaum, G., Roesch, M., Stalnaker, T., & Takahashi, Y. (2009). A new perspective on the role of the orbitofrontal cortex in adaptive behaviour. *Nature Reviews Neuroscience, 10,* 885–892.

Sokol-Hessner, P., Hsu, M., Curley, N., Delgado, M., Camerer, C., & Phelps, E. (2009). Thinking like a trader selectively reduces individuals' loss aversion. *Proceedings of the National Academy of Sciences of the United States of America, 106*(13), 5035–5040.

Sotres-Bayon, F., Cain, C., & LeDoux, J. (2006). Brain mechanisms of fear extinction: Historical perspectives on the contribution of prefrontal cortex. *Biological Psychiatry, 60*(4), 329–336.

Talmi, D., Seymour, B., Dayan, P., & Dolan, R. (2008). Human Pavlovian instrumental transfer. *The Journal of Neuroscience, 28*(2), 360–368.

Thaler, R. (1980). Toward a positive theory of consumer choice. *Journal of Economic Behavior & Organization, 1*(1), 39–60.

Tom, S. M., Fox, C. R., Trepel, C., & Poldrack, R. A. (2007). The neural basis of loss aversion in decision-making under risk. *Science, 315*(5811), 515–518.

Tovar, P. (2009). The effects of loss aversion on trade policy: Theory and evidence. *Journal of International Economics, 78*(1), 154–167.

Tremblay, L., & Schultz, W. (1999). Relative reward preference in primate orbitofrontal cortex. *Nature, 398*(6729), 661–663.

Tulving, E. (1973). Encoding specificity and retrieval processes in episodic memory. *Psychological Review, 80*(5), 352–373.

Tversky, A., & Kahneman, D. (1974). Judgment under uncertainty: Heuristics and biases. *Science, 185*(4157), 1124–1131.

Tversky, A., & Kahneman, D. (1981). The framing of decisions and the psychology of choice. *Science, 211*(4481), 453–458.

Tye, K., & Janak, P. (2007). Amygdala neurons differentially encode motivation and reinforcement. *The Journal of Neuroscience, 27*(15), 3937–3945.

von Neumann, J., & Morgestern, O. (1944). *Theory of games and economic behavior.* New Jersey: Princeton University Press.

Weber, B., Rangel, A., Wibral, M., & Falk, A. (2009). The medial prefrontal cortex exhibits money illusion. *Proceedings of the National Academy of Sciences of the United States of America, 106*(13), 5025–5028.

Williams, D., & Williams, H. (1969). Auto-maintenance in the pigeon: Sustained pecking despite contingent non-reinforcement. *Journal of the Experimental Analysis of Behavior, 12*(4), 511–520.

Yu, R., Mobbs, D., Seymour, B., & Calder, A. (2010). Insula and striatum mediate the default bias. *The Journal of Neuroscience, 30*(44), 14702–14707.

Preference Change through Choice

Petter Johansson[1]*, Lars Hall[2] and Nick Chater[3]

[1]Division of Psychology and Language Sciences,
University College London
[2]Lund University Cognitive Science, Lund University
[3]Behavioural Science Group, Warwick Business School,
University of Warwick

INTRODUCTION

In an unforgettable performance by the English comedian Eddie Izzard, he portrays the Spanish Inquisition as conducted by the wimpy Anglican Church. Playing a supremely domesticated inquisitor, he offers

*Corresponding author.

Neuroscience of Preference and Choice
DOI: 10.1016/B978-0-12-381431-9.00006-1

the accused heretics the intriguing choice between "Cake or Death?" Apart from the hilariousness of this scene, the social scientist can appreciate it as one of the last safe havens for neoclassical economics, rational choice, and expected utility theory, because people really and truly have stable and identifiable preferences to help them decide between cake and death. There is simply no amount of social-psychology shenanigans that could push this preference around (no matter how little Festinger would pay people for choosing the death sentence, it would not generate enough cognitive dissonance to sway anyone) (Festinger, 1957).

But as soon as we move away from cushy comedic inquisitions towards actual real-world decisions, the situation becomes considerably murkier. In another of his shows, Izzard delivers a brilliant caricature of people's vocational choices. "Vocation, you got to *go* for it, you can't just fall into it," he says. And then he illustrates his point with the fervent calls of the taxidermist and the beekeeper. "You gotta *want* to be a Taxidermist! Yes! I *want* to fill animals with sand! I want to get more sand into an animal than anyone has ever done before /.../ You got to *want* to be a beekeeper! I *want* to be a beekeeper! I *wanna* keep bees! Don't wanna let them get away; I *wanna* keep them! They have too much freedom." Here, much of the comedy lies in how difficult it is for us to imagine how, from all the imaginable options available, someone actually decides to become a taxidermist or beekeeper[1]. It is not that there is a reason to believe that stuffing sand into animals or stealing honey from bees are particularly unpleasant occupations to have (except perhaps as Izzard notes, the part where the beekeeper realizes: "oh my god, I'm covered in beeeeeeees"), but rather that it seems so unlikely that all those utility calculations line up properly at the many, many decision points that takes a person from a teensy toddler to a towering taxidermist. In contrast to the previous example, and barring ethical concerns, psychologists would have a field day framing and anchoring and frivolously fiddling with vocational trajectories like this.

In this chapter we look at examples of research showing how choice feedback effects may explain people's progression through a decision space, and we set out a number of desiderata for this type of research to meet in order to fully illuminate the impact of choices in real life.

Work by Ariely, Loewenstein and Prelec (2003, 2006) strongly suggests that arbitrary and irrelevant factors can not only influence participants in their assessment of the utility of different goods (such as when

---

[1] For all you closeted friends of the conjunction fallacy out there, we have to admit to ourselves that somehow it does not seem all that unlikely that a taxidermist would double as a beekeeper, but consulting the oracle of the Internet, we could only find reference to one such person; a William Jones Weeks, who lived in the hamlet of Yaphank in Longwood Community, Long Island, NY, and who was an inventor, scientist, horticulturalist, beekeeper, and taxidermist).

rumination on the digits of their social security number leads participants to create wildly different anchors for how much they are willing to pay for a bottle of wine), but that these factors can be maintained through longer decision trajectories, and creating a form of "coherent arbitrariness" (i.e. stable market patterns of revealed preferences) (Ariely, 2008). In the words of Norton and Ariely: "These results demonstrate a kind of self-herding, in which people observe their past behavior, infer some amount of utility and act in accordance with the inference of utility, despite the fact that this behavior can be based not on the initial choice driven by hedonic utility but on any host of trivial situational factors that impacted the first decision" (Ariely & Norton, 2008, p.14).

Recent research on the monetary valuation of pain has yielded similar conclusions (Kurniawan, Seymour, Vlaev et al., 2010; Vlaev, Seymour, Dolan & Chater, 2009). Given a fixed amount of "cash in hand," and two differing levels of pain to which they may be subjected, people will consistently pay more to avoid (or, in a motor task, take more trouble to avoid) the worse pain. But if the level of pains, or the amount of available money, is varied from one person to the next, the monetary "valuations" of any specific pain turn out to be wildly unstable, in highly predictable ways. I might be willing to pay just 20p to avoid a specific pain in one part of the experiment; but might cheerfully pay 40p to avoid the very same pain a few minutes later. One interpretation of these results, consonant with Ariely's explanation above, is that people have essentially no idea how to relate money and subjective experience; but they do want to be consistent, as far as possible. But such consistency cannot easily be enforced *across* contexts, perhaps because people have no *absolute* representation of subjective experiences (whether pains or other sensory inputs) (Chater & Vlaev, 2011; Stewart, Brown & Chater, 2005).

This, we surmise, is the first aspect to look out for in a framework to deal with feedback effects of choice: that it present itself as a serious contender to deal with both contextualized and realistically scaled time series of choices (even if the long-term effects are mainly implied).

But a limitation of the demonstrations described above is that it remains possible that the arbitrary factors might have fizzled if the starting state had concerned preferences of greater importance for the participants. In such a case, presumably the self-observation that propels self-herding of arbitrary preferences would instead work in concert with the original utility-based choice, making it even more resistant to contextual effects in further instances.

Thus, the second aspect to look out for in relation to feedback effects of choice is a demonstration of arbitrary contextual influences with real punching power. These are not uncommon. For example, in the case of hospital patients reporting on their suffering and life satisfaction, such a simple manipulation as giving out a high or a low frequency scale for rating the occurrence of physical symptoms markedly effects their judgments of health satisfaction, i.e. they "feel" much worse if the scale has a low frequency skew to make a particular symptom look very rare (the scale ranging from "never" to "more than twice a month"), and much better if it has the opposite skew (see Schwarz & Oyserman, 2001). This is so even if the low frequency scale in itself influences the patients to under-report the occurrence of the symptom in the first place. At the same time, if the patients' task instead concerns how much their symptoms bother them, then a high-frequency scale (which encourages higher estimates of symptom frequency) elicits reports of greater suffering (Schwarz & Oyserman, 2001). In this way, two very similar introspective reports (of general health satisfaction and subjective suffering) can be made to blatantly contradict each other. Perhaps even more striking is the result reported by Schwarz (1999). In this study participants rated the things that they deemed most important in their lives. When presented with a structured questionnaire format with a long list of alternatives, 61.5% reported that the most important thing for them was to care for and prepare their children for life. In an open-ended format only 4.6% reported similar sentiments. Thus, we find a gross, perhaps even disquieting, discrepancy in what the participants chose to report as being *the most important thing in their lives*, caused by nothing more than a switch between two common survey formats (for further examples, see Winkielman & Schwarz, 2001; Oishi, Schimmack & Colcombe, 2003). On the other hand, this remarkable series of studies made no attempt to track the potential reverberations of these snapshot judgments, or to relate them to economic variables, such as willingness to pay, or to even ask the participants to reason further about their statements. Therefore, this work does not address our first consideration, concerning the ramifications of choice of future beliefs, attitudes or behavior.

The third and perhaps most important dimension to get a grip on choice-feedback dynamics is to pay close attention to the relationship between what economists would call stated and revealed preferences, and what psychologists would call the attitude–behavior gap (Ajzen & Fishbein, 2000; Gross & Niman, 1975; Schelpler, 2010). However, as we see it, the best characterization of this problem comes neither from economists nor psychologists, but from philosophers, who would point out that the crucial thing lacking here is a basic theory of introspection and intentionality (Dennett, 1987). Decades of work on human decision-making, variously stressing either rational or irrational aspects of choice behavior (Krueger & Funder, 2004; Tversky & Kahneman, 1981), have yet to find a compelling method for studying the relationship between choice and introspection (Johansson, Hall, Sikström & Olsson, 2005). The greatest barrier for scientific research in this domain is simply the nature of subjective experience. How can researchers ever corroborate the reports of the participants involved, when they have no means of challenging them? As philosophers have long noted, incorrigibility is a mark of the mental (Rorty, 1970). Who are *they* (whether "they" are psychologists, economists or philosophers) to say what *my* reasons are?

The most troublesome thing about the dichotomy between stated and revealed preferences is not the fuss over whether actions speak louder than words (they do), but rather that talking is one of the most ubiquitous actions we engage in, and therefore one of the most obvious channels for the study of choice feedback effects. Thus, stated preferences are always prime and potent candidates to change and induce further preferences, particularly as these statements very seldom are allowed to stand undisputed in the social fray of our lives. As Dennett (1993) says: *"human beings have constructed a level that is composed of objects that are socially constructed, replicated, distributed, traded, endorsed..., rejected, ignored, obsessed about, refined, revised, attacked, advertised, discarded"* (p. 230). Here, we find both a glimmer of rationality in the distribution of information traveling between minds – in the asking, judging, revising, and clarifying of critical, communal discourse. But we also find the many social pressures and pitfalls where we unwittingly assume the warped views of parents and peers, and where we latch onto questionable authorities we think know better, or people that are just better known (when Penélope Cruz bats her nine inch L'Oreal-crusted eyelashes and says: *"because I'm worth it!"* we may think we are worth it too). Despite its obvious importance, this incessant discourse of decision-making is left out of the great majority of both economic and psychological models of preference change.

In summary, to forcefully approach the issue of choice feedback effects, we would like to see a focus on all three desiderata described above: (i) to clearly establish the effect as a plausible candidate for influence over multiple extended choice points, (ii) to aim for generalizability beyond preference free starting states by approaching contexts involving manifest preferences, and (iii) to realize the pivotal role individual and social discourse plays in shaping our preferences (whether consciously or not), and thus make an attempt to deal with both stated *and* revealed preferences in the same experimental paradigm.

CHOICE BLINDNESS

In an attempt to investigate self-knowledge and the representational nature of decisions and intentions, we recently introduced the phenomenon of choice blindness (e.g. Johansson et al., 2005). It is a choice paradigm inspired by techniques from the domain of close-up card-magic, which permits us to surreptitiously manipulate the relationship between choice and outcome that our participants experience. In Johansson et al.

(2005) participants were shown pairs of pictures of female faces, and were given the task of choosing which face in each pair they found most attractive. In addition, immediately after their choice, they were asked to verbally describe the reasons for choosing the way they did. Unknown to the participants, on certain trials, a double-card ploy was used to covertly exchange one face for the other. Thus, on these trials, the outcome of the choice became the opposite of what they intended.

From a common-sense perspective, it would seem that everyone would immediately notice such a radical change in the outcome of a choice. But that is not the case. The result showed that in the great majority of trials our participants were blind to the mismatch between choice and outcome, while nevertheless being prepared to offer introspectively derived reasons for why they chose the way they did. When analysing the reasons the participants gave it was also clear that they often confabulated their answers, as when they referred to unique features of the previously rejected face as being the reason for having made their choice (e.g. stating that *"I liked the earrings"* when the option they actually preferred did not have any). Additional analysis of the verbal reports in Johansson et al. (2005) as well as Johansson et al. (2006) also showed that very few differences could be found between cases where participants talked about a choice they actually made and those trials where the outcome had been reversed.

Choice blindness is a robust, replicable, and often dramatic effect. We have demonstrated it for attractiveness of abstract artistic patterns and for male and female faces, both when presented "by hand" as described above (Johansson et al., 2005, 2006), and when the alternatives are presented on a computer screen (Johansson, Hall & Sikström, 2008). In Hall, Johansson, Tärning et al. (2010) we examined if choice blindness would extend to choices made in more naturalistic settings, and to modalities other than vision. We set up a sample stand at a local supermarket, where we invited customers to participate in a blind test of two paired varieties of jam and tea. By using a concealed chamber in the jars, we could switch the content before asking the participants to sample again and motivate their choice. The results showed that no more than a third of all manipulation trials were detected by the participants. Even for such remarkably different tastes as spicy Cinnamon-Apple and bitter Grapefruit, or for the sweet smell of Mango and the pungent Pernod, less than half of all manipulation trials were detected.

Recently, we have also established the effect for multi-attribute choices (Johansson et al., in preparation (a)), for self-knowledge of personality traits (Johansson et al., in preparation (b)), and even for moral judgments involving hotly debated topics in the current political debate (Hall, Johansson & Strandberg (submitted)).

Choice blindness as an experimental design is the first to give experimental researchers the opportunity to systematically study how confabulatory reports are created and how they relate to standard or "truthful" reports about choice behavior. As a general method of investigation choice, blindness elevates inert hypothetical statements to powerful covert counterfactuals (i.e. from what do we think would have happened if we had chosen otherwise, to what actually happens when we get what we did not choose). However, perhaps the greatest potential of choice blindness as an experimental technique lies in the study of preference formation and preference change. Once the participants in a choice blindness experiment have accepted and argued for the opposite of their choice, they have already displayed (at least a stated) reversal of their initial preference. But in the recent studies, we have also started to look at what happens to that preference over time, e.g. when the manipulated choice enters the dynamics of preference formation, will the participants revert to their originally revealed preference, or will they prefer the option they were led to believe they liked?

CHOICE BLINDNESS AND PREFERENCE CHANGE FOR FACES

In psychology, an influential tradition has held that choices made influence future preferences; we come to prefer what we have chosen more, while the rejected alternative is liked even less (Brehm, 1956). This effect has been demonstrated for a wide range of choices (e.g. Gerard & White, 1983; Schultz & Léveillé, Lepper, 1999; Sharot, de Martino & Dolan, 2009), and for populations as different as amnesics (Lieberman, Ochsner, Gilbert & Schacter, 2001), young children (Egan, Santos & Bloom, 2007) and capuchin monkeys (Egan, Santos & Bloom, 2010).

But recent studies have questioned the free-choice paradigm (FCP), which is the main methodology used in these experiments (Chen, 2008; Chen & Risen, 2010). In its original form, the participants first rate a number of objects, then make a choice between two alternatives close in rank, and finally rate all the objects a second time. The typical finding is that the distance in rating between the chosen and the non-chosen object has increased when the rating is done at the end of the experiment, which has been interpreted as being due to the choice made between the two alternatives.

But Chen (2008) points out that this and all other versions of the free choice paradigm fall prey to a set of egregious statistical errors. In Brehm (1956), and in subsequent studies using the rating-based version of the FCP, the common procedure has been to remove all participants that are not consistent between the first rating and the choice – i.e. they first rate A over B, but then choose B over A in a direct binary choice between them (e.g. 21% of the participants in Brehm 1956 were removed for this reason). The remaining participants are then compared to a control condition in which only two ratings are performed without any intermediate choice. The underlying rationale for removing inconsistent participants is that there is no stable preference for the choice to influence. But this procedure unfortunately introduces a bias in the dataset. If the choice is seen as the "true" or revealed preference and the rating as an informative but less stable measure, removing the inconsistent participants effectively removes all participants that may have a weak preference for the chosen object. When comparing the difference between the first and second rating in the control and the choice condition it is thus not that surprising that a greater "spread" between the alternatives is found in the condition in which all the participants with weak preferences are removed.

The objection by Chen and Risen effectively undermines the entire tradition of research using the FCP. Given that the results of FCP research has been taken for granted for so long, and has been cited and relied upon in numerous other related studies, such a conclusion would have far reaching consequences.

In an effort to help fill this void, we recently adapted our choice blindness paradigm to incorporate a measure of preference change. If the participants in our experiments accept the reversed outcome of their choice and then also changed their future preferences in line with the manipulations made, it would serve as firm evidence that choices can indeed influence future preferences.

Using the same methodology as in Johansson et al. (2005), we let participants choose between two faces, and for some trials we reversed their choices using the card-trick. In the first version of the new experiment, to add critical time-series data, we allowed the participants to make a second round of choices using the same face pairs (Hall et al., in preparation (b)). First of all, the detection rate was as low as in Johansson et al. (2005), thus replicating the basic choice blindness effect with a different set of faces. The new measure of preference was choice consistency, i.e. to what extent the participants prefer the same face both the first and the second time they are presented with the pairs. For the non-manipulated trials, the choice consistency was very high (93%). However, in the manipulated trials consistency dropped as low as 56%, which demonstrates that choice blindness clearly influenced the participants preferences, as they were much more likely to pick the originally non-preferred faces the second time they evaluated a pair.

In addition to the first round of choices, the participants also rated the attractiveness of the chosen and the non-chosen picture directly after each verbal motivation. After the second round of choices, the pictures were presented one by one in a random sequence, and the participants rated them one more time. The difference in rating between the chosen and the non-chosen faces (i.e. a measure equivalent to the "spread" used in the free choice paradigm) also differed dramatically between the manipulated and the non-manipulated trials, both in the first and in the second rating. The originally chosen faces are thus rated much higher than the non-chosen faces for the non-manipulated trials (which makes perfect sense), while this relationship is more or less reversed for the manipulated trials (i.e. they rate the faces they were led to believe they liked higher, and the ones they thought they did not like lower). Of special interest here is the second rating, as the difference in spread indicates that the preference change is also present outside a pairwise comparison between the faces, and is stable enough to last long after the manipulated choice is performed.

Comparing with previous studies using the free choice paradigm, we avoid the statistical issues raised by Chen and Risen, and thereby show that making a choice can indeed influence future preferences for the chosen and the rejected alternative.

Very recently, another study remodelled and re-established the free choice paradigm itself. In Sharot, Velasques and Dolan (2010), the participants first rate a long list of names of holiday destinations, and are then asked to make choices between two destinations subliminally presented on the screen. The choice pairs are constructed from destinations with equal initial ratings, and they are presented for 2 msc followed by a mask. After the choice, the two masked alternatives are made visible accompanied by a star marking which alternative the participants had indicated as their choice. Finally, the participants rate all the alternatives one more time. The new feature is that there is actually nothing presented during the "subliminal" presentation – two non-words are displayed before the mask, which means that when the participants make their choice it is not based on any real information. Still, this procedure led to an increase in the rating of the "chosen" alternative (but no reduction in value for the "rejected" alternative). This new methodology is called "blind choice," which refers to the fact that the participants don't know what the alternatives are when the "choice" is made.

There are some important similarities and differences between this and our study. Through its elegant design, Sharot et al. (2010) prove that the choice *in itself* can have an impact on subsequent evaluations of the choice alternative. But this very design also makes sure there are no previous preferences involved in *making* the choice, as no choice was actually made. In contrast, when the preferences are changed in our study, they are reversed in relation to what the initial choice revealed was the participants preference, a result that indicates a much stronger potential for preference change as a result of prior choices.

Still, the two studies together clearly point at an interpretation of the choice effect as being due to the induced *belief* in preferring one option over the other. It has previously been argued that it is the comparison and evaluation of the two alternatives that drives the change in preference (e.g. Shafir, Simonson & Tversky, 1993), but in our studies, that cannot be the case, as the initial evaluative process ended in the opposite direction.

CHOICE BLINDNESS AND PREFERENCE CHANGE FOR RISKY CHOICES

In the study of risk and decision-making in both economics and psychology, an almost universally employed tool is hypothetical monetary gambles (e.g., Abdellaoui, 2000; Birnbaum, 2008; Brandstätter, Gigerenzer & Hertwig, 2006; Tversky & Wakker, 1995). By studying how people behave when faced with potential losses and gains under different probability conditions, models are constructed for behavior under risk and uncertainty (e.g., Gonzalez & Wu, 1999; Prelec, 1998; Wakker, 2004). This type of modeling gained even more recognition when Daniel Kahneman earned his Nobel Prize in Economics in 2002 for the invention and development of Prospect Theory (Kahneman & Tversky, 1979). And in recent years, the use of monetary gambles has become even more ubiquitous, with the advent of neuroeconomics as a separate and high-profile subfield of economics and cognitive science (e.g. Camerer, Loewenstein & Prelec, 2005; Glimcher, 2003; Kenning & Plassman, 2005).

One assumption that both economic and many psychological theories of decision-making often take for granted is that people's preferences for decisions under risk are stable constructs that stay constant over time and regardless of task (e.g., von Neumann & Morgenstern, 1947; Brandstätter et al., 2006; Tversky & Kahneman, 1992; Tversky & Koehler, 1994; but see Kusev, Van Schaik, Ayton et al., 2009; Stewart, Chater & Brown, 2006; Ungemach, Stewart & Reimers, 2011). For example, one of the main predictions of *Cumulative Prospect Theory* (Tversky & Kahneman, 1992) is the so-called four-fold pattern of risk attitudes: people will be risk averse (prefer the safe bet) for gains and risk seeking (prefer the risky option) for losses of moderate to high probability, and they will be risk seeking for gains and risk averse for losses of low probability. These risk profiles are sometimes thought of as analogous to personality characteristics; they are stable traits that can be used to predict how people will behave in the long-term.

The abundant use and reliance on monetary gambling scenarios, as well as the accompanying assumption of people's stability of preferences for decisions under risk, makes this a very interesting domain for the application of the method of choice blindness. Is it at all possible to change the outcome of a choice without people noticing when the alternatives are so detailed and explicit? And if the participants would fail to notice when their choices are being manipulated, what will happen to their general preferences over time? Will they change their risk profile from risk seeking to risk averse if this is the direction of the manipulation?

To help answer these questions, we recently constructed a choice blindness an experiment using hypothetical gambles (Kusev et al., in preparation (a)). The design closely follows the format of experiments in this tradition. The participants answer a number of gambling questions of the following type: *"what would you prefer: alternative (A) 45% of losing £100, or alternative (B) a certain loss of £50?"* The sums in alternative B are varied around the point of the expected value in alternative A (in this case £45). After the participants have completed their choices, they are presented with all the scenarios again with their previous choices highlighted. The task this time is to indicate if they want to confirm or reject their initial choice, and also to indicate how satisfied they are with what they have chosen. Finally, the scenarios are presented a third time, and the participants have to evaluate the gambles once more. When the participants are asked to confirm or reject their initial choices, some scenarios have the opposite of their first choice highlighted, thereby reversing their initial choices. These changes are consistently made in the opposite direction to participants' original preferences, e.g., from risk seeking to risk avoiding choices for all gambles with moderate and high probabilities of loss. The results showed that very few participants detected the manipulation in the second phase of the experiment, and when rating how satisfied they were with their previous choices, they were equally satisfied with their manipulated and non-manipulated choices (Kusev et al., in preparation (a)). In addition, we were able to demonstrate an overall significant change in their risk preferences for the repeated choice scenarios, and in some conditions even a complete preference reversal for some of the probability levels.

This result shows that choice blindness can affect not only esthetic preferences, but also supposedly more general preferences like risk aversion. Asymmetries and preference reversals for risk has been demonstrated many times before (see Lichtenstein & Slovic, 1971, 2006), but this is the first time it follows as a consequence of a manipulation of prior choices. As such, it adds to the accumulated evidence that people do in fact *not* have stable preferences for risk (e.g. Stewart, Chater, Stoff & Reimers, 2003; Stewart et al., 2006). Instead, a myriad of things can influence a risky choice, like task descriptions and presentation, memory of prior choices, complexity of the gamble, and computational skills (see Kusev, Tsaneva-Atanasova, van Schaik & Johansson, in preparation (b), for our attempt to integrate all these factors into one model of risky choice). In short, the conclusion must be that if it were the case that people had strong and enduring preferences for risk, it seems unlikely that they would accept a reversal of their choices, endorse and rate them as equally good as the choices in the non-manipulated trials, and then adjust their subsequent choices in line with the manipulations made.

CHOICE BLINDNESS AND PREFERENCE CHANGE FOR POLITICAL OPINION

The most salient and immovable ideological archetype across the political landscape in the EU and the US is the division between Socialists and Conservatives (left wing vs right wing). Despite a persistent trend towards diminishing and more flexible party affiliation among voters, partisanship across the left-right divide still holds an iron grip on the international western electorate, and has even shown evidence of further polarization in recent years. (For example, see Abramowitz & Saunders, 1998, 2008; Lewis-Beck, Norpath, Jacoby & Weisberg, 2008; Bafumi & Shapiro, 2009; Carsey & Layman, 2006; and Dodson, 2010 for analysis relating to the condition in the US; and Clarke, Sanders, Stewart & Whiteley, 2009; Kitschelt, 2010; Enyedi & Deegan-Krause, 2010; and Bornschier, 2010 for the EU perspective. See also Dalton, 2009; and Cwalina, Falkowski & Newman, 2010, for cross-cultural comparisons). Given this extraordinary stability of political opinion, we were very keen to investigate the potential impact of choice blindness for voter preferences during the final stretch of the 2010 general election in Sweden.

Despite a tradition of bipartisanship at the national assembly, at the level of voter identity the Swedish electorate is regarded as one of the most securely divided populations in the world, with a small number of voters that "float" or "swing" across partisan lines often having a profound impact on the outcome of the election (see, for example, Oscarsson & Holmberg, 2008 for a succinct analysis of the 2006 election). When we entered into the study, the tracking polls from commercial and government pollsters were polling the electorate at a mere 10% undecided between the two opposing (socialist-green and conservative-liberal) coalitions, with the common wisdom of the political scientist tagging very few additional voters as open for a coalition swing with only a few weeks left of the campaign (Petrocik, 2009; Holmberg & Oscarsson, 2004; Oscarsson, 2007).

To conduct the study, we approached people in the streets of the cities of Malmö and Lund and asked if they were willing to fill out an "election compass," a survey-format designed to establish which political coalition (or party) best fits the views of the respondent. The concept of election compasses is well known in Sweden; all major media outlets create their own online versions, often in collaboration with political scientists.

In our case, the questions concerned salient issues from the election campaign where the left- and the right-wing coalition held opposite positions (with a focus on traditional issues in the conservative-socialist divide, such as taxation and privatization).

The people who agreed to participate started by indicating their current voting intention for the election at the coalition-level, and then they proceeded to mark their opinion on our 12 statement election compass. After the participants had completed the survey, we asked them to explain and justify their stance on a few of the issues. However, at this point, we used a sleight-of-hand to alter the sum of the participants' answers in the opposite direction of the stated voting intention – i.e. from leaning socialist to conservative, or from conservative to socialist (see Figure 6.1). Thus, when the participants were asked to justify their opinion, their position had been reversed (for example, if they previously thought the tax on petroleum ought to be raised, the manipulated answering sheet now appeared to indicate that they had responded that it ought to be lowered).

Next is the crucial step of the experiment. After the participants had worked through the answers to the specific issues in the election compass, we overlaid a color coded semi-transparent correction template on their (manipulated) answering profile. In collaboration with the participants, we then tallied an aggregate "compass" score for the right and left wing side, indicating which political coalition they favored (Figure 6.1D). We then asked them to explain and comment on the summary score, and as the final step of the experiment, to once again indicate the direction and strength of their voting intention for the upcoming election.

Remarkably, the results showed that only 20% of the manipulated answers were noticed by the participants. Instead, they often volunteered coherent and elaborate arguments why they agreed with the reversed position. For the participants who did detect the changes, they almost invariably attributed the mismatch as a result of them having misunderstood the question the first time. In these cases, they were given the chance to once more express their attitude on the scale. The low rate of detection allowed us to move the aggregate scores of 91% of all participants across the partisan dividing line (i.e. only 9% of the participants detected enough manipulations to adjust their summary score back to the original coalition profile). Thus, we managed to create a situation where an overwhelming majority of the participants accepted a preference reversal across the socialist–conservative divide when aggregating the 12 campaign issues from the compass.

The critical concern now is whether this induced preference reversal managed to leap across the attitude–behavior gap, and impinge on real-world behavior (remember, this is immersed in a live campaign only a few weeks before the election, where stated voter intentions correlate

FIGURE 6.1 (A) The participant first indicates the direction and strength of his or her voting intention for the upcoming election, and then rates to what extent they agree with 12 statements selected to differentiate between the two political coalitions (e.g. concurring with "the tax on petroleum should be raised" would tally a point in agreement with the left-wing coalition, etc.). Meanwhile, the experimenter (left-most in the image) pretends to take notes, but instead monitors the markings of the participant, and creates an alternative answering profile favoring the opposite view. The new set of ratings is written on a slip of paper identical to the rating section on the questionnaire. (B) After completion, the participant hands over the questionnaire to the experimenter, who has hidden the alternative answer-slip under the notebook. The notebook has a non-permanent adhesive surface, and when the experimenter swipes it over the questionnaire it attaches and occludes the section containing the original ratings. (C) Next, the participant is confronted with the reversed answers, and asked to justify the manipulated opinions. (D) Then the experimenter covers the (manipulated) ratings of the participants with a color-coded, semi-transparent correction template, and sums up the results indicating which coalition the participant favors (8 right-wing vs 4 left-wing, or 2 right-wing vs 10 left-wing, etc.). Finally, the participant is asked to justify his or her aggregate position, and once again indicate the direction and strength of their voting intention for the upcoming election.

extremely well with actual voting, see Holmberg & Oscarsson, 2004). What we found was that, compared to the initial voting intention, 10% of our participants moved across the full ideological span, and switched their voting intention from firmly right-wing to firmly left-wing. A further 22% went from expressing unequivocal coalition support (left or right), to becoming entirely undecided, and 3% went from being undecided to having a clear voter intention. In addition, 10% of the participants recorded substantial movement in the manipulated direction along the confidence scale – moving from "absolutely sure" to "moderately sure." If we add to this that around 12% of participants were undecided both before and after the experiment (a figure roughly corresponding to the category of undecided voters in the traditional opinion polls), we end up with a figure of more than half of all participants being open for movement across the great partisan divide ("in play" as the pollsters would say), a figure dramatically different from the expectations of political scientist, pollsters, party campaign strategists, and not least the voters themselves.

In summary, we have demonstrated considerable levels of self-induced preference change through choice blindness for a highly charged and important domain of political life. As we see it, there is also a robustness to the results, as any potential experimental demand would line up more forcefully on the side of being consistent with the original voting intention (expressed on the exact same scale as the one at the end of the experiment), than being consistent with the compass score – which can always be discounted as not being representative of the interests of the participants, or the campaign focus, etc. It should also be noted that in no part of the experiment did we provide arguments in support or opposition to the expressed views of the participants – the participants did all the cognitive work themselves. This aspect is especially interesting, as a number of recent studies have emphasized how hard it is to influence peoples' voting intentions with "regular" social psychology tools, like framing (Druckman, 2004) and dissonance induction (Elinder, 2009), or indeed objective arguments, even when the political opinion held is based on factual misconceptions that are demonstrably false (Nyhan & Reifler, 2010).

DISCUSSION

Choice feedback effects have been a focus for psychologists for the better half of a century, and it is time to set some goals and standards for further studies in this domain. In the introduction to this chapter we listed three desiderata for such experiments: (i) to establish the effect as a plausible candidate for influence over multiple choice points, (ii) to aim for generalizability by involving manifest preferences, and (iii) to make an attempt to deal with both stated *and* revealed preferences in the same experimental paradigm. As we see it, the primary impetus for these criteria lies not in the promise of squeaky clean theoretical contrasts between utility and choice-based models (a real problem, admittedly), but rather in an attempt to marry the penetrating inventiveness of social-psychology and decision and judgment research with the crass real world demands of economics and consumer studies.

The studies described above use choice blindness as a novel tool to investigate preference change, and, in particular, consider the impact of modifying people's beliefs about what they previously chose. Echoing the results of the research on valuation described in the introduction (e.g., Ariely et al., 2003, 2006; Ariely & Norton, 2008; Vlaev et al., 2009), it appears that people's judgment about their preferences and opinions are unstable, and to some extent "arbitrary." Yet it appears that people also attempt to be consistent with what they believe to be their previously expressed attitudes or preferences: when they are misled about their previous responses, their subsequent responses are modified appropriately.

This presumed drive for consistency provides one possible explanation for the long-term stability of the artificially-induced preferences (for faces, for perceptions of risk, and even for political convictions). What is particularly interesting here is that different probes might have very different powers to promote consistency. For example, using our choice blindness paradigm one might contrast the preference induction arising from choice with verbal explanation, and/or choice with numerical ratings (of the mismatched alternative). Here, the question would be whether the naturalness and fluency of verbal explanation, or the vividness and "concreteness" of numerical ratings, would generate the greatest preference effect. The latter alternative would sit very well with the conjecture of Ariely et al. (2003) that it might be the quantification *as such* that leads participants to form further coherent preferences in the same domain of choice, and thus it might be some of the most quintessential features of economic choices that create the most troubling dynamic effects. Obviously, these results are still mere blips on the experimental radar, and the question remains whether they simply might be swamped or washed out in the competitive consumer and ideological landscape. As Simonson (2008) concludes: "*while the principles governing context, framing, and task effects may be general, the resulting 'preferences' often leave no trace and have little if any effect on subsequent decisions*" (p. 157; see also Yoon & Simonson, 2008). In our view, the attempt to trace such longitudinal and cross-contextual effects of various implicit influences, and to decide whether they represent the norm or just curious exceptions, is one of the most urgent and important quests to pursue for research on preference change.

In relation to the second goal, to use meaningful choices involving actual preferences, the choice blindness experiments we have presented here are more or less on the mark. Deciding which of two faces one finds more attractive, or which hypothetical gamble one prefers, are not the most exciting choices around. But they are simple and intuitive, and it is fair to assume that people have relatively stable guiding preferences for both domains. These are also domains in which the path to possible real-world tests and applications (for faces: the cosmetic and modeling industry, dating and mate selection, shifting cultural standards of beauty, etc., for risk: investment decisions, gambling, insurance, etc.) is relatively short. On the other hand, voting intentions measured at the cusp of the national election must be counted among the diamond league of dependent variables in psychology.

The final goal is also the most difficult. As we said in the introduction, there is currently no good model that can unravel the interplay between introspective reports and preference formation. For example, consumer psychologists and marketing researchers puzzled for some time over an effect they call "The Mere-Measurement Effect" (e.g. Morwitz & Fitzsimons, 2004; Williams, Fitzsimons & Block, 2004). True to its name, the mere-measurement effect concerns the fact that, by simply asking people about their intentions about one or another action (e.g. to buy a specific brand of automobile, to engage in a specific charity, to vote in an election), their behavior is likely to change (compared to a control group who receives no questions about their intentions). But why should the behavior of people change as a result of asking them what they intend to do? These are not questions that are framed to emphasize different benefits or drawbacks with a certain choice, or questions that entice participants into elaborating on reasons they had not thought about earlier, or some other cleverly designed technique of influence – they are just questions that ask people what they intend to do. In surface terms, Morwitz and Fitzsimons (2004) give the answer: "*somehow the act of measuring intentions affects consumers' thoughts associated with the behavior (e.g. representations of the behavior, attitudes towards the behavior, or thoughts associated with the act of purchasing). These altered thoughts in turn change the consumers subsequent purchase behavior*" (p. 4).

We agree, but to understand the basic process at the heart of the mere-measurement effect, the explanation needs to be framed yet more generally, and with reference to a discussion about introspection and intentionality. For all the intimate familiarity we have with everyday decision-making, it is very difficult to probe the representations underlying this process, or to determine what we can know about them from the "inside," by reflection and introspection (Nisbett & Wilson, 1977; Jack & Roepstorff, 2004; Johansson et al., 2005, 2006). Mere-measurement creates the effects it does because there is nothing mere about measurement (maybe the situation is not as paradoxical as in physics, but we see the litter of Schrödinger's cat all over our decision tasks). In fact, for Dennett (1987, 1991, 1996), the "mere-measurement effect" represents something fundamental about decision-making of the kind human beings engage in (and is a centerpiece in the explanation of why the notion of well-specified introspectively observable intentional states is so seductive). He writes: "*What creates the illusion of discrete, separate, individuatable beliefs [and preferences] is the fact that we talk about them: the fact that when we go to explain our anticipations, when we move from generating our own private expectations of what people are going to do, to telling others about these expectations we have, and explaining to them why we have them, we do it with language. What comes out, of course, are propositions. /.../ Then... it's only too easy to suppose that those sentences are not mere edited abstractions or distillations from, but are rather something like copies of or translations of the very states in the minds of the beings we're talking about"* (1991, pp. 88–89).

From an empirical perspective, choice blindness strongly suggests that there is no Archimedean point from which to observe and measure preferences. So, which side of the super sour apple does the economist and psychologist want to bite into? Does choice blindness demonstrate that only 20% of the opinions people held about campaign issues in the Swedish election involved real preferences? Or should we admit that preferences can often be blatantly reversed moments after the decision is made? As the discussion makes clear, we do not believe these questions are well formed, because the very concept of preference as an enduring state, which determines which choices we make, is itself under threat. Yet we do believe choice blindness is a uniquely positioned instrument to pry apart the relative impact of stated and revealed choice on future preferences and behavior.

Acknowledgment

The work by PJ was funded by a Marie Curie postdoc fellowship and the work by LH was funded by a grant from The Swedish Research Council.

References

Abdellaoui, M. (2000). Parameter free elicitation of utilities and probability weighting functions. *Management Science, 46*, 1497–1512.

Abramowitz, A. I., & Saunders, K. L. (1998). Ideological realignment in the US electorate. *Journal of Political, 60*(03), 634–652. doi:10.2307/2647642.

Abramowitz, A. I., & Saunders, K. L. (2008). Is polarization a myth? *Journal of Political, 70*(02), 542–555. doi:10.1017/S0022381608080493.

Ajzen, I., & Fishbein, M. (2000). Attitudes and the attitude-behavior relation: Reasoned and automatic processes. *European Review of Social Psychology, 11*(1), 1–33. Psychology Press doi: 10.1080/14792779943000116.

Ariely, D. (2008). *Predictably irrational.* London, UK: Harper Collins.

Ariely, D., & Norton, M. I. (2008). How actions create – not just reveal – preferences. *Trends in Cognitive Sciences, 12*(1), 13–16. doi: 10.1016/j.tics.2007.10.008.

Ariely, D., Loewenstein, G., & Prelec, D. (2003). "Coherent arbitrariness": Stable demand curves without stable preferences. *The Quarterly Journal of Economics, 118*(1), 73–105. doi: 10.1162/00335530360535153.

Ariely, D., Loewenstein, G., & Prelec, D. (2006). Tom Sawyer and the construction of value. *Journal of Economic Behavior & Organization, 60*(1), 1–10. doi: 10.1016/j.jebo.2004.10.003.

Bafumi, J., & Shapiro, R. Y. (2009). A new partisan voter. *Journal of Political, 71*(01), 1–24. doi:10.1017/S0022381608090014.

Birnbaum, M. H. (2008). New paradoxes of risky decision-making. *Psychological Review, 115*(2), 463–501. doi: 10.1037/0033–295X.115.2.463.

Bornschier, S. (2010). The new cultural divide and the two-dimensional political space in western europe. *West European Politics, 33*(3), 419–444. doi:10.1080/01402381003654387.

Brandstätter, E., Gigerenzer, G., & Hertwig, R. (2006). The priority heuristic: making choices without trade-offs. *Psychological Review, 113*(2), 409–432. doi: 10.1037/0033–295X.113.2.409.

Brehm, J. W. (1956). Postdecision changes in the desirability of alternatives. *Journal of Abnormal Psychology, 52*(3), 384–389.

Camerer, C. F., Loewenstein, G., & Prelec, D. (2005). Neuroeconomics: How neuroscience can inform economics. *Journal of Economic Literature, 43*(1), 9–64. American Economic Association. doi: 10.1257/0022051053737843.

Carsey, T. M., & Layman, G. C. (2006). Changing sides or changing minds? Party identification and policy preferences in the American electorate. *American Journal of Political Science, 50*(2), 464–477. doi:10.1111/j.1540–5907.2006.00196.

Chater, N., & Vlaev, I. (2011). The instability of value. In M. Delgado, E. A. Phelps & T. W. Robbins (Eds.), *Decision making: Attention and performance XXIII*. Oxford, UK: Oxford University Press.

Chen, M. K. (2008). *Rationalization and cognitive dissonance: Do choices affect or reflect preferences?* Working paper. New Haven, CT: Yale University.

Chen, M. K., & Risen, J. L. (2010). How choice affects and reflects preferences: Revisiting the free-choice paradigm. *Journal of Personality and Social Psychology, 99*, 573–594. doi: 10.1037/a0020217.

Clarke, H. D., Sanders, D., Stewart, M. C., & Whiteley, P. (2009). The American voter's British cousin. *Electoral Studies, 28*(4), 632–641. doi:10.1016/j.electstud.2009.05.019.

Cwalina, W., Falkowski, A., & Newman, B. (2010). Towards the development of a cross-cultural model of voter behavior Comparative analysis of Poland and the US. *European Journal of Marketing, 44*(3–4), 351–368. doi:10.1108/03090561011020462.

Dalton, R. J. (2009). Parties, partisanship, and democratic politics. *Perspectives on Politics, 7*(3), 628–629. doi:10.1017/S1537592709990600.

Dennett, D. C. (1987). *The intentional stance*. Cambridge, MA: MIT Press. doi: 10.2307/2026682.

Dennett, D. C. (1993). The message is: There is no medium. *Philosophy and Phenomenological Research, 53*(4), 919–931.

Dennett, D. C. (1991). *Consciousness explained*. Boston: Little, Brown & Company.

Dodson, K. (2010). The return of the American voter? Party polarization and voting behavior, 1988 to 2004. *Sociological Perspective, 53*(3), 443–449. doi:10.1525/sop.2010.53.3.443.

Druckman, J. (2004). Political preference formation: Competition, deliberation, and the (Ir) relevance of framing effects. *American Political Science Review, 98*(4), 671–686.

Egan, L. C., Santos, L. R., & Bloom, P. (2007). The origins of cognitive dissonance: evidence from children and monkeys. *Psychological Science, 18*(11), 978–983. doi: 10.1111/j.1467–9280.2007.02012.x.

Egan, L. C., Bloom, P., & Santos, L. R. (2010). Choice-induced preferences in the absence of choice: Evidence from a blind two choice paradigm with young children and capuchin monkeys. *Journal of Experimental Social Psychology, 46*(1), 204–207. doi: 10.1016/j.jesp.2009.08.014.

Elinder, M. (2009). *Correcting mistakes: Cognitive dissonance and political attitudes in Sweden and the United States* Working Paper Series. Uppsala University, Department of Economics 2009:12.

Enyedi, Z., & Deegan-Krause, K. (2010). Introduction: The structure of political competition in Western Europe. *West European Politics, 33*(3), 415. doi:10.1080/01402381003654254.

Festinger, L. (1957). *A theory of cognitive dissonance*. Stanford, CA: Stanford University Press.

Gerard, H. B., & White, G. L. (1983). Post-decisional reevaluation of choice alternatives. *Personality and Social Psychology Bulletin, 9*(3), 365–369. doi: 10.1177/0146167283093006.

Glimcher, P. W. (2003). *Decisions, uncertainty, and the brain: The science of neuroeconomics* (p. 395). Cambridge, MA: The MIT Press.

Gonzalez, R., & Wu, G. (1999). On the shape of the probability weighting function. *Cognitive Psychology, 38*(1), 129–166.

Gross, S. J., & Niman, C. M. (1975). Attitude–behavior consistency: A review. *Public Opinion Quarterly, 39*(3), 358. doi: 10.1086/268234.

Hall, L., Johansson, P. & Strandberg, T. (submitted). Lifting the Veil of Morality: Choice Blindness and Attitude Reversals on a Self-Transforming Survey.

Hall, L., Johansson, P., Tärning, B., & Sikström, S., Chater, N. (in preparation b). *Preference change and choice blindness.*

Hall, L., Johansson, P., Tärning, B., Sikström, S., & Deutgen, T. (2010). Magic at the market-place : Choice blindness for the taste of jam and the smell of tea. *Cognition, 117*(1), 54–61. doi: 10.1016/j.cognition.2010.06.010.

Holmberg, S., & Oscarsson, H. (2004). *Väljare. Svenskt väljarbeteende under 50 år.* Stockholm: Norstedts Juridik.

Jack, A. I., & Roepstorff, A. (2004). Trusting the subject II. *Journal of Consciousness Studies, 11*, 7–8.

Johansson, P., Hall, L., & Sikström, S. (2008). From change blindness to choice blindness. *Psychologia, 51*, 142–155.

Johansson, P., Hall, L., Sikström, S., & Olsson, A. (2005). Failure to detect mismatches between intention and outcome in a simple decision task. *Science, 310*, 116–119.

Johansson, P., Hall, L., Kusev, P., Aldrovandi, S., Yamaguchi, Y. & Watanabe, K. (in preparation a). *Choice blindness in multi attribute decision making.*

Johansson, P., Tentori, K., Harris, A., Hall, L., & Chater, N. (in preparation b). *Choice blindness for personality traits.*

Johansson, P., Hall, L., Sikström, S., Tärning, B., & Lind, A. (2006). How something can be said about telling more than we can know: On choice blindness and introspection. *Consciousness and Cognition, 15*, 673–692. doi: 10.1016/j.concog.2006.09.004.

Kahneman, D., & Tversky, A. (1979). Prospect theory: An analysis of decision under risk. *Econometrica, 47*(2), 263–291. The Econometric Society. doi: 10.2307/1914185.

Kenning, P., & Plassmann, H. (2005). Neuroeconomics: an overview from an economic per-spective. *Brain Research Bulletin, 67*(5), 343–354.

Kitschelt, H. (2010). The comparative analysis of electoral and partisan politics: A com-ment on a Special Issue of West European politics. *West European Politics, 33*(3), 659. doi:10.1080/01402381003654692.

Krueger, J. I., & Funder, D. C. (2004). Towards a balanced social psychology: causes, conse-quences, and cures for the problem-seeking approach to social behavior and cognition. *Behavioral and Brain Sciences, 27*(3), 313–327. discussion 328–376.

Kurniawan, I. T., Seymour, B., Vlaev, I., Trommershäuser, J., Dolan, R. J., & Chater, N. (2010). Pain relativity in motor control. *Psychological Science, 21*, 840–847.

Kusev, P., Johansson, P., Ayton, P., Hall, L., van Schaik, P., & Chater, N. (in preparation (a)). *Preference reversals: Memory and contextual biases with decision prospects.*

Kusev, P., Tsaneva-Atanasova, K., van Schaik, P., & Johansson, P. (in preparation (b)). *Relative theory of choice: preference change for risky choices.*

Kusev, P., Van Schaik, P., Ayton, P., Dent, J., & Chater, N. (2009). Exaggerated risk: pros-pect theory and probability weighting in risky choice. *Journal of Experimental Psychology. Learning, Memory, and Cognition, 35*(6), 1487–1505.

Lewis-Beck, M. S., Norpoth, H., Jacoby, W. G., & Weisberg, H. F. (2008). *The American voter revisited.* Ann Arbor, MI: University of Michigan Press.

Lichtenstein, S., & Slovic, P. (2006). The construction of preference: An overview. *The con-struction of preference* (pp. 1–40). New York, NY: Cambridge University Press.

Lichtenstein, Sarah, & Slovic, P. (1971). Reversals of preference between bids and choices in gambling decisions. *Journal of Experimental Psychology, 89*(1), 46–55.

Lieberman, M. D., Ochsner, K. N., Gilbert, D. T., & Schacter, D. L. (2001). Do amnesics exhibit cognitive dissonance reduction? The role of explicit memory and attention in attitude change. *Psychological Science: A Journal of the American Psychological Society/APS, 12*(2), 135–140.

Morwitz, V., & Fitzsimons, G. (2004). The mere-measurement effect: Why does measuring intentions change actual behavior? *Journal of Consumer Psychology, 14,* 64–74.

Neumann, J. von, & Morgenstern, O. (1947). *Theory of Games and Economic Behavior* (2nd ed.). Princeton, NJ: Princeton University Press.

Nisbett, R. E., & Wilson, T. D. (1977). Telling more than we can know: Verbal reports on mental processes. *Psychological Review, 84,* 231–259.

Nyhan, B., & Reifler, J. (2010). When corrections fail: The persistence of political misperceptions. *Political Behavior, 32*(2), 303–330. doi:10.1007/s11109–010–9112–2.

Oscarsson, H. (2007). A matter of fact? Knowledge effects on the vote in Swedish general elections, 1985–2002. *Scandinavian Political Studies, 30*(3), 301–322. doi: 10.1111/j.1467–9477.2007.00182.x.

Oscarsson, H., & Holmberg, S. (2008). *Regeringsskifte. Väljarna och valet 2006.* Stockholm: Norstedts.

Petrocik, J. R. (2009). Measuring party support: Leaners are not independents. *Electoral Studies, 28*(4), 562–572. doi:10.1016/j.electstud.2009.05.022.

Prelec, D. (1998). The probability weighting function. *Econometrica, 66*(3), 497–527. doi: 10.2307/2998573.

Rorty, R. (1970). Incorrigibility as the mark of the mental. *Journal of Philosophy, 67*(12), 399–424.

Schwarz, N., & Oyserman, D. (2001). Asking questions about behavior: Cognition, communication, and questionnaire construction. *American Journal of Evaluation, 22*(2), 127–160. doi: 10.1016/S1098–2140(01)00133–3.

Schwarz, N. (1999). Self-reports: How the questions shape the answers. *American Psychologist, 54*(2), 93–105. doi: 10.1037/0003–066X.54.2.93.

Shafir, E., Simonson, I., & Tversky, A. (1993). Reason-based choice. *Cognition, 49*(1–2), 11–36.

Sharot, T., De Martino, B., & Dolan, R. J. (2009). How choice reveals and shapes expected hedonic outcome. *Journal of Neuroscience, 29*(12), 3760–3765. doi: 10.1523/JNEUROSCI.4972–08.2009.

Sharot, T., Velasquez, C. M., & Dolan, R. J. (2010). Do decisions shape preference?: Evidence from blind choice. *Psychological Science, 21*(9), 1231–1235. (doi: 10.1177/0956797610379235)

Stewart, N., Chater, N., Stott, H. P., & Reimers, S. (2003). Prospect relativity: how choice options influence decision under risk. *Journal of Experimental Psychology. General, 132*(1), 23–46.

Stewart, N., Brown, G. D. A., & Chater, N. (2005). Absolute identification by relative judgment. *Psychological Review, 112,* 881–911.

Stewart, N., Chater, N., & Brown, G. D. A. (2006). Decision by sampling. *Cognitive Psychology, 53*(1), 1–26. doi: 10.1016/j.cogpsych.2005.10.003.

Tversky, A., & Kahneman, D. (1981). The framing of decisions and the psychology of choice. *Science, 211*(4481), 453–458.

Tversky, A., & Koehler, D. J. (1994). Support theory: A non-extensional representation of subjective probability. *Psychological Review, 101,* 547–567.

Tversky, A., & Wakker, P. (1995). Risk attitudes and decision weights. *Econometrica, 63*(6), 1255–1280.

Vlaev, I., Seymour, B., Dolan, R. J., & Chater, N. (2009). The price of pain and the value of suffering. *Psychological Science, 20,* 309–317.

Wakker, P. P. (2004). On the composition of risk preference and belief. *Psychological Review, 111*(1), 236–241. doi: 10.1037/0033–295X.111.1.236.

Williams, P., Fitzsimons, G., & Block, L. (2004). When consumers don't recognize 'benign' intention questions as persuasion attempts. *Journal of Consumer Research, 31,* 540–550.

Winkielman, P., & Schwarz, N. (2001). How pleasant was your childhood? *Psychological Science, 12*(2), 176–179.

Yoon, S.-O., & Simonson, I. (2008). Choice set configuration as a determinant of preference attribution and strength. *J Cons Res, 35,* 324–336.

II. CONTEXTUAL FACTORS

CHAPTER

7

Set-Size Effects and the Neural Representation of Value

Kenway Louie* and Paul W. Glimcher

Center for Neural Science, New York University

INTRODUCTION

Modern life is replete with choices that offer a diverse and often confounding array of options. Even simple decisions like purchasing a book or choosing an entree in a restaurant offer a wide array of options, but an extensive array of alternatives exists for far more consequential choices as well: choosing a university, selecting a retirement plan, deciding on

---

*Corresponding author.

Neuroscience of Preference and Choice
DOI: 10.1016/B978-0-12-381431-9.00007-3

143

the course of medical action. Contrary to popular and theoretical notions about the benefits of always having more options from which to choose, recent evidence suggests that having too many options can produce profoundly negative consequences. Large choice sets are associated with psychological costs, such as a decreased desire to choose and outcome dissatisfaction, and emerging evidence suggests that they affect the consistency and efficiency of choice behavior directly as well. This paradox of choice is particularly problematic for more traditional rational choice theories, which predict that larger choice sets should always lead to equivalent or better decisions.

Our goal in this review is to lay out a framework for understanding choice set effects at the neural level that may reconcile many previous unrelated observations about set size effects. To accomplish this goal we begin by reviewing laboratory and field results documenting the effects of choice set size on psychological outcomes and behavior. We then turn to a specific small-scale model for context-dependent decision-making used extensively in both the animal and human behavioral literature: the study of binary versus trinary choice. Finally, we review the most widely studied model system for decision-making in neuroscience, the primate vision-saccadic system, and explore how knowledge of the neural representation of valuation and choice in this system may shed light on the paradox of choice.

THE PARADOX OF CHOICE

It is widely believed that decision-makers are better off when given more choice options. One of the foundations for this belief is rational choice theory, which assumes that individuals hold stable rank-ordered preferences between all possible options – a requirement for any form of efficient maximization. Under this assumption, enlarging the choice set can only increase the likelihood of obtaining a better option according to the chooser's pre-existing preferences, a point made in Figure 7.1. Consider a chooser who is offered a choice between 6 options as in scenario 1. In this example, the darkness of the circles diagrammatically indicates the value of each of these stylized options, with the circled option being the best in this six-option choice set. Now consider what happens when 18 additional options are offered, all of lower value than the best option in the original set (scenario 2). Of course the chooser is, under these circumstances, still free to select the option she preferred before; in this case, the addition of alternatives has no effect on the welfare of the chooser. Finally, consider what happens in scenario 3 where one of the 18 new options is actually better than all of the original six. Under this condition the chooser can actually do better than in the original six-option choice set. This simple example highlights the fact that

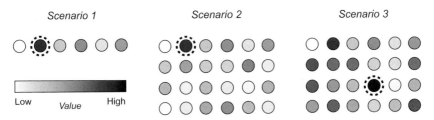

FIGURE 7.1   **Preference matching and the benefit of more choice.** Each scenario represents a choice between options with different values (indicated by the shading), with the highest valued option circled. The first six options are identical in all three scenarios. In the second scenario, the additional options are all of lower value than the best option in the original six. In the third scenario, the additional options include an alternative that is of higher value than any option in the first six alternatives. Thus, according to expected utility theory, decision-makers facing larger choice sets will realize a utility equal to or better than that from a smaller set.

having more options in principle cannot make one worse off, and it may indeed make one better off. More choices thus can only improve utility maximization by providing the opportunity for a higher utility outcome.

In a different vein, psychological theory and research suggest that more choice also leads to greater individual welfare via cognitive processes. A large body of literature shows that people prefer to make their own choices rather than have them externally dictated. Providing individuals with the ability to choose has been reported to result in increased intrinsic motivation and perception of self-control, and has been linked to higher life satisfaction and subjective evaluation of decision outcomes (Deci & Ryan, 1985; Taylor & Brown, 1988). In the marketing field, consumers have been shown to prefer choice variety and large product assortments, and stores that offer more options are conferred a competitive advantage (Arnold, Oum & Tigert, 1983; Chernev, 2003). Thus, research in economics, psychology, and marketing predicts that increasing the number of choice options will yield higher individual welfare, and that more choice should yield better choices and outcome satisfaction.

However, growing experimental evidence suggests that, contrary to existing theory, large choice sets can be detrimental, a phenomenon often referred to as "choice overload" or the "paradox of choice" Schwartz (2004). In particular, empirical evidence indicates that choice sets with large numbers of options are demotivating, for example driving individuals to default out of making a choice. An example of such choice set size effects is a seminal study conducted by Iyengar and Lepper (2000), who examined the effect of varying choice set size on empirical consumer behavior in an upscale grocery store. In a field experiment, they set up a tasting booth that allowed shoppers to sample from an array of exotic jams and tracked their subsequent purchasing behavior. To examine the effect of choice set size, the experiment exposed different groups of consumers to displays containing either a small (6) or large (24) number of different jams. Consistent with previous findings, the appearance of more choice was attractive, and more consumers who encountered the large choice set condition stopped at the display (60% versus 40%). However, despite this initial attractiveness, a much lower proportion of consumers who sampled the large choice set subsequently purchased one of the jams (3% versus 30%). Thus, the large choice set was demotivating in terms of choice, with a far higher percentage of consumers selecting the default option to not purchase a jam. This surprising result is the opposite of that predicted by classical theories of choice, where decision-makers with predetermined preference rankings should be more likely to find a suitable option in the larger choice set.

A number of subsequent or related studies now provide evidence for both psychological and empirical costs associated with increasing choice set size that may, at least in part, underlie this phenomenon. In laboratory settings, decision-makers facing extensive versus limited choice sets are more likely to defer decisions (Tversky & Shafir, 1992) and report greater dissatisfaction and regret with their choices (Iyengar & Lepper, 2000). Outside the laboratory, decision-makers facing more alternatives are less willing to purchase goods (Boatwright & Nunes, 2001), take loans (Bertrand, Karlan, Mallainathan et al., 2010), and participate in retirement plans (Iyengar, Jiang & Huberman, 2004). As shown in Figure 7.2, the participation rate of employees in 401(k) retirement plans falls

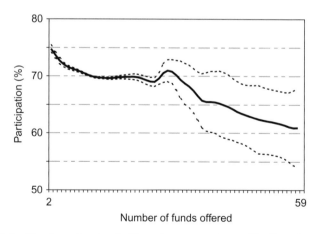

FIGURE 7.2   **The demotivational effect of extensive choice.** The line plots the empirical relationship between 401(k) plan participation rates and the number of different funds offered; data represents actual choices of clients of the Vanguard Group, an investment management company, in 2001. Dotted lines, 95% confidence intervals. Adapted from Iyengar et al. (2004).

significantly as the number of plans from which they could choose rises, indicating that the demotivating effect of large choice sets extends to major life decisions. Because contributions to such plans are tax-deferred and often matched by employers, the demotivating effect of set size represents a significant financial cost imposed on real decision-makers.

All of these observations indicate that extensive choice sets can decrease the motivation to choose, increase default option selection, decrease satisfaction, and produce negative psychological outcomes like disappointment and regret. But even worse, there is at least some evidence that large choice sets can systematically alter the choices subjects make in a way that prevents utility maximization (DeShazo & Fermo, 2002). As might be projected from the studies of opt-in/opt-out decisions mentioned above, DeShazo and Fermo found clear evidence that choices themselves become less consistent as choice set size increases. While the study can be criticized for having choosers select from hypothetical outcomes, this econometric analysis indicates that the effects of choice set size widely seen in opt-in tasks, like those studied by Iyengar and colleagues, can even manifest in situations where subjects choose from a set of well-defined alternatives.

How do large set sizes produce this decrement in choice consistency and a bias to opt out of choice situations? One possibility is that extended choice sets can overwhelm the limited information-processing capabilities of decision-makers (Malhotra, 1982). Alternatively, the addition of options can increase the similarity between alternatives and the conflict in making a choice, and make subsequent justification more difficult (Chernev, 2003; Sela, Berger & Liu, 2009). Expectations of consumers may also be important, as larger choice sets may raise initial expectations and drive post-choice dissatisfaction (Schwartz, 2000).

The choice overload literature focuses on the detrimental effects of very large choice sets, but larger set sizes may have multiple interacting consequences. While few studies have examined the effect of the number of alternatives in a parametric manner, there is some evidence in the behavioral literature that both purchasing behavior and reported satisfaction can behave as a non-monotonic function of set size (Reutskaja & Hogarth, 2009; Shah & Wolford, 2007). Satisfaction and purchasing both initially increase as the number of choices increases, peak at an intermediate size (~10 in these studies), and decline at larger values. This parametric inverted U-shape dependency on the number of alternatives suggests that multiple factors may play a role in producing the psychological and behavioral effects of choice set size. The suggestion is that the benefits of more choice (preference matching, attractiveness) predominate at smaller set sizes but are eventually outweighed by the costs (information load, conflict) at larger set sizes.

While these explanations are useful frameworks for examining the psychological impact of choice set effects, they cannot provide information about the interaction between multiple alternatives and the decision process itself. This requires an approach based in rational choice theory and normative decision-making, where choice itself is predicted from theory, documented in practice, and examined parametrically as a function of set size conditions. We turn next to evidence from animal and human behavioral studies for context-dependence in binary versus trinary choice, which may serve as a model for future studies of these more extensive set size effects.

CONTEXT-DEPENDENT CHOICE BEHAVIOR

How does the presence of more choice alternatives influence the decision-making process? Much of the work to date has focused on psychological descriptions of, and explanations for, the detrimental effects of extensive choice sets, but with the exception of the DeShazo and Fermo (2002) study, little is known about the effect of increasing set size on choice behavior itself. However, choice set size effects are a specific example of the broader phenomenon of context-dependent preferences and choice behavior, situations in which the term *context* refers to the other alternatives available in the choice set. In this literature, the current choice set is sometimes referred to as the *local context*, to distinguish it from choice sets experienced in the recent past, which is termed the *background context* and does not pertain directly to the effects considered here. While choice behavior in larger choice sets remains largely unexplored, there is a considerable body of work describing the effects of changing the choice set size from two (binary choice) to three (trinary choice). These context effects provide a valuable framework to examine the effect of the choice set itself on decision-making behavior, and so we turn next to this literature.

Neoclassical Explorations

Context-dependent decision-making violates specific assumptions required by economic rational choice theories. These requirements arise from the assumption that choice is driven by context-free, or absolute, valuations assigned to individual choice options by the decision-maker. Arrow captured this insight in his axiom of *regularity*, the notion that for a rational chooser the proportion of choices allocated to one of the options in a binary choice cannot be increased by the addition of a third option to the choice set (Arrow, 1963). Regularity represents a specific instantiation of the more broadly known Luce choice axiom, which states that the probability of choosing a given option is proportional to the value of that option divided by the total value of available options (Luce, 1959).

While these two formulations differ in some regards, they both effectively enforce the principle of *independence of irrelevant alternatives* (IIA) which asserts that the relative proportion of choice between two options should be unaffected by the addition or subtraction of other alternatives; if a subject picks option A twice as often as option B, that should be true whether or not an undesirable low valued alternative is available (Figure 7.3).

Contrary to the demands of rationality, however, human choice behavior has been observed to violate both regularity and Luce's choice axiom. Much of the existing literature on this point arises from the field of marketing and consumer behavior, where the implications of choice alternative effects hold considerable interest. For example, entrepreneurs have long believed that introducing a product to a marketplace consisting of two competing items would detrimentally draw consumers from the existing product closest to the new competition, and companies tend to diversify their own products in an effort to minimize cannibalization (Copulsky, 1976). This effect is summarized by the *similarity hypothesis*, according to which a new product draws shares disproportionately from the item most similar to it, a violation of proportionality (Tversky, 1972). A number of early studies examining choice behavior showed examples of such context-dependence in humans, with additional items drawing choice from similar options (Becker, Degroot & Marschak, 1963; Huber & Puto, 1983; Tversky & Russo, 1969).

Behavioral Explanations

A common informative framework for thinking about these kinds of context-dependent behavior is based on the idea of attributes. Consider a choice between two goods, for example cars, as illustrated in Figure 7.4. If we consider goods described fully by two attributes (here, stylishness and fuel economy), each choice can be described by its merits in the two attribute dimensions: the target (T) is stylish but has poor fuel economy while the competitor (C) is less stylish but more fuel efficient. According to the similarity hypothesis, the addition of an alternative near the target in this two-dimensional attribute space – for example, a stylish, fuel inefficient sports car – would draw choices (or market share) from the nearby target and disproportionately increase the preference for competitor over target.

This attribute-based approach to decisions provides a common interpretation for the source of comparative valuation mechanisms implied by context-dependent choice (Busemeyer & Townsend, 1993; Huber, Payne & Puto, 1982; Tversky, 1972; Tversky & Simonson, 1993; Wedell, 1991). The underlying hypothesis is that comparative valuation arises from the integration of ranking or preference information across different dimensions. An absolute valuation mechanism would combine the

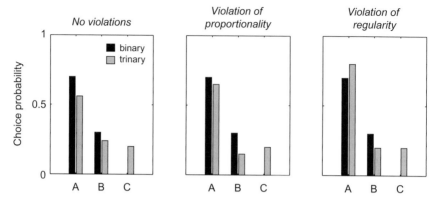

FIGURE 7.3 Examples of context-dependent violations of rationality. Graphs show the allocation of choices in a binary choice (black; options A, B) and a trinary choice (gray; options A, B, C). Left, no violations of rationality. The addition of the third option C proportionately decreases choices to A and B, such that the relative preference between A and B does not change. Middle, violation of proportionality. Addition of the third option C disproportionately draws choices from option B, increasing the relative preference for A versus B. Left, violation of regularity. Addition of the third option C increases the absolute proportion of choices for A, violating both regularity and proportionality.

multidimensional information about a given option (stylishness, fuel economy) into a single overall value, and compare these values across options when making a choice. Such a mechanism would not produce context-dependent choice behavior. A comparative valuation mechanism, however, might construct preferences rankings within a dimension and combine these rankings across dimensions. Such attribute-based models of information processing and choice have been proposed to explain various violations of rationality, for example Tversky's famous *Elimination-By-Aspects* model which sought to explain the similarity effect (Tversky, 1972).

However, the similarity effect describes only one of a number of well-documented violations of rationality and proportionality that depend on context. Much of the recent literature has focused on the effect on adding an *asymmetrically dominated alternative*, which produces behavior opposite to that predicted by the similarity hypothesis (Huber et al., 1982). An alternative is said to be *dominated* by another option if that option is equal or superior in all attribute dimensions, rendering the alternative clearly less attractive. In a typical trinary choice experiment of this kind, an asymmetrically dominated alternative (termed the decoy) is dominated by one option (the target) but not the other option (the competitor). Consider the decoy relative to the target and competitor options as

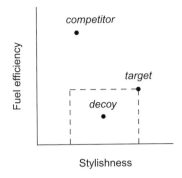

FIGURE 7.4 Attribute dimensions and context-dependent choice. Automobile choice options are positioned based on their qualities in two different attribute dimensions: fuel efficiency and stylishness. The target option and the competitor option both rate higher in only one of the attributes relative to the other option. The illustrated decoy is dominated by the target because it is worse than the target option in both attributes. The decoy is considered asymmetrically dominated because it is only dominated by the target and not the competitor (the dashed line outlines the space of all such asymmetrically dominated alternatives). Despite the fact that such a decoy should not be chosen (because it is dominated), its presence in the choice set can change preferences between the target and the competitor (typically towards the target), and is a specific example of a set size effect.

shown in Figure 7.4. The decoy is dominated by the target because it is both less fuel efficient and less stylish; however, like the target, it is more fuel efficient but less stylish than the competitor.

In this construction, the key variable is the relative preference between the target and competitor options in the absence (binary choice) or presence (trinary choice) of the asymmetrically dominated decoy. In contrast to the similarity effect, the asymmetric dominance effect predicts that the decoy will induce more choices to the target, to which it is more similar. Again, this is a violation of traditional neoclassical frameworks because it violates the principle of IIA. Violation or not, however, context-dependent choice when facing an asymmetrically dominated alternative is a robust finding in experimental and consumer research studies of human choice behavior (Doyle, O'Connor, Reynolds & Bottomley, 1999; Huber et al., 1982; Sedikides, Ariely & Olsen, 1999; Simonson & Tversky, 1992; Tversky & Simonson, 1993). This effect can produce violations of both proportionality and the similarity hypothesis, as the original option most similar to the decoy draws additional choices. Significantly, it can also produce violations of regularity, increasing the absolute preference of the target option. The asymmetric dominance effect represents a specific and robust example of what has been sometimes called the *attraction effect*, which generalizes to additional options near the target in attribute space, including those that are not dominated (relatively inferior). As for the similarity effect, several attribute-based choice models have been proposed to explain the context-dependent attraction effect (Simonson & Tversky, 1992; Wedell, 1991).

Were the attribute-based approaches, similarity effects and attraction effects not confusing enough, researchers have also documented what is known as the *compromise effect* (Simonson & Tversky, 1992). In this construction, the introduction of an extreme alternative (e.g. a car that has better fuel economy but less stylishness than the target) increases the probability of choosing the option that is now a compromise between the extreme options. All of these effects, and the hypotheses that seek to explain them, share one key property however, and this is that they all describe robust context-dependent violations of IIA. Context-dependent violations of IIA thus seem to be a key feature of human choice behavior, although a unified explanation for these violations has so far eluded scholars.

Emerging evidence from the animal literature suggests that these types of context-dependence are not limited to humans. Animals, like humans, often show a preference for larger choice set sizes in a variety of behaviors, a prime example being mate choice (Hutchinson, 2005). However, multiple studies have demonstrated context-dependent violations of rationality in species ranging from insects to birds (Bateson, 2002; Bateson, Healy & Hurly, 2003; Royle, Lindstrom & Metcalfe, 2008; Shafir, Waite & Smith, 2002). In one of the first demonstrations of this type, Sharoni Shafir and colleagues (2002) examined the asymmetric dominance effect in foraging preferences of both honeybees (*A. mellifera*) and gray jays (a member of the crow family; *P. canadensis*). As in human experiments, Shafir constructed options that differed in two dimensions, for example presenting jays with alternatives that varied in the magnitude of reward and the effort required. When the jays were tested in binary (target, competitor) and trinary (target, competitor, decoy) choices, they exhibited an attraction effect and chose the target with a higher probability in the presence of the decoy, violating both Luce's choice rule and regularity.

Evidence for context-dependent preferences in animals suggests that such behavior may arise independent of higher-order explanations for such behavior in humans such as reasoning or justifying ones preferences. A recent study showed that even a unicellular slime mold (*P. polycephalum*), choosing between food sources that varied in food content and illumination, can exhibit violations of regularity and proportionality (i.e. an asymmetric dominance effect) similar to those observed in humans (Latty & Beekman, 2011). While this example of comparative valuation, in a species taxonomically distant from typical laboratory organisms and lacking a central nervous system, may arise via a different mechanism than that underlying human violations of rationality, it nonetheless suggests that context-dependent behavior may be a characteristic of biological decision-making. In addition to choice set effects, violations of transitivity in animals also support the presence of comparative valuation mechanisms (Shafir, 1994; Waite, 2001).

Such mechanisms may have evolved despite their violations of rationality for evolutionary optimization reasons (Glimcher, 2011; Hutchinson & Gigerenzer, 2005; Shafir, 1994). For example, comparative valuation mechanisms may be more computationally efficient than absolute valuation mechanisms, while still generating behavior sufficiently close to optimal or rational in the organism's natural environment. Tversky was amongst the first to raise this point when he suggested that comparisons within a given attribute dimension may be easier than comparisons across different dimensions, effectively making a comparative valuation system more efficient (Simon, 1955; Tversky, 1969). Such systems, one might hypothesize, would have been favored by natural selection for the advantages they conferred in the evolutionary environment, but could generate apparent violations of rationality when tested in specific conditions today.

THE NEURAL BASIS OF DECISION-MAKING

We have reviewed above some of the psychological effects of extensive choice sets, behavioral evidence for contextual effects in simple choice set manipulations, and proposed psychological models of such context-dependent decision-making. How is this suite of effects related to the neural mechanisms that instantiate the choice process? We briefly review below the current state of knowledge about a standard model neurophysiological system for decision-making, the primate visuo-saccadic system, and describe recent experiments relevant to the influence of the choice set on the underlying neural representation of choice, which might shed some light on the sources of these violations of IIA.

The processes of visual sensation and eye movement generation are intricately intertwined in primates. Because the region of the retina with the highest acuity is small and located centrally, primates constantly perform fast eye movements, termed saccades, to foveate different, relevant parts of the visual scene (Carpenter, 1988; Yarbus, 1967). Thus, we are continually making decisions about where to look next, a process that is driven by both the salient properties of the visual scene (a class of variables termed bottom-up factors) and active cognitive control (a class of variables termed top-down factors). In its entirety, saccade selection involves the processing of incoming sensory information, incorporation of internal information such as prior knowledge about the environment and state, comparison of different actions, selection of a single option, and the subsequent activation of the proper motor pathways, providing a model system in which to examine the complete sensory to motor transformation underlying decision-making.

Over the past half century, neuroanatomists and electrophysiologists have mapped out the bulk of the anatomical pathway that transforms visual information into a saccadic eye movement. Almost all of this system encodes information in a *topographic* manner, with neighboring neurons representing adjacent regions of visual space. This topographic representation is present in both sensory visual areas and saccadic motor areas: visual cortical neurons respond to stimuli in a circumscribed portion of the visual field (the *receptive field*), and oculomotor neurons are selective for specific eye movements (the *response field*). Importantly, the intervening brain regions that are neither purely sensory nor purely motor display both sensory and motor properties, and their sensory receptive fields and motor response fields generally coincide. Based on both anatomical connectivity and physiological findings, these cortical areas, including portions of the parietal and frontal lobes, are believed to be critically involved in action selection and oculomotor decision-making (Glimcher, 2003).

The visual information necessary for saccade selection arrives via the extremely well-studied early visual system, passing from the retina to the lateral geniculate nucleus of the thalamus to primary visual cortex (V1). From V1 onwards, there is a general division of processing into two anatomically and functionally separable streams of information processing, a dorsal stream representing spatial location (often termed the *where pathway*) and a ventral stream representing object identification (*what pathway*) (Goodale & Milner, 1992; Mishkin & Ungerleider, 1982). Consistent with visual and motor planning in a spatial reference frame, the brain areas involved in visuo-saccadic decision-making are located in the dorsal stream. Visual information passes from V1 through extra-striate cortical regions, including the medial temporal area (MT) and the middle superior temporal area (MST), to the lateral intraparietal area (LIP) of the posterior parietal cortex (Blatt, Andersen & Stoner, 1990). Importantly, neurons in area LIP are responsive to visual stimulus input but also display strong topographically organized responses prior to saccade execution (Barash, Bracewell, Fogassi et al., 1991a, 1991b; Mountcastle, Lynch, Georgopoulos et al., 1975), suggesting the parietal cortex encodes, at least in part, information necessary for decision-making. The posterior parietal cortex may be a general site for such visuomotor transformation, as similar activity exists in neighboring posterior parietal areas for other action modalities, for example reach-related activity in the parietal reach region (Andersen & Buneo, 2002; Mountcastle et al., 1975; Snyder, Batista & Andersen, 1997).

Area LIP in turn has direct projections to brain areas responsible for saccade control and execution. Outgoing connections from LIP target both the frontal eye fields (FEF), a subdivision of premotor cortex in the frontal cortex, and the superior colliculus (SC), a midbrain region with direct outputs to brainstem oculomotor nuclei (Blatt et al., 1990). FEF outputs in turn target the SC, and also directly innervate oculomotor centers in the brainstem. Neurons in FEF, like those in LIP, have both sensory and motor characteristics, with strong spatially selective responses before and during saccades. FEF is closely linked to the control of saccades: activity in FEF neurons specifies whether and when saccades are initiated (Hanes, Patterson & Schall, 1998; Hanes & Schall, 1996), and inactivation of FEF disrupts saccade generation (Dias, Kiesau & Segraves, 1995; Schiller, True & Conway, 1979). Finally, the intermediate layers of the SC, to which both LIP and FEF project, also contain a topographic, retinotopically-organized motor map of saccadic endpoints, and activity in this structure is required for normal saccade generation (Lee, Rohrer & Sparks, 1988; Schiller et al., 1979). Collicular neurons exhibit strong phasic bursts of action potentials just before saccades into the response field, but can also display slow, low-frequency activity prior to the saccade related to more cognitive aspects of motor preparation and decision (Dorris & Munoz, 1998; Glimcher & Sparks, 1992).

Numerous studies have shown that the intermediate cortical areas of this visuo-saccadic pathway display decision-related neural activity, most commonly examined in perceptual choice tasks. The goal of such studies is typically to examine how neural circuits represent the subjective information that decision-makers use to make a choice, particularly in the cases where the sensory information itself is ambiguous. In one landmark study, Newsome and his colleagues examined how the activity of neurons in the extrastriate area MT, early in the visuo-saccadic pathway, correspond to behavioral choice (Newsome, Britten & Movshon, 1989). MT neurons are motion sensitive, with responses tuned to visual stimuli moving at a particular speed and direction, and these neurons are believed to encode in their instantaneous firing rates the strength of motion in their preferred direction. In the Newsome study, monkeys were presented moving dot stimuli, where some dots moved coherently in a single direction while the remainder moved randomly, and asked to report the direction of motion with a saccade; behavioral performance in this task is a monotonic function of the coherence of the random dots. Significantly, when motion stimuli are placed in the receptive field, the reliability and sensitivity of most MT cells matched or exceeded the psychophysical performance of the monkeys, suggesting that the decision-making (in this task) could be driven by a small number of these neurons.

Much additional decision research has focused on LIP, which receives direct input from MT. Unlike the purely sensory responses in MT, LIP activation has a strong, selective motor response for saccades into the response field (Barash et al., 1991b; Gnadt & Andersen, 1988). Examined in the motion discrimination task, LIP responses are markedly different than those in MT and indicative of a further stage in decision-making (Roitman & Shadlen, 2002; Shadlen & Newsome, 1996). In these experiments, the motion stimulus cue was presented centrally while monkeys held fixation; based on the perceived motion in the cue, the monkeys had to choose between two targets, one of which was placed in the recorded LIP neuron's response field. Because their activity is selective for particular actions (i.e. a saccade to the response field), LIP neurons are always indicative of choice by the time of action implementation, analogous to a motor neuron for right arm movement always active for a reach to a target on the right. The more relevant and informative issue is how such activity evolves prior to, and culminating in, action selection.

In the motion discrimination task, LIP neurons do not code motion itself, but rather how that motion informs the upcoming required saccade. If the motion stimulus indicated that a saccade into the recorded LIP neuron's response field would yield a reward, spiking activity increased over time, consistent with the idea that LIP is accumulating evidence about the value of a particular action. Furthermore, the rate of increase was dependent on the coherence of the motion signal; if there was stronger evidence for a rightward saccade, the activity of LIP neurons with a response field to the right increased more rapidly. Importantly, such activity is dissociable from sensory information; when there is no coherent motion in the stimulus (when the dots are entirely random, and there is no sensory information dictating the correct choice), LIP activity nonetheless indicates the upcoming saccadic choice the monkey will make (Shadlen & Newsome, 1996). In a reaction-time version of the task where monkeys were permitted to respond as soon as a decision was made, LIP activity under different coherence conditions – despite widely varying rates of activity increase – consistently reached a common threshold firing rate prior to saccade initiation. Thus, neural activity in LIP provides a physiological marker for the dynamics and evolution of the decision process. These initial results led to the idea that decision-related neural circuits, LIP being a well-studied example, represent the necessary and sufficient information to drive behavioral choice, a quantity termed a *decision variable* (Platt & Glimcher, 1999). Subsequent experiments showed that LIP neurons are modulated by the task-defined relevant behavioral information across a number of different paradigms, such as color, temporal information, and the accumulation of sensory evidence (Leon & Shadlen, 2003; Toth & Assad, 2002; Yang & Shadlen, 2007). The integration of such relevant information is captured by the concept of *subjective value* (Dorris & Glimcher, 2004), and the representation of value information is critical to understanding the function of neural decision circuits, a topic we explore further below.

While activity in the parietal cortex represents much of the sensorimotor information processing linked to choice, neurophysiological evidence suggests that frontal circuits are involved as well. FEF is likely more closely aligned to saccadic initiation than LIP: acute inactivation of FEF produces more severe saccade deficits (Dias et al., 1995; Dias & Segraves, 1999; Li, Mazzoni & Andersen, 1999; Wardak, Olivier & Duhamel, 2002), and FEF activity is more tightly coupled to saccadic reaction times (Hanes & Schall, 1996). However, evidence suggests that FEF activity also represents earlier stages in information processing, beyond just the initiation of an eye movement. In pop-out visual search tasks, where a single differently colored *oddball* target appears with a number of identically colored distractors, FEF neurons reliably distinguish targets from distractors appearing in the response field (Thompson, Bichot & Schall, 1997; Thompson, Hanes, Bichot & Schall, 1996). This activity is not simply visual, as FEF neurons do not initially differentiate target and distractor; however, it is more aligned with visual selection than simple motor preparation, since neural discrimination of target and distractor occurs at a relatively constant time after stimulus presentation, despite variable latencies in saccade generation. These studies suggest that FEF activity, like that in LIP, may represent the evolving decision variables guiding choice. Consistent with such a representation, microstimulation of FEF neurons while monkeys performed the random-dot motion discrimination task evoked eye movements that deviated in the direction of the monkey's intended saccade, with the magnitude of deviation reflecting the accumulated sensory evidence (Gold & Shadlen, 2000).

The evidence outlined above suggests that an interconnected network of brain areas carries the neural activity that evolves over the course of the decision process. We note here that this circuit does not exist in isolation, and a number of these regions are heavily interconnected with other brain areas that may influence the choice process (for example, reciprocal connections between FEF and other frontal cortical areas like dorsolateral prefrontal cortex and the supplementary eye fields, and between LIP and other posterior parietal areas). Nevertheless, the examination of decision-related activity in LIP and FEF provides a valuable framework to examine the influence of the choice set on neural representations.

NEURAL ACTIVITY AND THE EFFECT OF ALTERNATIVES

Given this knowledge of some of the neural structures underlying decision-making, what are the potential neurophysiological correlates of choice set size effects? Unfortunately, little is known about the effects of set size and value-guided choice in these decision circuits. However,

Feature search   Conjunction search

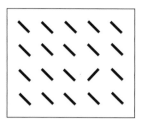 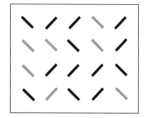

FIGURE 7.5   **Visual search arrays.** In the left panel, the goal is to identify the location of a target (tilted to the right) amongst a number of distractors (tilted to the left). In this example of feature search, the target differs from distractors by a single feature such as size, color, or orientation. In the right panel, the target is defined by more than a single stimulus feature (gray, tilted to the right), an example of conjunction search.

extensive findings exist regarding the neurophysiological effects of set size in a closely related process known as *visual search*, where a subject must detect a target stimulus against a background of other distractor stimuli (Figure 7.5). While visual search involves sensory information processing, it has also been effectively used to examine the selection process in areas like FEF and LIP. In some paradigms, the visual difference between the target and the distractors is a unique stimulus feature, for example color or orientation, leading to rapid detection thought to be driven by bottom-up processing in early visual areas. In other paradigms, the difference between target and distractors is a combination of features (e.g. green-square target and green-round, red-round, red-square distractors), and may involve top-down integrative processing in higher cortical areas (Treisman & Gelade, 1980). Many such studies are framed in terms of target selection, where the identified stimulus is also the location of a rewarded action (typically, a saccade).

In visual search tasks, psychophysical studies show a significant dependence of behavior on the number of possible targets. Classically, the time it takes to choose is a monotonic increasing function of the number of possible choice alternatives, an idea formalized as Hick's Law (Hick, 1952; Teichner & Krebs, 1974). In the human visual search literature, this reaction time *set-size effect* is much more significant when stimuli are more complex, a finding often interpreted as evidence for the involvement of cognitive, top-down processes related to covert attention or decision (Carrasco & Yeshurun, 1998; Wolfe, 1994). In particular, set-size effects arise when the target must be distinguished based on the conjunction of a number of stimulus features (a conjunctive search), reminiscent of the multi-attribute nature of options in context-dependent choice effects, such as in asymmetric dominance. Importantly, performance in visual search can decrement as the number of distractors increases, indicating that the set-size effect can affect accuracy as well as speed (Desimone & Duncan, 1995; Treisman & Gelade, 1980), analogous to proportionality and regularity violations in the context-dependent choice literature. While many attentional theories hypothesize that these behavioral set-size effects reflect a limited capacity of information processing, independent process models rooted in signal detection theory (and akin to neoclassical economic and Bayesian models) suggest that such effects can arise from stochastic features of the decision process itself (Palmer, 1995; Palmer, Verghese & Pavel, 2000).

How are these effects of set size on visual search reflected in neural activity? Recent studies have identified neurophysiological correlates of the set-size effect in FEF neurons (Cohen, Heitz, Woodman & Schall, 2009; Lee & Keller, 2008), a class of cells previously known to be active during visual search and target selection and which discriminate (in their activity) between the presence of a target versus a distractor in their response fields. Cohen and colleagues (2009) specifically examined how FEF activity varies with set size in an "oddball" detection task. In this experiment, monkeys searched for a target ("L") amongst a varying number (1, 3, or 7) of distractors ("T") and were rewarded for a subsequent saccade to the correct location of the target (Figure 7.6). Consistent with a classic behavioral set-size effect, reaction times were significantly longer when more distractors were presented; error rates were low and weak functions of set size, suggesting that in this task the monkeys traded improved accuracy for slower speed. When FEF activity was examined, neural target selection times – the duration before neuronal activity differentiated target and distractor – increased with set size. This effect mirrors the increase in behavioral reaction times, suggesting that saccades cannot be initiated until FEF neurons locate the target.

Furthermore, set sizes had an additional effect on neural activity, with peak firing rates decreasing as set sizes increased. Similar number-dependent, stimulus-driven suppressive effects, originally described in the superior colliculus (Basso & Wurtz, 1997, 1998), have been observed in multiple decision-related brain areas including FEF (Lee & Keller, 2008; Schall, Hanes, Thompson & King, 1995) and the posterior parietal cortex (Balan, Oristaglio, Schneider & Gottlieb, 2008; Constantinidis & Steinmetz, 2001). Different possible mechanisms have been proposed to explain this suppressive phenomenon. Suppression may be bottom-up, driven by local suppressive cortical interactions in early visual areas and dependent only on the stimulus configuration (Schall et al., 1995). Such suppressive surround interactions, where stimulus-driven activity is inhibited by the presence of stimuli outside the receptive field, are common in the early visual pathway and may play a critical role here. Alternatively, suppression may also be driven by top-down, cognitive

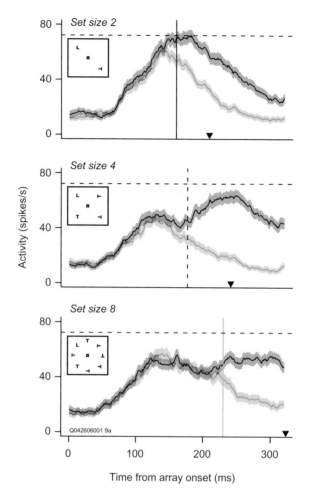

FIGURE 7.6   **Set-size effects on FEF activity during visual search.** Monkeys viewed an array containing a target stimulus (L) and various numbers of distractor stimuli (variably rotated T), and were rewarded for identifying and making an eye movement to the target location. Graphs present the activity of a single example FEF neuron presented with a target and 1, 3, or 7 distractors (dark trace, target in response field; light trace, distractor in response field). There are two primary effects of set size on neural activity. First, there is an effect on *timing*: the time at which neuronal activity distinguishes between a target and a distractor in its response field (vertical bar) increases as set size increases. Second, there is an effect on *firing rates*: peak firing rates are lower at larger set sizes. The first effect matches visual search behavior, where reaction times increase at larger set sizes (triangles, median reaction times). The second effect, mean rate suppression, may be related to choice accuracy and, in value paradigms, context-dependent choice. Adapted from Cohen et al. (2009), with permission.

influences reflecting knowledge of the task contingencies, such as choice uncertainty (Basso & Wurtz, 1998).

The effects of choice set size on neural activity during visual search demonstrates that, at least in this kind of perceptual task, the number of alternatives explicitly modulates neural activity in decision circuits. Moreover, the specificity of both neural and behavioral set-size effects to more complicated search paradigms (conjunctive versus feature search) implicates the higher order cognitive processes over simple bottom-up sensory mechanisms. Most visual search studies frame these effects in terms of covert attention, but such attentional processes are likely intricately linked to the decision process itself (Palmer, 1994; Palmer et al., 2000). Furthermore, the dependence of search set-size effects on conjunctive stimuli suggests a functional similarity with context-dependent preferences, which typically emerge when choice options are close to subjective equivalence, for example generating stochastic choice. This similarity suggests that set-size dependent modulation of the neural representation of option values may underlie behavioral set-size effects during more traditional decision-making tasks.

SET-SIZE EFFECTS AND THE NEURAL REPRESENTATION OF SUBJECTIVE VALUE

One critical distinction between the visual search and context-dependent preference literatures is the emphasis on value. The paradox of choice is that the empirical dependence of preference on choice context violates traditional, long-standing assumptions of how value representations are constructed; specifically, context-dependent choice implies a comparative rather than absolute valuation mechanism. In the search field, emphasis is placed on the visual or cognitive mechanisms that enable target identification, and visual search is viewed in terms of sensory processing. However, value is implicitly included in search paradigms, as trials which are rewarded require the correct identification of the search target. One might thus hypothesize that the examination of value representation during decision-making could provide a bridge between set-size effects in visual search and context-dependent preferences and choice.

How is valuation represented in the neural circuits underlying decision? One critical finding was the demonstration that LIP neurons encoding saccades are modulated by the value of the associated action (Platt & Glimcher, 1999). Platt and Glimcher recorded LIP neurons while monkeys performed a simple oculomotor decision task, in which monkeys made saccades to two different targets (Figure 7.7). Across blocks, the authors manipulated either the probability or the magnitude of

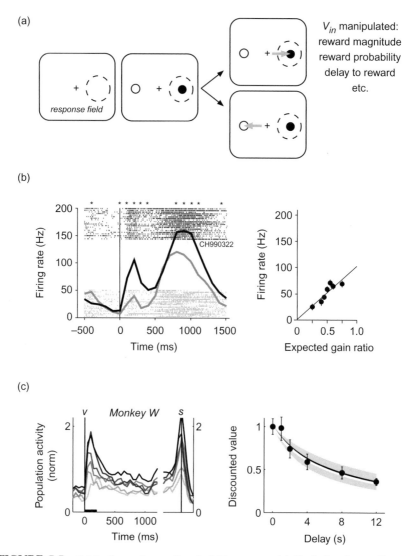

FIGURE 7.7 **Subjective value coding in LIP neurons.** (a) Typical task paradigm for examining LIP activity. LIP neurons respond to visual stimulus presentation in a constrained region of the visual field (the *response field*); saccadic activity typically is selective for eye movements to the same region of visual space. LIP neurons are known to represent the value of a saccade to the response field (V_{in}), which is manipulated by varying parameters of the associated reward such as magnitude, probability, and delay to reinforcement. (b) LIP represents value modulated by reward magnitude. Left, average firing rate of an example LIP neuron under two different conditions of reward magnitude (gray, low; black, high). Right, mean firing rate as a function of expected gain ratio $V_{in}/(V_{in}+V_{out})$ during the early visual period after stimulus onset. Data are from trials with identical stimuli and motor actions (saccade to the response field), indicating that

reward associated with the targets. When they examined trials in which the monkeys made a saccade into the response field, neuronal firing rates were strongly modulated by the expected value of the eye movement, despite the fact that the sensory stimuli and required motor action were identical in all trials. Subsequent studies reinforced the notion that LIP neurons carry a robust subjective value signal, whether value is determined by local fluctuations in a dynamic foraging task (Sugrue, Corrado & Newsome, 2004), equilibrium constraints in strategic games (Dorris & Glimcher, 2004; Seo, Barraclough & Lee, 2009), or temporal information in a delay discounting task (Louie & Glimcher, 2010). Value representation in the parietal cortex serves as a general framework to explain many of the previously observed modulations of LIP activity, such as the accumulation of sensory evidence (Shadlen & Newsome, 1996; Yang & Shadlen, 2007), since such behaviorally relevant parameters inform the subject about expected reward.

Since value information is a fundamental element of the decision process, it is important to understand how set-size manipulation affects the neural representation of value. Economic theories do not require that value be represented in specific units, and thus do not distinguish between representations coding *absolute value*, where action values are modulated strictly by the value of the target option, and those coding *relative value*, where action values are normalized to the value of all available options. Neural systems, however, must instantiate value representation in biophysical parameters, such as neuronal firing rates. Growing evidence suggests that the representation of value in decision circuits is expressed in relative terms, in a manner explicitly dependent on the available choice set. For example, doubling the values of two saccade targets produces little modulation in the activity of an LIP neuron representing a saccade to one of the targets (Dorris & Glimcher, 2004), despite the fact that these neurons are sensitive to the value of the response field target. This suggests that a neuron in LIP represents the value of a saccade into its response field (V_{in}) in a relative rather than absolute manner, dependent on both available target values – a form of nonlinear suppression. The classic assumption in the decision-making literature is that value is represented as the quantity $V_{in}/(V_{in}+V_{out})$, known as the expected gain ratio or fractional value (Herrnstein, 1961); studies have found that

neural activity is modulated by value. Adapted from Platt and Glimcher (1999). (c) LIP represents value modulated by delay to reinforcement. Left, average population LIP activity under conditions of fixed reward and varying delays (darker shades correspond to longer delays). Right, average LIP activity (black) and behavioral subjective value (white) plotted as a function of delay, demonstrating a precise neurometric-psychometric match for subjective value. Adapted from Louie and Glimcher (2010).

LIP value coding is closer to fractional than absolute, whether value is explicitly controlled by the experimenter or determined by local behavior in a dynamic foraging task (Platt & Glimcher, 1999; Sugrue et al., 2004).

How is this relative value representation constructed? In recent work in our laboratory, we have explicitly examined the effect of the choice context on value coding and find that LIP neurons instantiate a specific relative value representation driven by divisive normalization, a gain control algorithm ubiquitous in early sensory cortices (Louie, Grattan & Glimcher, 2011). Importantly, these results are analogous to the suppressive effect of increasing numbers of targets on firing rates observed in visual search experiments, but incorporate value as well as number of alternatives into the gain control mechanism. To precisely examine the influence of alternative option values and quantify the nature of this modulation, we recorded LIP neurons in a three-target task, where we systematically varied the number of saccade targets and their values, enabling us to test the divisive normalization model from visual search in the context of value representation. Monkeys fixated a central cue and were presented with either one, two, or three targets, each of which was associated with a different magnitude of water reward. After target presentation, monkeys were subsequently instructed to select one of the presented targets: a medium reward target situated in the response field (RF) or either the small and or large reward targets placed outside the RF, typically in the opposite hemifield. Each trial consisted of one of seven possible target arrays, presented randomly and with equal probability (three single targets, three dual targets, and one triple target trial). Each randomized target array provided a unique combination of value associated with the target in the RF and values available outside the RF, allowing us to quantify the relationship between target value (V_{in}) and value context (V_{out}).

We found two primary effects of the choice set, or value context, on LIP responses. First, consistent with qualitative reports in two target tasks, activity elicited by target onset in the RF is modulated by the value of the alternatives, with larger V_{out} magnitudes leading to greater suppression. Second, activity when no RF target is present is suppressed in a context-dependent manner, with larger V_{out} values driving activity further below baseline activity levels. Analogous to extra-classical modulation in early visual cortex, both of these effects are driven by the value of targets that themselves do not drive the recorded neuron. Significantly, when we performed a quantitative model comparison with other possible relative reward representations including fractional value, $V_{in}/(V_{in} + V_{out})$, and differential value, $V_{in} - V_{out}$, contextual value modulation was clearly best explained by a divisive normalization based model.

What are the implications of a relative value representation on the choice process itself? One promising framework for understanding the effects of set size on both visual search and value-guided decision-making is based on signal detection theory (Green & Swets, 1966). The goal of a decision system is to select the highest valued option or action for implementation, based on internal, subjective estimates of the corresponding action values. However, these neural representations of value may be uncertain, with multiple possible sources of noise. First, there may be uncertainty in the learned or projected value itself, for example when an animal possesses incomplete knowledge of the potential reward. Second, neural activity is inherently noisy, with cortical neurons exhibiting a variance that scales approximately with the mean firing rate (Shadlen & Newsome, 1998; Tolhurst, Movshon & Dean, 1983). In contrast to the much more regular activity of peripheral neurons, such response variability is one of the dominant features of neurons in the cortex. Critically, the presence of noise transforms value from a single quantity into a probability distribution, a notion incorporated into economic theory as random utility models (McFadden, 1974).

We propose that choice set effects on the decision related neural activity may influence behavior when combined with noisy representations of value. Consider Figure 7.8 (top panel), which depicts a subject choosing between two high-valued options (solid and dashed black lines). Each probability distribution depicts the possible activity of a neuron representing the value of an option; this distribution can be equivalently interpreted as the firing rate of a single neuron over many repetitions, or the activity of a population of identical neurons on a single trial. As a simple assumption for this example, each probability distribution has a mean firing rate proportional to the option's fractional value and a fixed variance. The subsequent panels depict the addition of either 1, 10, or 20 additional low-valued alternatives to the same choice set. Because the neurons represent relative rather than absolute value, the presence of additional alternatives decreases the mean firing rate of every distribution as the normalization term includes the larger choice set. This decreases the discriminability between the two high-valued options, evident as increasing overlap between the two highest valued distributions as set size increases. The implementation of a simple decision rule, for example drawing a sample from each probability distribution and selecting the option with the maximum firing rate, predicts a violation of both proportionality and regularity. The relative proportions of choices allocated to best option (solid) compared to the second best option (dashed) decreases, violating proportionality; furthermore, if the irrelevant alternatives are sufficiently low in value, the choices lost by the best option will be captured by the second best option, violating regularity.

This simple model suggests that the now well-documented relative representation of value in decision circuits can lead to at least some of the context-dependent choice effects reviewed above, but many additional details regarding the neurophysiology of choice and value representation

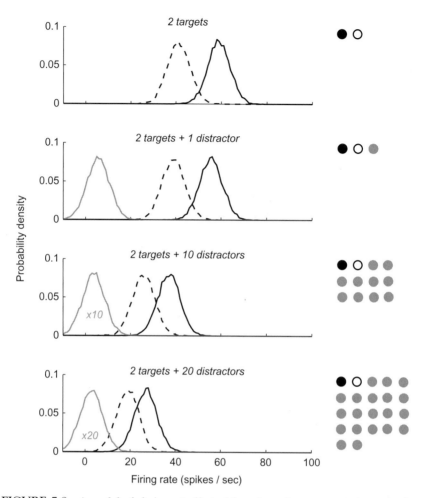

FIGURE 7.8 **A model of choice set effects driven by noisy representations of value.** Hypothetical distributions of neural activity representing the values of two high-valued options (solid and dashed black lines) in the absence or presence of different numbers of low-valued distractors (gray line; 0, 1, 10, or 20 distractors). In this simple model, the distributions of neural activity have a fixed variance and a mean firing rate proportional to $V_{in}/(V_{in} + V_{out})$. As the number of alternatives in the choice set increases, the value gain control implemented by the relative value representation decreases the distance between the neural activity distributions and decreases the discriminability between the two high-valued options. In this example, because the additional low-valued options capture very few additional choices, the result is a violation of regularity and proportionality as the absolute preference for the second best option increases.

still need to be integrated to fully understand the link between neural representation and behavioral set-size effects. First, the actual probability distributions of neural value representations are not fully known. Cortical neurons classically have response variances that scale with mean firing rates, which implies that variances (and uncertainty) shrink as additional alternatives are included in the choice set; however, in a simple model, this scaling of variance does not offset the discriminability problem, since the width of normal distributions depend on the standard deviation rather than the variance. Furthermore, such scaling does not address other, additive sources of noise for which there is growing evidence in the neurobiological literature. A full understanding of population coding across cortical areas will involve a better understanding of a number of additional factors, such as population averaging and correlated noise across the neurons (Glimcher, 2001; Shadlen & Newsome, 1998). Second, neither the location nor the mechanism of the decision rule itself is yet well understood, though biophysical models of simple choice have been proposed (Wang, 2008). Regardless of further work, however, it is clear in both the visual search and the value-guided decision literature that the composition of the choice set has a quantifiable effect on neural firing rates. A more thorough understanding of the neural circuits of value and decision-making may provide a critical step to linking the size of the choice set to its behavioral consequences.

CONCLUSION

The decision process is vital to organisms that live in a dynamic environment and require a behavioral repertoire beyond simple stimulus-response reflexes. Many behavioral and neurophysiological decision studies have focused on two alternative choices, for obvious reasons of simplicity and tractability. However, there is now unambiguous evidence that the size of the choice set a biological decision-maker faces influences his or her efficiency as a chooser. Large choice sets robustly lead decision-makers to opt-out of important choices at significant economic and personal cost. In fact, sophisticated economic analyses now indicate that choosers become internally inconsistent in their choices, failing to maximize their welfare as choice sets grow in size – observations that are both troubling and puzzling, troubling because of the costs they impose on choosers and puzzling because we do not yet have good explanatory models for these costly behavioral phenomena.

Phenomena like these are, however, not new. There has been clear psychological evidence that the context in which a decision is made influences the consistency of a decision-maker's choices for almost half a century. A range of context effects have been documented and a host of psychological theories have been proposed to explain these effects. One feature that these many theories lack, however, is unified mechanisms for explaining context effects in general, and set-size effects in particular.

Recent studies of the neurobiological mechanism of choice may, however, offer an opportunity to mechanistically explain the sources of set-size effects in choice. Basic neurobiological studies clearly indicate that the representation of option value in the nervous system is highly sensitive to context. Studies of visual search, for example, have now widely documented non-linear effects of context and set size on the value representations that guide neural mechanisms for decision-making. Incorporating these existing neurobiological models of set size and context into existing choice models may prove a valuable strategy for reconciling many behavioral observations around a single conceptual framework. Indeed, studies of cortical firing-rate variability, the neural representation of value, and the structure of cortical circuits may even allow one to reconcile violations of traditional rational choice theory as examples of optimization under constraint. The hope thus is that a detailed understanding of the mechanism of choice and the constraints faced by the human nervous system can provide both deep and broad explanations of these many context-dependent phenomena that have thus far escaped normative synthesis.

In this chapter, we have reviewed the effects of changing set size on both choice behavior and the response of neurons in primate visuo-saccadic decision system. For the most part, these fields have remained largely separate, though increasing interest in interdisciplinary approaches bridging neuroscience, psychology, and economics have begun to bridge the gap. It is important to emphasize that while the simple model of primate visuo-saccadic decision-making affords a framework for beginning to examine the neurophysiology of set-size effects, it does not comprehensively describe the strategies monkeys may take in more complicated choice situations, or humans likely employ in the natural environment. We propose above that uncertainty in value representation coupled with a context-dependent, relative neural coding of value can produce apparent violations of rationality, even without incorporating the effect of multiple attributes. However, attribute dimensions play a significant role in examples and models of context-dependent decision-making, such as for the asymmetric dominance effect, and many of the psychological costs of extensive set sizes may arise from difficulty in combining information across different attribute dimensions. Going forward, more comprehensive neural and computational models will have to incorporate not just how value is represented but how it is constructed as well.

References

Andersen, R. A., & Buneo, C. A. (2002). Intentional maps in posterior parietal cortex. *Annual Review of Neuroscience, 25*, 189–220.

Arnold, S. J., Oum, T. H., & Tigert, D. J. (1983). Determinant attributes in retail patronage – seasonal, temporal, regional, and international comparisons. *Journal of Marketing Research, 20*(2), 149–157.

Arrow, K. J. (1963). *Social choice and individual values* (2d ed.). New York: Wiley.

Balan, P. F., Oristaglio, J., Schneider, D. M., & Gottlieb, J. (2008). Neuronal correlates of the set-size effect in monkey lateral intraparietal area. *PLoS Biology, 6*(7), e158.

Barash, S., Bracewell, R. M., Fogassi, L., Gnadt, J. W., & Andersen, R. A. (1991). Saccade-related activity in the lateral intraparietal area. I. Temporal properties; comparison with area 7a. *Journal of Neurophysiology, 66*(3), 1095–1108.

Barash, S., Bracewell, R. M., Fogassi, L., Gnadt, J. W., & Andersen, R. A. (1991). Saccade-related activity in the lateral intraparietal area. II. Spatial properties. *Journal of Neurophysiology, 66*(3), 1109–1124.

Basso, M. A., & Wurtz, R. H. (1997). Modulation of neuronal activity by target uncertainty. *Nature, 389*(6646), 66–69.

Basso, M. A., & Wurtz, R. H. (1998). Modulation of neuronal activity in superior colliculus by changes in target probability. *Journal of Neuroscience, 18*(18), 7519–7534.

Bateson, M. (2002). Context-dependent foraging choices in risk-sensitive starlings. *Animal Behaviour, 64*, 251–260.

Bateson, M., Healy, S. D., & Hurly, T. A. (2003). Context-dependent foraging decisions in rufous hummingbirds. *Proceedings of the Royal Society of London B: Biological Sciences, 270*(1521), 1271–1276.

Becker, G. M., Degroot, M. H., & Marschak, J. (1963). Probabilities of choices among very similar objects – an experiment to decide between two models. *Behavioral Science, 8*(4), 306–311.

Bertrand, M., Karlan, D., Mullainathan, S., Shafir, E., & Zinman, J. (2010). What's advertising content worth? Evidence from a consumer credit marketing field experiment. *Quarterly Journal of Economics, 125*(1), 263–306.

Blatt, G. J., Andersen, R. A., & Stoner, G. R. (1990). Visual receptive field organization and cortico-cortical connections of the lateral intraparietal area (area LIP) in the macaque. *Journal of comparative neurology, 299*(4), 421–445.

Boatwright, P., & Nunes, J. C. (2001). Reducing assortment: An attribute-based approach. *Journal of Marketing, 65*(3), 50–63.

Busemeyer, J. R., & Townsend, J. T. (1993). Decision field theory: A dynamic-cognitive approach to decision making in an uncertain environment. *Psychological Review, 100*(3), 432–459.

Carpenter, R. H. S. (1988). *Movements of the eyes* (2nd ed.). London: Pion.

Carrasco, M., & Yeshurun, Y. (1998). The contribution of covert attention to the set-size and eccentricity effects in visual search. *Journal of Experimental Psychology: Human Percept Perform, 24*(2), 673–692.

Chernev, A. (2003). Product assortment and individual decision processes. *Journal of Personality and Social Psychology, 85*(1), 151–162.

Cohen, J. Y., Heitz, R. P., Woodman, G. F., & Schall, J. D. (2009). Neural basis of the set-size effect in frontal eye field: Timing of attention during visual search. *Journal of Neurophysiology, 101*(4), 1699–1704.

Constantinidis, C., & Steinmetz, M. A. (2001). Neuronal responses in area 7a to multiple-stimulus displays: I. neurons encode the location of the salient stimulus. *Cerebral Cortex, 11*(7), 581–591.

Copulsky, W. (1976). Cannibalism in marketplace. *Journal of Marketing, 40*(4), 103–105.

Deci, E. L., & Ryan, R. M. (1985). *Intrinsic motivation and self-determination in human behavior*. New York: Plenum.

DeShazo, J. R., & Fermo, G. (2002). Designing choice sets for stated preference methods: The effects of complexity on choice consistency. *Journal of Environmental Economics and Management*, 44(1), 123–143.

Desimone, R., & Duncan, J. (1995). Neural mechanisms of selective visual attention. *Annual Review of Neuroscience*, 18, 193–222.

Dias, E. C., Kiesau, M., & Segraves, M. A. (1995). Acute activation and inactivation of macaque frontal eye field with GABA-related drugs. *Journal of Neurophysiology*, 74(6), 2744–2748.

Dias, E. C., & Segraves, M. A. (1999). Muscimol-induced inactivation of monkey frontal eye field: Effects on visually and memory-guided saccades. *Journal of Neurophysiology*, 81(5), 2191–2214.

Dorris, M. C., & Glimcher, P. W. (2004). Activity in posterior parietal cortex is correlated with the relative subjective desirability of action. *Neuron*, 44(2), 365–378.

Dorris, M. C., & Munoz, D. P. (1998). Saccadic probability influences motor preparation signals and time to saccadic initiation. *Journal of Neuroscience*, 18(17), 7015–7026.

Doyle, J. R., O'Connor, D. J., Reynolds, G. M., & Bottomley, P. A. (1999). The robustness of the asymmetrically dominated effect: Buying frames, phantom alternatives, and in-store purchases. *Psychology and Marketing*, 16(3), 225–243.

Glimcher, P. W. (2003). The neurobiology of visual-saccadic decision making. *Annual Review of Neuroscience*, 26, 133–179.

Glimcher, P. W. (2011). *Foundations of neuroeconomic analysis*. New York: Oxford University Press.

Glimcher, P. W., & Sparks, D. L. (1992). Movement selection in advance of action in the superior colliculus. *Nature*, 355(6360), 542–545.

Gnadt, J. W., & Andersen, R. A. (1988). Memory related motor planning activity in posterior parietal cortex of macaque. *Experimental Brain Research*, 70(1), 216–220.

Gold, J. I., & Shadlen, M. N. (2000). Representation of a perceptual decision in developing oculomotor commands. *Nature*, 404(6776), 390–394.

Goodale, M. A., & Milner, A. D. (1992). Separate visual pathways for perception and action. *Trends in Neurosciences*, 15(1), 20–25.

Green, D. M., & Swets, J. A. (1966). *Signal detection theory and psychophysics*. New York: Wiley.

Hanes, D. P., Patterson, W. F., 2nd, & Schall, J. D. (1998). Role of frontal eye fields in countermanding saccades: Visual, movement, and fixation activity. *Journal of Neurophysiology*, 79(2), 817–834.

Hanes, D. P., & Schall, J. D. (1996). Neural control of voluntary movement initiation. *Science*, 274(5286), 427–430.

Herrnstein, R. J. (1961). Relative and absolute strength of response as a function of frequency of reinforcement. *Journal of the Experimental Analysis of Behavior*, 4, 267–272.

Hick, W. E. (1952). On the rate of gain of information. *Quarterly Journal of Experimental Psychology*, 4, 11–26.

Huber, J., Payne, J. W., & Puto, C. (1982). Adding asymmetrically dominated alternatives – violations of regularity and the similarity hypothesis. *Journal of Consumer Research*, 9(1), 90–98.

Huber, J., & Puto, C. (1983). Market boundaries and product choice – illustrating attraction and substitution effects. *Journal of Consumer Research*, 10(1), 31–44.

Hutchinson, J. M. (2005). Is more choice always desirable? Evidence and arguments from leks, food selection, and environmental enrichment. *Biological Reviews*, 1, 73–92.

Hutchinson, J. M., & Gigerenzer, G. (2005). Simple heuristics and rules of thumb: Where psychologists and behavioural biologists might meet. *Behavioural Processes*, 69(2), 97–124.

Iyengar, S. S., Jiang, W., & Huberman, G. (2004). How much choice is too much: Determinants of individual contributions in 401 K retirement plans. In O. S. Mitchell

& S. Utkus (Eds.), *Pension design and structure: New lessons from behavioral finance* (pp. 83–97). Oxford: Oxford University Press.

Iyengar, S. S., & Lepper, M. R. (2000). When choice is demotivating: Can one desire too much of a good thing? *Journal of Personality and Social Psychology, 79*(6), 995–1006.

Latty, T, & Beekman, M. (2011). Irrational decision-making in an amoeboid organism: Transitivity and context-dependent preferences. *Proceedings of the Royal Society B: Biological Sciences, 278*(1703), 301–302.

Lee, C., Rohrer, W. H., & Sparks, D. L. (1988). Population coding of saccadic eye movements by neurons in the superior colliculus. *Nature, 332*(6162), 357–360.

Lee, K. M., & Keller, E. L. (2008). Neural activity in the frontal eye fields modulated by the number of alternatives in target choice. *Journal of Neuroscience, 28*(9), 2242–2251.

Leon, M. I., & Shadlen, M. N. (2003). Representation of time by neurons in the posterior parietal cortex of the macaque. *Neuron, 38*(2), 317–327.

Li, C. S., Mazzoni, P., & Andersen, R. A. (1999). Effect of reversible inactivation of macaque lateral intraparietal area on visual and memory saccades. *Journal of Neurophysiology, 81*(4), 1827–1838.

Louie, K., & Glimcher, P. W. (2010). Separating value from choice: Delay discounting activity in the lateral intraparietal area. *Journal of Neuroscience, 30*(16), 5498–5507.

Louie, K., Grattan, L. E., & Glimcher, P. G. (2011). Reward value-based gain control: Divisive normalization in parietal cortex. *The Journal of Neuroscience, 31*(29), 10627–10639.

Luce, R. D. (1959). *Individual choice behavior: A theoretical analysis.* New York: Wiley.

Malhotra, N. K. (1982). Information load and consumer decision-making. *Journal of Consumer Research, 8*(4), 419–430.

McFadden, D. (1974). Conditional logit analysis of qualitative choice behavior. In P. Zarembka (Ed.), *Frontiers of econometrics.* New York: Academic Press.

Mishkin, M., & Ungerleider, L. G. (1982). Contribution of striate inputs to the visuospatial functions of parieto-preoccipital cortex in monkeys. *Behavioural Brain Research, 6*(1), 57–77.

Mountcastle, V. B., Lynch, J. C., Georgopoulos, A., Sakata, H., & Acuna, C. (1975). Posterior parietal association cortex of the monkey: Command functions for operations within extrapersonal space. *Journal of Neurophysiology, 38*(4), 871–908.

Newsome, W. T., Britten, K. H., & Movshon, J. A. (1989). Neuronal correlates of a perceptual decision. *Nature, 341*(6237), 52–54.

Palmer, J. (1994). Set-size effects in visual search: The effect of attention is independent of the stimulus for simple tasks. *Vision Research, 34*(13), 1703–1721.

Palmer, J. (1995). Attention in visual search – distinguishing four causes of a set-size effect. *Current Directions in Psychological Science, 4*(4), 118–123.

Palmer, J., Verghese, P., & Pavel, M. (2000). The psychophysics of visual search. *Vision Research, 40*(10–12), 1227–1268.

Platt, M. L., & Glimcher, P. W. (1999). Neural correlates of decision variables in parietal cortex. *Nature, 400*(6741), 233–238.

Reutskaja, E., & Hogarth, R. M. (2009). Satisfaction in choice as a function of the number of alternatives: When "goods satiate". *Psychology and Marketing, 26*(3), 197–203.

Roitman, J. D., & Shadlen, M. N. (2002). Response of neurons in the lateral intraparietal area during a combined visual discrimination reaction time task. *Journal of Neuroscience, 22*(21), 9475–9489.

Royle, N. J., Lindstrom, J., & Metcalfe, N. B. (2008). Context-dependent mate choice in relation to social composition in green swordtails Xiphophorus helleri. *Behavioral Ecology, 19*(5), 998–1005.

Schall, J. D., Hanes, D. P., Thompson, K. G., & King, D. J. (1995). Saccade target selection in frontal eye field of macaque. I. Visual and premovement activation. *Journal of Neuroscience, 15*(10), 6905–6918.

Schiller, P. H., True, S. D., & Conway, J. L. (1979). Effects of frontal eye field and superior colliculus ablations on eye movements. *Science, 206*(4418), 590–592.

Schwartz, B. (2000). Self-determination – the tyranny of freedom. *American Psychologist,* 55(1), 79–88.

Schwartz, B. (2004). *The paradox of choice: Why more is less* (1st ed.). New York: ECCO.

Sedikides, C., Ariely, D., & Olsen, N. (1999). Contextual and procedural determinants of partner selection: Of asymmetric dominance and prominence. *Social Cognition, 17*(2), 118–139.

Sela, A., Berger, J., & Liu, W. (2009). Variety, vice, and virtue: How assortment size influences option choice. *Journal of Consumer Research, 35*(6), 941–951.

Seo, H., Barraclough, D. J., & Lee, D. (2009). Lateral intraparietal cortex and reinforcement learning during a mixed-strategy game. *Journal of Neuroscience, 29*(22), 7278–7289.

Shadlen, M. N., & Newsome, W. T. (1996). Motion perception: Seeing and deciding. *Proceedings of the National Academy of Sciences of the United States of America, 93*(2), 628–633.

Shadlen, M. N., & Newsome, W. T. (1998). The variable discharge of cortical neurons: Implications for connectivity, computation, and information coding. *Journal of Neuroscience, 18*(10), 3870–3896.

Shafir, S. (1994). Intransitivity of preferences in honeybees – support for comparative evaluation of foraging options. *Animal Behaviour, 48*(1), 55–67.

Shafir, S., Waite, T. A., & Smith, B. H. (2002). Context-dependent violations of rational choice in honeybees (Apis mellifera) and gray jays (Perisoreus canadensis). *Behavioral Ecology and Sociobiology, 51*(2), 180–187.

Shah, A. M., & Wolford, G. (2007). Buying behavior as a function of parametric variation of number of choices. *Psychological Science, 18*(5), 369–370.

Simon, H. A. (1955). A behavioral model of rational choice. *Quarterly Journal of Economics, 69*(1), 99–118.

Simonson, I., & Tversky, A. (1992). Choice in context – tradeoff contrast and extremeness aversion. *Journal of Marketing Research, 29*(3), 281–295.

Snyder, L. H., Batista, A. P., & Andersen, R. A. (1997). Coding of intention in the posterior parietal cortex. *Nature, 386*(6621), 167–170.

Sugrue, L. P., Corrado, G. S., & Newsome, W. T. (2004). Matching behavior and the representation of value in the parietal cortex. *Science, 304*(5678), 1782–1787.

Taylor, S. E., & Brown, J. D. (1988). Illusion and well-being: A social psychological perspective on mental health. *Psychological Bulletin, 103*(2), 193–210.

Teichner, W. H., & Krebs, M. J. (1974). Laws of visual choice reaction time. *Psychological Review, 81*(1), 75–98.

Thompson, K. G., Bichot, N. P., & Schall, J. D. (1997). Dissociation of visual discrimination from saccade programming in macaque frontal eye field. *Journal of Neurophysiology, 77*(2), 1046–1050.

Thompson, K. G., Hanes, D. P., Bichot, N. P., & Schall, J. D. (1996). Perceptual and motor processing stages identified in the activity of macaque frontal eye field neurons during visual search. *Journal of Neurophysiology, 76*(6), 4040–4055.

Tolhurst, D. J., Movshon, J. A., & Dean, A. F. (1983). The statistical reliability of signals in single neurons in cat and monkey visual cortex. *Vision Research, 23*(8), 775–785.

Toth, L. J., & Assad, J. A. (2002). Dynamic coding of behaviourally relevant stimuli in parietal cortex. *Nature, 415*(6868), 165–168.

Treisman, A. M., & Gelade, G. (1980). A feature-integration theory of attention. *Cognitive Psychology, 12*(1), 97–136.

Tversky, A. (1969). Intransitivity of preferences. *Psychological Review, 76*(1), 31–48.

Tversky, A. (1972). Elimination by aspects – a theory of choice. *Psychological Review, 79*(4), 281–299.

Tversky, A., & Russo, J. E. (1969). Substitutability and similarity in binary choices. *Journal of Mathematical Psychology, 6*(1), 1–12.

Tversky, A., & Shafir, E. (1992). Choice under conflict – the dynamics of deferred decision. *Psychological Science*, *3*(6), 358–361.

Tversky, A., & Simonson, I. (1993). Context-dependent preferences. *Management Science*, *39*(10), 1179–1189.

Waite, T. A. (2001). Intransitive preferences in hoarding gray jays (Perisoreus canadensis). *Behavioral Ecology and Sociobiology*, *50*(2), 116–121.

Wang, X. J. (2008). Decision making in recurrent neuronal circuits. *Neuron*, *60*(2), 215–234.

Wardak, C., Olivier, E., & Duhamel, J. R. (2002). Saccadic target selection deficits after lateral intraparietal area inactivation in monkeys. *Journal of Neuroscience*, *22*(22), 9877–9884.

Wedell, D. H. (1991). Distinguishing among models of contextually induced preference reversals. *Journal of Experimental Psychology: Learn*, *17*(4), 767–778.

Wolfe, J. M. (1994). Guided search 2.0 – a revised model of visual search. *Psychonomic Bulletin & Review*, *1*(2), 202–238.

Yang, T., & Shadlen, M. N. (2007). Probabilistic reasoning by neurons. *Nature*, *447*(7148), 1075–1080.

Yarbus, A. L. (1967). *Eye movements and vision*. New York: Plenum Press.

SOCIAL FACTORS

CHAPTER

8

Social Factors and Preference Change

Daniel Campbell-Meiklejohn and Chris D. Frith
Center for Functionally Integrative Neuroscience, Aarhus University and
Wellcome Centre for Neuroimaging at University College London

Neuroscience of Preference and Choice
DOI: 10.1016/B978-0-12-381431-9.00008-5

INTRODUCTION

Humans are intensely social animals. Much of our behavior is enacted in the presence, and during interactions, with others. We should not be surprised at the extent to which others influence our decisions and values. As Adam Smith suggested (1759), obtaining money (i.e. resources) may not be the prime force that drives human behavior. Rather, our principle motivation is to be noticed by our fellows (i.e. to have a reputation) and acquiring wealth is just one way to enhance our reputation. In this chapter we review evidence for social motivations, considering implicit social processes that alter our individual behavior (without our awareness) and explicit social factors that play a crucial role in enabling our collaborations with others to achieve more than the sum of the individuals involved.

THE PRESENCE OF OTHERS

Simply by watching other creatures, animals (including humans) can obtain useful information such as the location of food, the presence of predators, and the physical characteristics of the creature being observed. In consequence, the creature being observed may modify its behavior to influence the observer (Earley, 2010). In the presence of bystanders, individuals might behave in a way that exaggerates their physical strength or improves their reputation for being cooperative. These effects of an audience on behavior are widely reported (Tennie, Frith & Frith, 2010).

The Effects of Being Watched

Audience Effects in Non-Human Animals

The three examples we list here give a flavor of the wide range of species that show audience effects and the wide range of contexts in which these effects occur. (1) On hearing the calls of other frogs, male green frogs lower the pitch of their calls. This has the effect of exaggerating their size, since larger frogs make deeper calls (Bee, Perrill & Owen, 2000). This lowering of the call's pitch is more likely to be made by a small frog on hearing a large frog. (2) Male domestic chickens make more predator-elicited alarm calls in the presence of other chickens, but not in the presence of other species (Karakashian, Gyger & Marler, 1988). (3) Reef cleaner fish feed more cooperatively (i.e. they refrain from biting their clients) when observed by other potential host clients (Bshary & Grutter, 2006).

The frogs and fish in these examples appear to be changing their behavior in order to enhance their "reputation" thereby giving themselves an immediate advantage, but this does not seem to explain the case with chickens and their alarm calls. In this latter instance the change in behavior seems more directly prosocial since the calls provide an advantage for the listener, but a disadvantage for the caller (at least in the short-term) since the calls potentially run the risk of the unwanted attention of a predator (see Kokolakis, Smith & Evans, 2010). This can be seen as an example of kin selection where the call is made if there is an advantage for related, but not for unrelated, individuals. These two effects of an audience, to increase prosocial behavior and to work harder to enhance the reputation of the self, are also widely observed in humans (Tennie et al., 2010).

Audience Effects in Humans

Ten-month-old infants spontaneously smile at their mothers, an effect that is greatly enhanced when the mother is looking at the infant (Jones, Collins & Hong, 1991). A similar audience effect is seen in adults. On seeing someone in pain, people often respond by making a facial expression of pain and this response is greatly enhanced if the person in pain is looking at the responder (Bavelas, Black, Lemery & Mullett, 1986). As we saw from the example of the frogs trying to make themselves appear larger, there are many attributes in addition to being cooperative or empathetic, which can be used to send signals to an audience. One example comes from a study of Gardner and Steinberg (2005) who found that adolescents take more risks in the presence of their peers.

We can speculate about the function of these behaviors. The response of the infant to the gaze of the mother might be a signal, reflecting an implicit desire to increase a reputation for cooperativeness. Likewise, my enhanced response to a person in pain is a signal to that person that I am an empathic individual. Direct experimental evidence that audience effects increase prosocial behavior comes from an ingenious real-world study by Bateson, Nettle and Roberts (2006). The authors monitored the amount of money placed in an honesty box in the university coffee room. By simply placing a photograph of a pair of human eyes above the box, the amount of money contributed was significantly enhanced. Presumably this effect is largely automatic and unconscious, since a moment's thought would lead a person to the conclusion that the photograph can have no effect on one's reputation for cooperativeness.

The Neural Basis of Audience Effects

We know little about the neural basis of these audience effects concerning reputation. However, a related audience effect has been observed in the brain's so-called mirror system. Many studies have revealed the strong tendency to mirror the behavior and expressions of those we observe, in terms of covert neural activity. For example, seeing people move their limbs elicits neural activity in the motor system of the observer, even though the observer does not actually move (Rizzolatti & Craighero,

2004). There is preliminary evidence that this tendency to mirror the behavior of others is enhanced by an audience effect. Kilner and colleagues measured the suppression of oscillatory activity associated with action and also with the observation of action and found suppression associated with action observation was greatly enhanced when the actor was looking at the observer (Kilner, Marchant & Frith, 2006).

Reputation

A major finding from studies in which volunteers play economic games is that people do not always behave "rationally" in order to maximize their gains. For example, in the ultimatum game the responder will turn down "unfair" offers, thereby foregoing a certain gain. Likewise, the proposer does not offer the smallest possible amount. This might be because the proposer reflects on the mental state of the responder and suspects that an unfair offer will be refused. However, this problem does not arise in the dictator game, since the responder does not have the option to refuse the offer. So why do proposers in this game typically not make the smallest possible offer? One explanation is that the proposer does not wish to acquire a reputation for being mean. Consistent with this idea is the observation of an audience effect: if proposers can be persuaded that they are unobserved, then they will make significantly smaller offers (Hoffman, McCabe & Smith, 1996). However, these low offers are still larger than they need to be. This observation is consistent with the idea that we have an automatic and largely unconscious tendency to guard our reputation.

While not relevant in one-shot games, reputation for cooperativeness has a direct effect on gains in sequential games. Milinski and colleagues have shown how cooperation rapidly declines when players switch from games where they can be identified to those where they are anonymous and cannot acquire a reputation for good or bad behavior (Semmann, Krambeck & Milinski, 2004). In situations where a reputation can be acquired on the basis of past behavior, a reputation for being cooperative has two obvious advantages. First, a cooperative person is more likely to receive investments and rewards (Wedekind & Milinski, 2000) since others can anticipate these will be returned. Second, the cooperative person is more likely to be chosen as a partner in collaborative projects (Sylwester & Roberts, 2010). In contrast, the person with a bad reputation for cooperation is more likely to be either punished (Fehr & Gachter, 2002) or avoided (Panchanathan & Boyd, 2004).

The Neural Basis of our Reputation

As suggested by Adam Smith, acquiring a good reputation is as rewarding as receipt of money. Izuma, Saito & Sadato (2010a) scanned volunteers while they chose whether or not to donate money to charity. When such donations are made in the presence of observers, social reward (i.e. reputation) can be gained, but not when identical donations are made anonymously. In this study, donation rates were enhanced by the presence of observers. In addition, enhanced activity in the striatum was observed when a high social reward was anticipated (donation in public) and when monetary gain was anticipated, but without any social cost (no donation in the absence of observers). It appears that an improved reputation activates our brain's basic reward system in the same way as gaining money and other valuable resources.

However, representing our own reputation is not as simple as representing the value of some object or another person. Our reputation is, after all, determined exclusively by other people. To have insight into my own reputation requires knowledge about what other people think about me. In other words, thinking about our own reputation involves thinking about how we are represented in the minds of others. Many studies point to the involvement of medial prefrontal cortex and adjacent paracingulate cortex in representing the mental states of other people and of the self. On this basis, it seems plausible that this region might be engaged by reputation representation since this involves simultaneously representing both the self and others (Amodio & Frith, 2006).

This idea has been confirmed in a host of studies. The requirement for participants to rate themselves in relation to social norms of behavior has been shown to elicit enhanced activity in mPFC. This activity is greatly enhanced if such a rating is carried out in the presence of an audience (Izuma, Saito & Sadato, 2010b) where one assumes that participants are more concerned about the impact their ratings have on their reputation.

Bengtsson, Lau and Passingham (2009) conducted a study where participants were led to believe that the task they were performing provided a measure of their cognitive abilities. In this context, the occurrence of errors could lead to a lowering of reputation and the occurrence of errors was associated with increased activity (measured with fMRI) in anterior paracingulate cortex. Similar results were obtained by priming people to think about their reputation for cleverness (Bengtsson, Dolan & Passingham, 2010) where, after such priming, a response to errors was enhanced in anterior paracingulate cortex. Similar results have been obtained using EEG measurements of error related negativity, which is believed to have its origin in ACC. Errors in the performance of a racial stereotype inhibition task reveals underlying prejudice and has the potential to lower reputation, particularly in people sensitive to external pressures to respond without prejudice. When performing this stereotype inhibition task in public, participants showed enhanced amplitude of the error-perception component of the error-related potential linked to activity in rostral ACC (Amodio, Kubota, Harmon-Jones & Devine, 2006).

The concept of self-esteem is closely linked with reputation, since a person with high self-esteem is generally confident of having a good reputation. In keeping with this, people with low self-esteem are more sensitive to social feedback (i.e. whether or not they are liked by others) and show enhanced activity in ventral striatum and in mPFC in response to positive social feedback (Somerville, Kelley & Heatherton, 2010).

Conclusions

The mere presence of others alters our behavior, such that we become more likely to make prosocial rather than selfish decisions. Much, but not all, of this effect can be explained in terms of reputation. In a highly interactive society it is an advantage to have a good reputation for being cooperative and trustworthy. However, only when others observe us, does behavior have the potential to raise or lower our reputation. The performance of acts that enhance our reputation activates the brain's reward system (including ventral striatum). However, when we think about our reputation, we need to think about how we are perceived by other people by implicitly asking, for example, *"If I do this what will he or she think of me?"* Such thoughts are associated with enhanced activity in mPFC and paracingulate cortex, components of a putative mentalizing system.

The Effects of Being Imitated

Sensitivity to Imitation

When engaged in a face-to-face interaction, people have a strong and automatic tendency to imitate one another (Chartrand & van Baaren, 2009). In the laboratory this can be revealed by interference effects on one's own action when observing the actions of others (e.g. Liepelt, Ullsperger, Obst et al., 2009). It is widely reported that there is activation in the relevant facial muscles when observing the facial expressions of another, even when the presentation is subliminal (Dimberg, Thunberg & Elmehed, 2000). In a more naturalistic setting, this mutual imitation is known as the *chameleon effect* (Chartrand & Bargh, 1999) consisting of unconscious mimicry of postures, mannerisms, and facial expressions. This mimicry facilitates the smoothness in our interactions and increases liking between partners. Empathic individuals show a greater chameleon effect. However, imitation of actions (Gutsell & Inzlicht, 2010) and facial expressions (Avenanti, Sirigu & Aglioti, 2010; Xu, Zuo, Wang & Han, 2009) does not seem to occur for interactions with members of an out-group.

Increasing Preference for the Imitator

An increased liking for someone who (covertly) imitates us is also observed in monkeys who show greater affiliative behavior towards humans who imitate them (Paukner, Suomi, Visalberghi & Ferrari, 2009). Preschool children find imitation reinforcing (Miller & Morris, 1974), showing an increase in the probability of performing an action that has been imitated, and decreased probability of performing an action that has been counter-imitated (Fouts, Waldner & Watson, 1976). In addition, preference increases for neutral stimuli that are associated with imitation (Parton & Priefert, 1975). Direct interaction is not needed for these effects and identical effects are observed in a child who observes another child being imitated (Thelen, Lada, Lasoski et al., 1980).

These effects are not restricted to children. Adults are also influenced by and prefer people who imitate them (e.g. Chartrand & Bargh, 1999). Furthermore, these effects are harnessed to increase liking. For example, people with a need for affiliation, after having been ostracized, show increased non-conscious behavioral mimicry (Lakin & Chartrand, 2003; Lakin, Chartrand & Arkin, 2008). Similar effects are seen in children primed by seeing a video in which one shape is ostracized by a group of other shapes (Over & Carpenter, 2009).

Imitation Increases Prosocial Behavior

The effects of being imitated are not restricted to feelings of affiliation towards the person doing the imitation. After people have been mimicked, they become generally more prosocial in their behavior. Such behavior is evident, as in an increased disposition to help people other than the imitator (van Baaren, Holland, Kawakami & van Knippenberg, 2004), to give more money to charity (Stel, Van Baaren & Vonk, 2008), and to vote for more prosocial political parties (Stel & Harinck, 2010). These effects are not restricted to the person being imitated and extend even to people who have been instructed to imitate (Stel & Vonk, 2010).

It is widely believed that these effects are unconsciously mediated. Except when they have been explicitly instructed by the experimenter, imitators are not aware they are imitating and great care is taken to make sure that person being imitated is not aware that they are being imitated. It has been suggested that high levels of discomfort are experienced by people who are aware that they are being imitated (see Bailenson, Yee, Patel & Beall, 2008). Therefore, affiliative and prosocial effects of being imitated do not occur when we are aware that we are being imitated. These apparently dramatic differences in the effects of covert and overt imitation on behavior are of considerable interest and deserving of more in-depth research.

Neural Basis of Imitation Effects

These effects of imitation on behavior involve at least three mechanisms. First, there is the basic process by which we translate what we see into what we do. This likely depends upon the action mirror system in the brain involving STS and parietal cortex. This was confirmed in an early study involving overt imitation in both directions (Decety,

Chaminade, Grezes & Meltzoff, 2002) and in many subsequent studies of action observation (Rizzolatti & Craighero, 2004). Second is the mechanism through which being imitated is rewarding. Schilbach and colleagues used an ingenious paradigm in which participants' eye movements were recorded and used to control the position of the eyes of an avatar (Schilbach, Wilms, Eickhoff et al., 2010). When the gaze of the avatar was contingent on the gaze of the participant (i.e. when the participant was imitated), activity was seen in the ventral striatum, the major reward area of the brain. In another study, participants were asked to adopt an expression of happiness or sadness and then shown a face with either the congruent or the incongruent expression. When the face was congruent, whether happy or sad, there was increased activity in medial orbitofrontal cortex (mOFC) and ventromedial prefrontal cortex (vmPFC), brain regions associated with positive feelings and reward processing (Kühn, Müller, van der Leij et al., 2010a). A problem with these studies is that the imitation was overt, whereas, as we have noted above, the most striking effects occur with covert imitation. In a third study, participants observed an interaction in which one partner either imitated the gestures of the other or not. Behavioral ratings confirmed that this indirect observation of imitation did indeed increase liking for the imitator. Here also observing imitation was associated with increased activity in mOFC and vmPFC (Kühn, Müller, van Baaren et al., 2010b). In this experiment the attention of the participants was not drawn to the occurrence of imitation in the interaction, since they were performing an irrelevant visual attention task. An informal debriefing after scanning suggested that very few participants were aware of the mimicry (Simone Kühn, personal communication) rendering it likely that these regions, associated with pleasure and reward, are activated by covert imitation.

Third is a mechanism through which being (covertly) imitated increases prosocial behavior in general. Hints about this mechanism come from a behavioral study suggesting that affective empathy mediates the relationship between imitation and prosocial behavior (Stel et al., 2008). Our interpretation of this observation is that imitation alters the balance between self-oriented and group-oriented behavior towards a group orientation.

THE BEHAVIOR AND EXPERIENCE OF OTHERS

By observing the behavior of others we can learn about our environment, each other and ourselves. This remarkable ability not only saves time and effort but also allows knowledge to accumulate over generations, enabling advances in technology and culture. Other people's fortunes and misfortunes can also profoundly affect us: tears at weddings, joyous cheers when our favorite teams score, hidden feelings of envy, and the guilty pleasure of a competitor's loss, are but a few examples of how we respond to other people's fate.

Such responses form through a complex interaction between cognition and context. In the next two sections we explore what neuroscience has told us about our ability to learn from others, to be fair, to trust, and to effectively use social information as well as conform. We also explore how neuroscience has begun to isolate processes behind our responses to the experience of others and how these responses are realized within the architecture of the brain.

Gains and Losses of Others

My experience of another person's experience depends on our similarities; how likeable the other is, and the extent to which my outcome depends on theirs.

Similarity

When research participants watched a gambler play a roulette game where the outcome was either money or an electric shock, those who believed the gambler was similar to themselves, tended to have stronger emotional and physiological reactions to the gambler's outcomes (Krebs, 1975). When subsequently choosing between being selfish or helping the gambler at a cost, a participant who reacted strongly to the gambler's fate was also more likely help. This finding suggested that similarity plays an important role in vicarious reward and altruistic behavior.

In a related study, but with neuroimaging, participants were asked to observe videos of gamblers winning or losing money and rate how satisfied they were with each outcome (Mobbs, Yu, Meyer et al., 2009). Participants felt more rewarded and manifested more ventral striatum activity (within the area activated when winning oneself) if a more likeable gambler won relative to when a less likeable gambler won. To the extent that participants believed the gambler was similar to themselves, their vmPFC and ventral anterior cingulate cortex (vACC) were more active in the same contrast. Finally, watching a similar gambler win relative to a dissimilar gambler win produced a stronger relationship between vACC and ventral striatum activity. So, not only did the positive outcome of a similar and likeable other activate more low-level reward circuitry than a win of a dissimilar other, a stronger link was made to the frontal cortex, where the authors speculated participants engendered self-relevant positive emotion (Mobbs et al., 2009; Moran, Macrae, Heatherton et al., 2006).

An intuitive reason to be rewarded by the positive outcomes to others is that they may be working towards a goal from which we too could benefit. One, perhaps implicit, goal is to produce more people like us.

A highly influential theory to explain why we might feel more invested in outcomes of those we see as similar to ourselves is Hamilton's theory of "kin-selection" (Hamilton, 1964). This theory proposes an evolutionary mechanism by which animals (e.g. Sherman, 1977) and humans (e.g. Burnstein, Crandall & Kitayama, 1994; Essock-Vitale & McGuire, 1985; Kruger, 2003) have a hard-wired disposition to assist close relations in preference to distant ones. It suggests that helping our close relatives, even at a cost to ourselves, may have evolved to increase the likelihood that future generations carry our own genes. Close relatives have more of our genes than distant relatives. Since we cannot see each other's genomes, we can infer our relation to others by observing the traits that we share. The relationship between traits and kinship is understood from a very young age (Springer, 1992). The relationship between vicarious reward and similarity fits well within this evolutionary framework. If we experience reward from the positive outcomes of those who share our traits, we will be more likely to bring about positive outcomes for them in the future, and thus increase chances that future generations will have the same genes as ourselves, including the genes that encourage us help similar others.

Competition

How we interpret another person's fortune is highly dependent on its relationship to our own. If I observed my sister playing the world-championship game of chess, I would likely be most happy if she was supremely victorious over her opponent. When her opponent turned out to be me, however, my preference would most likely change. The anticipated vicarious reward from my sister's victory would now be competing with my own desire for a quick checkmate.

If another's goal does not compete with my own status, their rewards can be experienced as vicariously rewarding to me. For example, Harbaugh, Mayr and Burghart (2007) observed neural activity as participants donated to a local charity (which participants presumably assumed had goals that did not significantly compete with their own). Satisfaction and activation of a neural reward system increased in proportion to the donation to recipients, even when donations were forced. These effects also predicted how much the participant donated voluntarily. Thus, when the recipient's payoff did not compete with the goals or status of the donor, the donor was happy to incur a cost for the recipient's benefit.

However, if a recipient has similar goals, ability and status to one's own, satisfaction from vicarious reward can compete with a disposition to experience envy and Schadenfreude (pleasure derived from someone else's misfortune). In a neuroimaging study, Takahashi and colleagues asked participants to imagine themselves as a protagonist in a story and, prior to scanning, read a detailed description of the protagonist's goals, abilities and social status (Takahashi, Kato, Matsuura et al., 2009). While scanned with fMRI, participants learned about the goals, abilities and status of other characters and discovered a series of unfortunate events that befell these characters; reading about characters with similar aspirations to the protagonist, but more ability and more status, evoked envy in participants that correlated with activity in cingulate cortex. The more a character was envied, the more pleasure was reported when bad things happened to them. This pleasure correlated with activity in the participant's ventral striatum.

Although both the aforementioned studies were concerned with the effects of similarity, the results differ from those of Mobbs and colleagues (2009), who reported that vicarious reward increased with similarity; whereas Takahashi and colleagues (2009) reported increasing Schadenfreude. The critical difference might be that the latter experiment was structured so that participants could compare outcomes of others to those of their own lives. This allowed for comparisons and competition of social status and reward at the time of evaluating outcomes of others. By contrast, Mobbs and colleagues (2009) did not allow for such a comparison because participants played the novel game themselves only after seeing the other people play. In summary, similarity can promote vicarious reward and empathy in response to the outcomes of others, but these feelings can be supplanted by concerns about one's self. It remains an interesting and useful avenue for future research to explore how competition between vicarious reward and envy is resolved.

Social Yardsticks

Contrasting our abilities, actions and experiences with those of others is known in psychology as social comparison (Festinger, 1954). This social process has a considerable effect on our subjective value of outcomes and is a necessary component of envy and perceived fairness. Supporting its effect on value, social comparison has been shown to modify the neural response to a reward. With neuroimaging, Fliessbach and colleagues showed that a ventral striatum blood oxygenation level dependent (BOLD) response to reward can be modulated by knowledge of reward provided to another participant (Fliessbach, Weber, Trautner et al., 2007). The very same reward generated a stronger response if another participant received less, and a weaker response if the other participant received more. Other researchers have looked at an interaction between performance and social hierarchies. Zink and colleagues asked participants to play a video game in parallel with two other players (Zink, Tong, Chen et al., 2008). Participants could either be the best player, the average player, or the worst player and were told their rankings. The participant's neural response to performance feedback depended on how others performed. When the person ranked as inferior performed better than the participant (a lowering of status for the participant), there was a stronger BOLD response in the anterior insula cortex, which is often associated with negative outcomes (Eisenberger, Lieberman & Williams, 2003; Singer, Seymour, O'Doherty et al., 2004). In contrast, when the participant performed better than the superior player, activity was seen in the dorsal striatum and other regions. These results support the theory of Adam Smith (1759) – that changes of social status may be as motivating as money or food.

Fairness and Equity

Intricately linked to social comparison, perceived fairness also shapes our subjective valuation of rewards. Preference for equity is an aversion to receiving either less or more than others. Behavioral evidence about a person's equity preferences (a willingness to decrease their own payoff to improve those who are worse off) is mixed. Some studies indicate that the better-off participant is motivated to reduce a gap in wealth, while other studies suggest that, at least in some cultures, people will pay to maintain or even increase their relative status (e.g. Fehr & Schmidt, 1999; Herrmann, Thoni & Gachter, 2008; Loewenstein, Thompson & Bazerman, 1989). This makes it likely that there are other factors determining equity preferences, such as perceived competition, fear of punishment, and reputation (i.e. an audience).

Tricomi and colleagues used fMRI to examine neural responses relating to equity (Tricomi, Rangel, Camerer & O'Doherty, 2010). First, they endowed one participant of a pair with $50 and then asked both participants to rate different divisions of future money between the other participant and themselves. In both participants, BOLD signals in vmPFC and ventral striatum were greater for divisions that favored the participant without an initial endowment, a finding the authors interpreted as evidence for an underlying preference for equity. Oddly, reported preferences by the participants did not follow suit such that the participant with an endowment still reported a preference for receiving the higher reward. This inconsistency could reflect a difference between an implicit preference for equity and perhaps more explicit processes of social comparison. Another interpretation is that neural activity was driven by preferences for efficiency rather than equity. Individuals with more relative wealth may experience less utility for new rewards insofar as a reward to them would be less efficient (have less utility) than a reward to a relatively less wealthy participant.

An earlier study by Hsu and colleagues examined equity and efficiency preferences in the absence of self-comparison (Hsu, Anen & Quartz, 2008). The experimenters pitted efficiency (maximizing the overall number of meals) against fairness (distribution of meals between needy orphans). Weight given to the overall utility (combining equity and efficiency) of the distribution correlated with activity in the ventral striatum. Weight given to equity of distribution was associated with enhanced BOLD activity in anterior insula cortex. Given the anterior insula's association with emotion and empathy (Singer, Critchley & Preuschoff, 2009) the researchers concluded that actions derived from perceived unfairness might be rooted in emotional processing. Supporting Hsu and colleagues' (2008) interpretation, rejection of offers in response to unfairness in the ultimatum game is also associated with enhanced activity in anterior insula (Sanfey, Rilling, Aronson et al., 2003).

Finally, in an fMRI study, Haruno & Frith (2010) contrasted people with prosocial preferences (preferring to maximizing joint gain) to those with individualistic preferences (preferring to maximize gain for self) (Van Lange, Otten, De Bruin & Joireman, 1997). In prosocials, the absolute difference of reward between the self and other activated amygdala, an activation that predicted their dislike for the arrangement. This effect distinguished prosocials from individualists and was unaffected by cognitive load. The result supports a theory that automatic emotional processing in the amygdala lies at the core of prosocial value orientations in addition to more strategic processes involving the vmPFC (Tricomi et al., 2010), anterior insula (Hsu et al., 2008; Sanfey et al., 2003) and ventral striatum (Tricomi et al., 2010).

In summary, self-similarity may provide a strong basis for how we react to the outcomes of others. If competition evokes a comparison of another's outcome to our own, however, our response will be determined by complex array of motivations. These include affiliative (e.g. vicarious reward and fairness) and competitive (e.g. envy, status changes, schadenfreude, and reputation) desires. Some of these influences are likely to be processed implicitly, while others may involve more explicit cognition.

Learning about the World through the Values of Others

Many animals learn their values from each other and this form of learning enables us to choose valuable resources without incurring a cost of exploration through trial and error. Even without direct communication, we can still infer the value of our choices simply by observing other people's choices and rewards. This ability is very cost-effective, it is no surprise that it observed across a wide range of species (Laland, 2004).

Tracking the Rewards of Others

Burke and colleagues used fMRI to study how the brain tracks information about the actions and rewards of other people (Burke, Tobler, Baddeley & Schultz, 2010). The authors asked participants to make a decision in three conditions where they could see either: their outcomes;

their outcomes and the choices of others; or their outcomes, other's choices, and other's outcomes. With increasing amounts of social information, participants made more correct choices. BOLD activity increased within participants' dorsolateral prefrontal cortex (DLPFC) and correlated with the absolute (unsigned) prediction error generated by the choices of others. In contrast, prediction errors generated by other people's rewards correlated with activity in vmPFC.

Tracking the Fears of Others

Just as we learn from what others like, we also learn from their dislikes. Many species, from cows (Munksgaard, DePassillé, Rushen et al., 2001), cats (John, Chesler, Bartlett & Victor, 1968) and primates (Mineka, Davidson, Cook & Keir, 1984) learn what to fear by observing others. Primates learn particularly well from fearful facial expressions (Mineka et al., 1984; Vaughan & Lanzetta, 1980). An observed fearful expression could act as an unconditioned stimulus that can be enhanced by empathy and mentalizing and is paired with a conditioned stimulus (i.e. the object which scared the conspecific) through the same circuit as fear-learning from direct experience (for review see Olsson & Phelps, 2007).

Trustworthiness: Tracking the Reputation of Others

Reputation is a very important attribute of the people with whom we interact, especially when it comes to their reputation for being cooperative. We quickly learn to avoid people who have a bad reputation and approach others whom we can trust. When we trust an individual, we believe they will act in our best interests. They have a good reputation of being trustworthy and we are likely to cooperate with them (McAllister, 1995). The same mechanisms through which we learn to approach or avoid objects can be applied to learning about the trustworthiness of others.

Learning about Trustworthiness from Interactions

We can learn about other people's trustworthiness by interacting with them. When someone is more cooperative than we expect, their reputation goes up; when they behave worse than we expect, their reputation goes down. As with other kinds of error-based learning, these neural signals about trust can be observed in the ventral striatum (Delgado, Frank & Phelps, 2005). As a result of such learning, the sight of someone who has acquired a reputation elicits activity in regions associated with the representation of value, such as the amygdala (Singer, Kiebel, Winston et al., 2004). In two seminal studies, Rilling and colleagues observed that reward-associated areas of the brain responded to mutually beneficial outcomes in a prisoner's dilemma game, but only when both players cooperate (Rilling, Gutman, Zeh et al., 2002; Rilling, Sanfey, Aronson et al., 2004).

The ventral striatum signal may have been a sign of reinforcement, since participants with stronger responses in this region were more likely to contribute again. This result was not as strong when the participants played with a computer, leading the authors to conclude that the signal was caused by a uniquely social phenomenon, mutual trust.

In a more recent exploration, Behrens and colleagues investigated how trust might be learned in an experiment where trust was modeled as a prediction about the reliability of another player's advice (Behrens, Hunt, Woolrich & Rushworth, 2008). Participants in this neuroimaging study were provided with advice before making a choice for an uncertain reward. Neural activity tracked prediction errors of trustworthiness of advice in a network of regions, including dorsomedial prefrontal cortex, middle temporal gyrus, and the temporoparietal junction (TPJ), areas previously implicated in tracking the intentions of others (Frith & Frith, 2006). In addition, anterior cingulate cortex gyrus (ACCg) activity produced a signal correlating with volatility of the advice. With volatile advice, participants should give more weight to recent advice and thereby learn about it at a faster rate. The more that the ACCg of a participant varied its activity with advice volatility, the more that participant allowed the advice to affect their choices and anticipation of a reward, reflected in a signal of the vmPFC. Thus, while a network of areas tracked the fidelity of advice, ACCg response volatility predicted the impact of that advice on valuation. In parallel, the impact of reward history alone correlated with volatility tracking in the ACC sulcus. These results suggested that trust is learned by similar computational mechanisms as learning about non-social features of the environment, but instantiated in slightly different anatomy (Behrens, Woolrich, Walton & Rushworth, 2007).

Learning about Trustworthiness from Gossip

In addition to direct interactions with people, we also learn about the reputation, and hence the value of other people, from verbal reports about their trustworthiness (gossip, see Sommerfeld, Krambeck, Semmann & Milinski, 2007). Delgado and colleagues (2005) investigated this phenomenon with a "trust game." In this game, participants chose to give their money to a trustee, where it was multiplied by three. The trustee could then decide to keep all the money or give some of it back. Without a reputation of the trustee, a participant would learn whether or not to invest from trial to trial (e.g. Rilling et al., 2002). However in this study, the experimenters provided participants with descriptions of the trustee's reputation before the game. Trustees were "good," "neutral" or "bad." Trustees of different moral characters returned money at the same rate, yet participants were still persistently more likely to invest with the "good" trustee. Despite the influence of reputation on decisions, post-experiment ratings of trustworthiness of the "good" trustee ended up similar to that of other moral characters. This suggested that conscious and implicit representations of trust and choice-relevant representations of trust are not the same thing. However, what was particularly intriguing was that participant reliance on prior reputation for determining their choices was also evident in the pattern of brain activity. The ventral striatum, which clearly tracked the neutral trustee behavior, did not respond to the actions of the good trustee and responded only weakly to the actions of the bad one. Overall, a strong prior reputation for trustworthiness seems to inhibit implicit learning about it in a way that would affect choice.

Conformity

Conformity is the act of matching attitudes, beliefs, and behaviors to what individuals perceive is normal of their society or social group. It is necessarily preceded by conflict: I must initially have a different value to that of a group if I am to change my value to match. Neuroimaging studies have explored how the human brain responds to a conflict with group values and whether this response can predict how much we will change our opinion. We asked participants to rate a list of 20 songs, which they wanted but did not own, for desirability (Campbell-Meiklejohn, Bach, Roepstorff et al., 2010). While scanned with fMRI, participants learned whether or not two expert reviewers preferred their song or another song. After each review, participants received a token for the song they preferred or for the alternative and, after scanning, they again rated their songs for desirability. For each participant, a measure of social influence was obtained by observing to what extent the reviews predicted changes in song desirability. Agreement between the reviewers and the participant activated ventral striatum, as did receiving a token for the song that the participant preferred. This suggested that agreement with others is rewarding. Intriguingly, the level of activation associated with the opinions of the reviewers predicted how much the participant would subsequently conform. This activity, associated with reviewer opinions that conflicted with their own, was observed in several regions including temporoparietal junction and rostral anterior cingulate cortex (rACC). The right temporoparietal junction has been shown to monitor others' choices (Behrens et al., 2008; Hampton, Bossaerts & O'Doherty, 2008), while activity in rACC reflects the extent to which conflict, or other negative information, alters values and corresponding behavior (Jocham, Neumann, Klein et al., 2009; Kerns, Cohen, MacDonald et al., 2004).

Conflict of opinion with a large group over the attractiveness of faces also produces rACC activity that predicts subsequent conformity (Klucharev, Hytonen, Rijpkema et al., 2009). Moreover, both rACC and anterior insula cortex responses to popular opinion have been found to predict conformity to that opinion in adolescents (Berns, Capra,

Moore & Noussair, 2010). It seems that the degree to which the brain responds to social conflict, reliably predicts how much influence social conflict will have on values.

Other studies have investigated where in the brain social influence on value can be observed. Two studies, one of perceptual decision-making and one of value-based decision-making, suggest that conformity affects the very earliest stages of cognition. In the perceptual study, participants were asked to decide if two objects were rotated versions of each other, after hearing the answers of other people (Berns, Chappelow, Zink et al., 2005). Rather than exerting influence through higher-order reinterpretation of bottom-up information in the frontal cortex, evidence for the influence of other people's opinions was found within a network of regions more associated with mental rotation itself. The story for value is similar. In our own study, the influence of two expert reviews on the value of an object altered the basic ventral striatum signals of value that occurred as the object was received (Campbell-Meiklejohn et al., 2010). Just how the opinions of others can so rapidly affect the very basic processes of perception and valuation remains a deeply interesting question.

Open questions about conformity remain. We do not know if social influence on ventral striatum responses to receiving objects is due to change in self-esteem (i.e. reputation) associated with owning that object, updating of the object's worth based on increased demand, or on inferences about the quality of the object based on expert opinion. It is likely that all the factors that influence the value of an object are combined into a single signal of value in the human reward system that increases its desirability. The mechanism of this integration is not known.

THE MINDS OF OTHERS

A basic requirement for successful interaction with others is to be able to predict what they are going to do next. Fundamental concepts in game theory, such as the Nash equilibrium and backward induction, are based on the assumption that I will choose my actions on the basis of what I expect my partner will do, and that she will likewise be thinking about what I am going to do. This implies that a player must view the game from the other player's perspective (Singer & Fehr, 2005). Thus our decisions are determined, in part, by inferences about other minds: what our partners and competitors know, believe and desire.

What Will He Think of My Offer?

Evidence for the influence of mentalizing on decision-making is provided by a comparison of the ultimatum and dictator games. In both games the proposer offers a portion of an endowment to a responder. However, in the ultimatum game the responder can turn down the offer, in which case neither player gets any money. A "rational self-interested" responder should accept low offers since a small amount of money is better than none. In practice, however, most responders turn down offers of less than about a third (Camerer & Thaler, 1995) where the responder consider a small offer as "unfair." If the proposer wants his or her offer to be accepted, then he or she must take into account the responder's point of view in relation to small offers. Such considerations do not apply in the dictator game since the responder has no choice but to accept what is given. As might be expected, therefore, offers made in the dictator game are significantly lower where typically, they are about half those made in the ultimatum game (Forsythe, Horowitz, Savin & Sefton, 1994).

It is even more important to think about the intentions of one's partner in games involving trust and reciprocity, such as prisoners' dilemma, with many repeated trials. The best strategy in such games is usually *tit-for-tat* (Axelrod & Hamilton, 1981) in which the partner is effectively rewarded for being cooperative and punished for defecting. However, this strategy only works when there is a close relationship between the player's intention and the behavioral outcome. In real life, as a result of performance errors and other kinds of noise, this relationship can break down. For example, if someone fails to respond to an email it need not indicate a lack of cooperation and may simply be a local network outage. When noise is added to a trust and reciprocity game, so that generous intentions are not always correctly transmitted to the partner, the *tit-for-tat* strategy no longer elicits the highest degree of cooperation, since the partner may too frequently be punished despite good intentions. In the presence of noise, a higher degree of cooperation is achieved if the responder behaves slightly more cooperatively than the actor did on the previous trial (i.e. *tit-for-tat* plus 1) (Van Lange, Ouwerkerk & Tazelaar, 2002). In the case with noise in the system, players are assuming, correctly, that their partner's intentions are more cooperative than is evident in their apparent behavior. This demonstrates the advantages of tracking intentions, rather than behavior, when we make decisions in a social interaction.

A related strategy for repairing broken cooperation is *coaxing*, that is responding to a low offer, with a large repayment. Such behavior signals to the partner that the actor is worthy of trust (King-Casas, Sharp, Lomax-Bream et al., 2008). In normal interactions, coaxing behavior repairs cooperation and pays off for the coaxer in subsequent rounds. However, when one of the partners has a diagnosis of borderline personality disorder (BPD), cooperation breaks down after a few rounds. At this stage the partner with BPD is significantly less likely to engage in coaxing behavior and consequently cooperation does not get repaired

(King-Casas et al., 2008). We might speculate that, after experiencing intended, or even unintended, punishment the person with BPD concludes that a partner has bad intentions and will not be influenced by coaxing. Consistent with this idea, a person with BPD does not appear to take account of the behavior of her partner. This is evident in the observation that activity in anterior insula does not show the normal pattern of increasing with the perceived unfairness (i.e. lowness) of offers (King-Casas et al., 2008).

What Does He Think I Will Think of His Offer?

In the example of coaxing, player A is trying to alter what player B thinks about him or her. This goes beyond the representation of the mental state of another. Player A is representing player B's representation of A. This is the first step in a recursion that, in principle, could go on indefinitely.

Hampton and colleagues computationally modeled this kind of thinking in an analysis of behavior in the *inspector game* (Hampton et al., 2008). The *influence* model has a specific component concerning the player's estimate of the influence his or her behavior will have on their opponent. In other words, the player estimates what the opponent estimates the player will do next. This model provides a good fit to the behavior of players, and a significantly better fit than *fictive play* in which the player simply estimates what the opponent will do next on the basis of the opponent's previous choices.

John Maynard Keynes' *beauty contest* is another example of a competitive game, involving many players, in which the players need to think about what the players think the other players are going to do (Keynes, 1936). Each player has to choose a number between 1 and 100 and the winning number is the one nearest to some fraction (1/2 for example) of the mean number chosen. To win you clearly have to choose a number that is lower than what most other people choose, but you have to bear in mind that the other players may be adopting the same strategy. Players who don't think at all (level 0) might choose 50 in which case you should choose 25. However, if the other players assume that 50 will be chosen (level 1) then you should choose 12, and so on through recursion. If everyone makes such a super-rational analysis, then all will choose the lowest possible number. When this game is played in real life (e.g. a newspaper competition) people choose numbers much higher than 0, such that, on average players operate at around the level of one recursion, so that numbers associated with 2 levels of recursion are likely to win (e.g. Schou, 2005).

Jean-Jacques Rousseau's *Stag and Rabbit* hunt is an example of collaborative game that requires each player to think about the other in a recursive fashion (Rousseau, 1984). If a player hunts for a rabbit he will catch it and makes a small gain, representing the safe, *risk dominant* option. However, if both players hunt the stag they will catch it and achieve a much greater gain. This is the best *payoff dominant* option. The worst outcome occurs if you hunt the stag and your partner hunts the rabbit, since you will achieve no gain. The problem then in this game is how to achieve cooperation. You will only choose to hunt the stag if you believe your partner is going to hunt the stag. However, you know that your partner will only hunt the stag if he or she believes that you are going to hunt the stag. Once again we enter a potentially infinite recursion. Yoshida and colleagues have shown that, to respond optimally in this game, a player needs to estimate his or her partner's level of recursion and then play at one level higher (Yoshida, Dolan & Friston, 2008). This is a very similar strategy to what we have just described for the beauty contest game, and assumes that each player has an upper limit of recursion.

Neural Correlates of Recursive Mentalizing

Scanning studies of the three games discussed above have given remarkably consistent results. The greater the extent to which a player used the recursive *Influence Model* in the Inspector game, the greater the activity seen in medial prefrontal cortex (Hampton et al., 2008). The greater the level of recursion used by a player in the beauty contest, the greater the activity in mPFC (Coricelli & Nagel, 2009). These observations are consistent with the many studies showing that mPFC is active when people have to think about mental states, whether their own or others (for a review see e.g. Amodio & Frith, 2006). The analysis of the Stag and Rabbit hunt as developed by Yoshida and colleagues provides a more sophisticated account of the computational processes involved. In addition to estimating the level of recursion of his or her partner, a player must also compute the degree of certainty about this estimate. A brain imaging implementation of the paradigm suggested that DLPFC encoded the depth of recursion being used, an index of executive sophistication. In contrast, mPFC was involved in encoding the uncertainty of the inference of the partner's strategy, with greater activity being associated with greater uncertainty (Yoshida, Dziobek, Kliemann et al., 2010). Does greater activity reflect more processing to resolve uncertainty or, alternatively, reflect high level predictions errors concerning the strategy of the other? These studies show the way forward towards a much more exact account of the role of this region which is persistently implicated in mentalizing.

Decisions Made by Interacting Minds

The Stag and Rabbit hunt is an example of a collaborative interaction where two people working together can achieve more than one person working alone. A more trivial example would be two people moving an item of furniture, too heavy for one person to lift. The problem in both examples concerns how to achieve coordination, rather than what to do once coordination has been achieved. In other words, these problems are concerned with a decision whether or not to collaborate, rather than the decisions that are made once the collaboration has commenced. In this final section we consider the processes whereby collaborating groups make better decisions than even the best single individual within the group.

How Group Decisions Can Be Better Than Individual Decisions

Eusocial insects, such as ants and bees, can make decisions as a group. This is the case, for example, when choosing a new nest site. As with all decisions, this process will have three stages: (1) collecting relevant information, (2) deciding when enough information has been collected to make a sufficiently accurate decision (speed-accuracy trade-off), and finally, (3) implementing the decision. In the case of an individual decision-maker, these stages can be captured by a simple learning algorithm. Through the collection of information, values can be assigned to the various possible options. Once there is one option with a sufficiently higher value than the others, then this option can be implemented. The problem for a group decision is to collate the information that has been collected by different individuals. In human societies this is often achieved explicitly by voting. The option with the most individuals voting for it is then chosen. In the case of ants and bees, the same effect is achieved by the kind of automatic processes we have discussed above, such as learning by observation and mimicry. Individual scouts discover potential nest sites and integrate multiple properties of these sites into assessments of their quality. In the case of bees, the scouts then return to the swarm and indicate the desirability of the site by the vigor and length of their dance (Visscher, 2007). The more vigorous dances attract more scouts to investigate the site. In this way discovered sites compete for a limited pool of nest-site scouts, and the most desirable site is revealed by the increasing number of scouts who visit it. Once a sufficient majority have chosen one particular sight (*Quorum sensing*), information gathering ceases and the whole swarm decamps to the chosen site.

In the case of ants, a scout indicates the value of the site by the speed of recruiting another ant to visit that site (*tandem running*). Here, the most desirable site attracts the largest number of scouts. When a sufficiently large number of ants have arrived at one site, information gathering ceases (*Quorum sensing*) and, rather than bringing additional ants to the selected site by tandem running, the ants simply carry ants from the old nest – which is three times faster (Franks, Dechaume-Moncharmont, Hanmore & Reynolds, 2009).

In these examples, the advantage of the group over the individual decision arises because a much larger space can be explored to find the best site for a new nest. In other words, more information can be gathered by many people than by a single person. In the course of evolution, various mechanisms have emerged for collating information from many individuals to facilitate a group decision.

The likelihood that similar processes occur in humans was recognized by Galton and led to the idea of the wisdom of crowds (Surowiecki, 2004). More recently, it has been shown that humans also reach consensus decisions without verbal communication or obvious signaling (Dyer, Ioannou, Morrell et al., 2008). However, the more obvious human parallel is "market judgment" which supposedly allows goods and services to be valued faster, or more accurately, by groups (Surowiecki, 2004). Yet, unlike bees and ants, whose group decisions reliably lead to the choice of the best nest site, human joint decisions sometimes go spectacularly wrong, as demonstrated by the recent banking crisis.

There are many reasons why group decisions fail to be better than individual decisions, and perhaps the most important is a lack of independence between the individuals involved. Another possibility, which we enlarge on in the next section, concerns the problem of the extent to which the signals about the value of options genuinely reflect their true value. When a scout bee dances vigorously to indicate the value of a potential nesting site, that vigor does not directly indicate the value of the site. It would be more accurate to say that it reflects the bee's belief about the site or the bee's confidence in its own report.

Since other scouts go out to check the potential nest site, this possible divergence between belief and actual value is likely to be minimized. However, the potential for divergence between belief and value is not minimized in the case of the stock market. In fact this is the very point Keynes was making through the idea of the beauty contest. Who is the most beautiful is not as important as *who people think* is the most beautiful. It is all too often the case that money can be made by predicting what people believe whether or not this belief converges with reality.

Explicit Information Sharing

We recently developed a paradigm for studying in detail how people share information in order to make a group decision that is better than the best individual within the group. Pairs of volunteers worked together to detect a weak signal in a standard psychophysical task requiring a forced choice between two options (Bahrami, Olsen, Latham et al., 2010). When the pair disagreed about the answer they discussed the problem until they came up with a joint answer. Our aim was to discover how the information from the two individuals was integrated.

Combining information from different sources is ubiquitous within the single brain. For example, the understanding of speech is greatly enhanced when we can see the person speaking as well as hearing them (Summerfield, 1992). Recently, it has been shown that humans integrate information from vision and touch in a statistically optimum fashion (Ernst & Banks, 2002). This can be achieved by weighting the information in the signals from the two sources on the basis of the reliability (variance) of the signals. To achieve this when signals come from two individuals (rather than from two sensory modalities within the same brain), the partners need to communicate an estimate of the reliability of their signal to each other (confidence sharing).

We found that, as long as the pair were not too dissimilar in their ability to detect signals on their own, their joint decision was significantly better than the better member of the pair, and was statistically optimum in terms of the performance predicted by our confidence sharing model. We also found that the discussion that led up to the joint decision was crucial for reaching this optimum integration, while the presence of feedback as to whether the joint decision was right or not made little difference. We assume the discussion that occurred before making the joint decision involved the partners explaining to each other how well they had seen the signal, so that a final decision is based on who had "seen it better." Such an interaction clearly depends upon introspection about sensations (metacognition) and the ability to convey and match these introspections using a common metric. Confidence sharing seems to be a relatively rare case where explicit representations play a critical role in social influences on decision-making.

But does confidence sharing need to be explicit? The vigor of the waggle dance of a scout bee not only indicates the value of the potential nest site, but also the confidence of the scout. So what advantage does explicit human confidence sharing have over what bees can do? This is a key question for future research into group decision-making. For the moment, one obvious advantage is flexibility. The explicit confidence sharing that may be unique to humans can probably be applied to any kind of signal and in any domain. The content of ant and bee communication is predetermined and extremely restricted.

As yet we know almost nothing about the neural basis of confidence sharing, but the paradigm we describe readily lends itself to future studies in which brain imaging and pharmacological manipulations can be used to explore this question. However, computational models and neural processes concerning metacognition, a critical component of confidence sharing, are now beginning to emerge (Cleeremans, Timmermans & Pasquali, 2007; Fleming, Weil, Nagy et al., 2010).

CONCLUSIONS

We are all deeply embedded in a social context, and this embedding impacts on our values and our decisions. Most of these effects occur without our awareness. The presence of, and covert imitation by, similar others is rewarding and can be the basis for association learning. The presence of others, and imitation by others, also tips the balance towards prosocial rather than self-interested behavior. We get vicarious pleasure when others are rewarded, particularly if they are similar to us. We find the agreement of others about our values rewarding and we tend to alter our values to conform to the values of others. These effects seem to be driven by basic reward learning systems in the brain.

When interacting with others we track their intentions, rather than their behavior, in order to optimize our own decisions. We also recognize that others will be tracking our intentions, and consequently we make decisions that enhance our reputation for cooperation (or whatever behavior is relevant to the group aims). These processes are largely explicit and require deliberate thought.

Human social interactions necessitate recursive representations, since, while I am trying to predict what you will do, you are trying to predict what I will do. Thus, I need to represent what you are going to do, what you think I am going to do, and so on. In competitive situations we need to go at least one level deeper in a recursion to get a competitive edge. In collaborative situations we need some degree of recursion to be sure that the aims and means of our collaborations are common knowledge.

When decisions are made by a group, there are mechanisms available for determining who has the most reliable and appropriate information relevant to the decision, and for determining when no further information is needed. It seems likely that the optimal information sharing needed for making good group decisions critically depends on explicit metacognitive processes. While we still know very little about the mechanisms through which social factors influence decision-making, one thing is clear. None of our decisions and values can escape the intensely social milieu in which we are all embedded.

Acknowledgments

DCM & CDF are supported by the Danish National Research Foundation, the Danish Council for Independent Research and the Lundbeck Foundation.

References

Amodio, D. M., & Frith, C. D. (2006). Meeting of minds: The medial frontal cortex and social cognition. *Nature Reviews Neuroscience, 7*(4), 268–277.

Amodio, D. M., Kubota, J. T., Harmon-Jones, E., & Devine, P. G. (2006). Alternative mechanisms for regulating racial responses according to internal vs. external cues. *Social Cognitive and Affective Neuroscience, 1*(1), 26–36.

Avenanti, A., Sirigu, A., & Aglioti, S. M. (2010). Racial bias reduces empathic sensorimotor resonance with other-race pain. *Current Biology, 20*(11), 1018–1022.

Axelrod, R., & Hamilton, W. D. (1981). The evolution of cooperation. *Science, 211*(4489), 1390–1396.

Bahrami, B., Olsen, K., Latham, P. E., Roepstorff, A., Rees, G., & Frith, C. D. (2010). Optimally interacting minds. *Science, 329*(5995), 1081–1085.

Bailenson, J. N., Yee, N., Patel, K., & Beall, A. C. (2008). Detecting digital chameleons. Article. *Computers in Human Behavior, 24*(1), 66–87.

Bateson, M., Nettle, D., & Roberts, G. (2006). Cues of being watched enhance cooperation in a real-world setting. *Biology Letters, 2*(3), 412–414.

Bavelas, J. B., Black, A., Lemery, C. R., & Mullett, J. (1986). I show how you feel – motor mimicry as a communicative act. *Journal of Personality and Social Psychology, 50*(2), 322–329.

Bee, M. A., Perrill, S. A., & Owen, P. C. (2000). Male green frogs lower the pitch of acoustic signals in defense of territories: A possible dishonest signal of size? *Behavioral Ecology, 11*(2), 169–177.

Behrens, T. E., Hunt, L. T., Woolrich, M. W., & Rushworth, M. F. (2008). Associative learning of social value. *Nature, 456*(7219), 245–249.

Behrens, T. E., Woolrich, M. W., Walton, M. E., & Rushworth, M. F. (2007). Learning the value of information in an uncertain world. *Nature Neuroscience, 10*(9), 1214–1221.

Bengtsson, S. L., Lau, H. C., & Passingham, R. E. (2009). Motivation to do well enhances responses to errors and self-monitoring. *Cerebral Cortex, 19*(4), 797–804.

Bengtsson, S. L., Dolan, R. J., & Passingham, R. E. (2010). Priming for self-esteem influences the monitoring of one's own performance. *Social Cognitive and Affective Neuroscience.* doi:10.1093/scan/nsq048.

Berns, G., Chappelow, J., Zink, C., Pagnoni, G., Martin-Skurski, M., & Richards, J. (2005). Neurobiological correlates of social conformity and independence during mental rotation. *Biological Psychiatry, 58*(3), 245–253.

Berns, G., Capra, C., Moore, S., & Noussair, C. (2010). Neural mechanisms of the influence of popularity on adolescent ratings of music. *Neuroimage, 49*(3), 2687–2696.

Bshary, R., & Grutter, A. (2006). Image scoring and cooperation in a cleaner fish mutualism. *Nature, 441*(7096), 975–978.

Burke, C. J., Tobler, P. N., Baddeley, M., & Schultz, W. (2010). Neural mechanisms of observational learning. *Proceedings of the National Academy of Sciences of the United States of America, 107*(32), 14431–14436.

Burnstein, E., Crandall, C., & Kitayama, S. (1994). Some neo-Darwinian decision rules for altruism: Weighing cues for inclusive fitness as a function of the biological importance of the decision. *Journal of Personality and Social Psychology, 67*(5), 773–789.

Camerer, C., & Thaler, R. H. (1995). Ultimatums, dictators and manners. *The Journal of Economic Perspectives, 9*(2), 209–219.

Campbell-Meiklejohn, D. K., Bach, D. R., Roepstorff, A., Dolan, R. J., & Frith, C. D. (2010). How the opinion of others affects our valuation of objects. *Current Biology, 20*(13), 1165–1170.

Chartrand, T. L., & Bargh, J. A. (1999). The chameleon effect: The perception-behavior link and social interaction. *Journal of Personality and Social Psychology, 76*(6), 893–910.

Chartrand, T. L., & van Baaren, R. (2009). Human mimicry. *Advances in Experimental Social Psychology, 41*, 219–274.

Cleeremans, A., Timmermans, B., & Pasquali, A. (2007). Consciousness and metarepresentation: A computational sketch. *Neural Networks, 20*(9), 1032–1039.

Coricelli, G., & Nagel, R. (2009). Neural correlates of depth of strategic reasoning in medial prefrontal cortex. *Proceedings of the National Academy of Sciences of the United States of America, 106*(23), 9163–9168.

Decety, J., Chaminade, T., Grezes, J., & Meltzoff, A. N. (2002). A PET exploration of the neural mechanisms involved in reciprocal imitation. *Neuroimage, 15*(1), 265–272.

Delgado, M. R., Frank, R. H., & Phelps, E. A. (2005). Perceptions of moral character modulate the neural systems of reward during the trust game. *Nature Neuroscience, 8*(11), 1611–1618.

Dimberg, U., Thunberg, M., & Elmehed, K. (2000). Unconscious facial reactions to emotional facial expressions. *Psychological Science, 11*(1), 86–89.

Dyer, J. R. G., Ioannou, C. C., Morrell, L. J., Croft, D. P., Couzin, I. D., Waters, D. A., et al. (2008). Consensus decision making in human crowds. Article. *Animal Behaviour, 75*, 461–470.

Earley, R. L. (2010). Social eavesdropping and the evolution of conditional cooperation and cheating strategies. *Philosophical Transactions of the Royal Society of London. Series B, Biological Sciences, 365*(1553), 2675–2686.

Eisenberger, N., Lieberman, M., & Williams, K. (2003). Does rejection hurt? An FMRI study of social exclusion. *Science, 302*(5643), 290–292.

Ernst, M. O., & Banks, M. S. (2002). Humans integrate visual and haptic information in a statistically optimal fashion. *Nature, 415*(6870), 429–433.

Essock-Vitale, S., & McGuire, M. (1985). Women's lives viewed from an evolutionary perspective. II. Patterns of helping. *Ethology and Sociobiology, 6*(3), 155–173.

Fehr, E., & Gachter, S. (2002). Altruistic punishment in humans. *Nature, 415*(6868), 137–140.

Fehr, E., & Schmidt, K. M. (1999). A theory of fairness, competition, and cooperation. *The Quarterly Journal of Economics, 114*(3), 817–868.

Festinger, L. (1954). A theory of social comparison processes. *Human Relations, 7*, 117–140.

Fleming, S. M., Weil, R. S., Nagy, Z., Dolan, R. J., & Rees, G. (2010). Relating introspective accuracy to individual differences in brain structure. *Science, 329*(5998), 1541–1543.

Fliessbach, K., Weber, B., Trautner, P., Dohmen, T., Sunde, U., Elger, C. E., et al. (2007). Social comparison affects reward-related brain activity in the human ventral striatum. *Science, 318*(5854), 1305–1308.

Forsythe, R., Horowitz, J. L., Savin, N. E., & Sefton, M. (1994). Fairness in simple bargaining experiments. *Games and Economic Behavior, 6*(3), 347–369.

Fouts, G. T., Waldner, D. N., & Watson, M. W. (1976). Effects of being imitated and counterimitated on the behavior of preschool children. *Child Development, 47*(1), 172–177.

Franks, N. R., Dechaume-Moncharmont, F. X., Hanmore, E., & Reynolds, J. K. (2009). Speed versus accuracy in decision-making ants: Expediting politics and policy implementation. *Philosophical Transactions of the Royal Society of London. Series B, Biological Sciences, 364*(1518), 845–852.

Frith, C. D., & Frith, U. (2006). The neural basis of mentalizing. *Neuron, 50*(4), 531–534.

Gardner, M., & Steinberg, L. (2005). Peer influence on risk taking, risk preference, and risky decision making in adolescence and adulthood: An experimental study. *Developmental Psychology, 41*(4), 625–635.

Gutsell, J. N., & Inzlicht, M. (2010). Empathy constrained: Prejudice predicts reduced mental simulation of actions during observation of outgroups. *Journal of Experimental Social Psychology, 46*(5), 841–845. doi: 10.1016/j.jesp.2010.03.011.

Hamilton, W. D. (1964). The genetical evolution of social behavior. I. *Journal of Theoretical Biology, 7*(1), 1–16.

Hampton, A. N., Bossaerts, P., & O'Doherty, J. P. (2008). Neural correlates of mentalizing-related computations during strategic interactions in humans. *Proceedings of the National Academy of Sciences of the United States of America, 105*(18), 6741–6746.

Harbaugh, W. T., Mayr, U., & Burghart, D. R. (2007). Neural responses to taxation and voluntary giving reveal motives for charitable donations. *Science, 316*(5831), 1622–1625.

Haruno, M., & Frith, C. (2010). Activity in the amygdala elicited by unfair divisions predicts social value orientation. *Nature Neuroscience, 13*(2), 160–161.

Herrmann, B., Thoni, C., & Gachter, S. (2008). Antisocial punishment across societies. *Science, 319*(5868), 1362–1367.

Hoffman, E., McCabe, K., & Smith, V. L. (1996). Social distance and other-regarding behavior in dictator games. *The American Economic Review, 86*(3), 653–660.

Hsu, M., Anen, C., & Quartz, S. (2008). The right and the good: Distributive justice and neural encoding of equity and efficiency. *Science, 320*(5879), 1092–1095.

Izuma, K., Saito, D. N., & Sadato, N. (2010). Processing of the incentive for social approval in the ventral striatum during charitable donation. *Journal of Cognitive Neuroscience, 22*(4), 621–631.

Izuma, K., Saito, D. N., & Sadato, N. (2010). The roles of the medial prefrontal cortex and striatum in reputation processing. *Social Neuroscience, 5*(2), 133–147.

Jocham, G., Neumann, J., Klein, T. A., Danielmeier, C., & Ullsperger, M. (2009). Adaptive coding of action values in the human rostral cingulate zone. *The Journal of Neuroscience, 29*(23), 7489–7496.

John, E. R., Chesler, P., Bartlett, F., & Victor, I. (1968). Observation learning in cats. *Science, 159*(3822), 1489–1491.

Jones, S. S., Collins, K., & Hong, H. W. (1991). An audience effect on smile production in 10-month-old infants. *Psychological Science, 2*(1), 45–49.

Karakashian, S. J., Gyger, M., & Marler, P. (1988). Audience effects on alarm calling in chickens (*Gallus gallus*). *Journal of Comparative Psychology, 102*(2), 129–135.

Kerns, J. G., Cohen, J. D., MacDonald, A. W., III, Cho, R. Y., Stenger, V. A., & Carter, C. S. (2004). Anterior cingulate conflict monitoring and adjustments in control. *Science, 303*(5660), 1023–1026.

Keynes, J. M. (1936). *General theory of employment interest and money.* London: Macmillan & Co.

Kilner, J. M., Marchant, J. L., & Frith, C. D. (2006). Modulation of the mirror system by social relevance. *Social Cognitive and Affective Neuroscience, 1*(2), 143–148.

King-Casas, B., Sharp, C., Lomax-Bream, L., Lohrenz, T., Fonagy, P., & Montague, P. R. (2008). The rupture and repair of cooperation in borderline personality disorder. *Science, 321*(5890), 806–810.

Klucharev, V., Hytonen, K., Rijpkema, M., Smidts, A., & Fernandez, G. (2009). Reinforcement learning signal predicts social conformity. *Neuron, 61*(1), 140–151.

Kokolakis, A., Smith, C. L., & Evans, C. S. (2010). Aerial alarm calling by male fowl (*Gallus gallus*) reveals subtle new mechanisms of risk management. *Animal Behavior, 79*(6), 1373–1380. doi: 10.1016/j.anbehav.2010.03.013.

Krebs, D. (1975). Empathy and altruism. *Journal of Personality and Social Psychology, 32*(6), 1134–1146.

Kruger, D. J. (2003). Evolution and altruism: Combining psychological mediators with naturally selected tendencies. *Evolution and Human Behavior, 24*, 118–125.

Kühn, S., Müller, B. C., van der Leij, A., Dijksterhuis, A., Brass, M., & van Baaren, R. B. (2010). Neural correlates of emotional synchrony. *Social Cognitive and Affective Neuroscience.* doi: 10.1093/scan/nsq044.

Kühn, S., Müller, B. C., van Baaren, R. B., Wietzker, A., Dijksterhuis, A., & Brass, M. (2010). Why do I like you when you behave like me? Neural mechanisms mediating positive consequences of observing someone being imitated. *Social Neuroscience, 5*(4), 384–392.

Lakin, J. L., & Chartrand, T. L. (2003). Using nonconscious behavioral mimicry to create affiliation and rapport. *Psychological Science, 14*(4), 334–339.

Lakin, J. L., Chartrand, T. L., & Arkin, R. M. (2008). I am too just like you – Nonconscious mimicry as an automatic behavioral response to social exclusion. *Psychological Science, 19*(8), 816–822.

Laland, K. N. (2004). Social learning strategies. *Learning & Behavior, 32*(1), 4–14.

Liepelt, R., Ullsperger, M., Obst, K., Spengler, S., von Cramon, D. Y., & Brass, M. (2009). Contextual movement constraints of others modulate motor preparation in the observer. *Neuropsychologia, 47*(1), 268–275.

Loewenstein, G. F., Thompson, L., & Bazerman, M. H. (1989). Social utility and decision making in interpersonal contexts. *Journal of Personality and Social Psychology, 57*(3), 426–441.

McAllister, D. (1995). Affect- and cognition-based trust as foundations for interpersonal cooperation in organizations. *Academy of Management Journal, 38*, 24–59.

Miller, R. S., & Morris, W. N. (1974). The effects of being imitated on children's responses in a marble-dropping task. *Child Development, 45*(4), 1103–1107.

Mineka, S., Davidson, M., Cook, M., & Keir, R. (1984). Observational conditioning of snake fear in rhesus monkeys. *Journal of Abnormal Psychology, 93*(4), 355–372.

Mobbs, D., Yu, R., Meyer, M., Passamonti, L., Seymour, B., Calder, A., et al. (2009). A key role for similarity in vicarious reward. *Science, 324*(5929), 900.

Moran, J. M., Macrae, C. N., Heatherton, T. F., Wyland, C. L., & Kelley, W. M. (2006). Neuroanatomical evidence for distinct cognitive and affective components of self. *Journal of Cognitive Neuroscience, 18*(9), 1586–1594.

Munksgaard, L., DePassillé, A. M., Rushen, J., Herskin, M. S., & Kristensen, A. M. (2001). Dairy cows' fear of people: Social learning, milk yield and behavior at milking. *Applied Animal Behaviour Science, 73*(1), 15–26.

Olsson, A., & Phelps, E. (2007). Social learning of fear. *Nature Neuroscience, 10*(9), 1095–1102.

Over, H., & Carpenter, M. (2009). Priming third-party ostracism increases affiliative imitation in children. *Developmental Science, 12*(3), F1–F8.

Panchanathan, K., & Boyd, R. (2004). Indirect reciprocity can stabilize cooperation without the second-order free rider problem. *Nature, 432*(7016), 499–502.

Parton, D. A., & Priefert, M. J. (1975). The value of being imitated. *Journal of Experimental Child Psychology, 20*(2), 286–295. doi: 10.1016/0022–0965(75)90104–6.

Paukner, A., Suomi, S. J., Visalberghi, E., & Ferrari, P. F. (2009). Capuchin monkeys display affiliation toward humans who imitate them. *Science, 325*(5942), 880–883.

Rilling, J., Gutman, D., Zeh, T., Pagnoni, G., Berns, G., & Kilts, C. (2002). A neural basis for social cooperation. *Neuron, 35*(2), 395–405.

Rilling, J., Sanfey, A., Aronson, J., Nystrom, L., & Cohen, J. (2004). The neural correlates of theory of mind within interpersonal interactions. *Neuroimage, 22*(4), 1694–1703.

Rizzolatti, G., & Craighero, L. (2004). The mirror–neuron system. *Annual Review of Neuroscience, 27*, 169–192.

Rousseau, J.-J. (1984) *A discourse on inequality* (M. Cranston, Trans.). London: Penguin (original work published 1755).

Sanfey, A., Rilling, J., Aronson, J., Nystrom, L., & Cohen, J. (2003). The neural basis of economic decision-making in the Ultimatum Game. *Science, 300*(5626), 1755–1758.

Schilbach, L., Wilms, M., Eickhoff, S. B., Romanzetti, S., Tepest, R., Bente, G., et al. (2010). Minds made for sharing: Initiating joint attention recruits reward-related neurocircuitry. *Journal of Cognitive Neuroscience, 22*(12), 2702–2715.

Schou, A. (2005, September 22nd). Gæt-et-tal konkurrence afslører at vi er irrationelle. *Politiken.*

Semmann, D., Krambeck, H. J., & Milinski, M. (2004). Strategic investment in reputation. *Behavioral Ecology and Sociobiology, 56*(3), 248–252.

Sherman, P. W. (1977). Nepotism and the evolution of alarm calls. *Science*, *197*(4310), 1246–1253.

Singer, T., Critchley, H., & Preuschoff, K. (2009). A common role of insula in feelings, empathy and uncertainty. *Trends in Cognitive Sciences*, *13*(8), 334–340.

Singer, T., & Fehr, E. (2005). The neuroeconomics of mind reading and empathy. *The American Economic Review*, *95*(2), 340–345.

Singer, T., Kiebel, S., Winston, J., Dolan, R., & Frith, C. (2004). Brain responses to the acquired moral status of faces. *Neuron*, *41*(4), 653–662.

Singer, T., Seymour, B., O'Doherty, J., Kaube, H., Dolan, R., & Frith, C. (2004). Empathy for pain involves the affective but not sensory components of pain. *Science*, *303*(5661), 1157–1162.

Smith, A. (1759). *The theory of moral sentiments*. Indianapolis: Liberty Classics. 1982

Somerville, L., Kelley, W., & Heatherton, T. (2010). Self-esteem modulates medial prefrontal cortical responses to evaluative social feedback. *Cerebral Cortex*, *20*(12), 3005–3013.

Sommerfeld, R. D., Krambeck, H. J., Semmann, D., & Milinski, M. (2007). Gossip as an alternative for direct observation in games of indirect reciprocity. *Proceedings of the National Academy of Sciences of the United States of America*, *104*(44), 17435–17440.

Springer, K. (1992). Children's awareness of the biological implications of kinship. *Child Development*, *63*(4), 950–959.

Stel, M., Van Baaren, R. B., & Vonk, R. (2008). Effects of mimicking: Acting prosocially by being emotionally moved. *European Journal of Social Psychology*, *38*(6), 965–976.

Stel, M., & Harinck, F. (2010). Being mimicked makes you a prosocial voter. *Experimental Psychology*, *58*, 79–84.

Stel, M., & Vonk, R. (2010). Mimicry in social interaction: Benefits for mimickers, mimickees, and their interaction. *British Journal of Psychology*, *101*(Pt 2), 311–323.

Summerfield, Q. (1992). Lipreading and audio-visual speech perception. *Philosophical Transactions of the Royal Society of London. Series B, Biological Sciences*, *335*(1273), 71–78.

Surowiecki, J. (2004). *The wisdom of crowds: Why the many are smarter than the few and how collective wisdom shapes business, economies, societies and nations*. New York: Doubleday.

Sylwester, K., & Roberts, G. (2010). Cooperators benefit through reputation-based partner choice in economic games. *Biology Letters*, *6*, 659–662. doi: 10.1098/rsbl.2010.0209.

Takahashi, H., Kato, M., Matsuura, M., Mobbs, D., Suhara, T., & Okubo, Y. (2009). When your gain is my pain and your pain is my gain: Neural correlates of envy and schadenfreude. *Science*, *323*(5916), 937–939.

Tennie, C., Frith, U., & Frith, C. D. (2010). Reputation management in the age of the world-wide web. *Trends in Cognitive Sciences*, *14*(11), 482–488.

Thelen, M. H., Lada, S. T., Lasoski, M. C., Paul, S. C., Kirkland, K. D., & Roberts, M. C. (1980). On being imitated: Effects on models and observers. *Journal of Experimental Child Psychology*, *29*(1), 50–59.

Tricomi, E., Rangel, A., Camerer, C. F., & O'Doherty, J. P. (2010). Neural evidence for inequality-averse social preferences. *Nature*, *463*(7284), 1089–1091.

van Baaren, R. B., Holland, R. W., Kawakami, K., & van Knippenberg, A. (2004). Mimicry and prosocial behavior. *Psychological Science*, *15*(1), 71–74.

Van Lange, P., Otten, W., De Bruin, E., & Joireman, J. (1997). Development of prosocial, individualistic, and competitive orientations: Theory and preliminary evidence. *Journal of Personality and Social Psychology*, *73*(4), 733–746.

Van Lange, P., Ouwerkerk, J., & Tazelaar, M. (2002). How to overcome the detrimental effects of noise in social interaction: The benefits of generosity. *Journal of Personality and Social Psychology*, *82*(5), 768–780.

Vaughan, K. B., & Lanzetta, J. T. (1980). Vicarious instigation and conditioning of facial expressive and autonomic responses to a model's expressive display of pain. *Journal of Personality and Social Psychology*, *38*(6), 909–923.

Visscher, P. K. (2007). Group decision making in nest-site selection among social insects. *Annual Review of Entomology, 52*, 255–275.

Wedekind, C., & Milinski, M. (2000). Cooperation through image scoring in humans. *Science, 288*(5467), 850–852.

Xu, X., Zuo, X., Wang, X., & Han, S. (2009). Do you feel my pain? Racial group membership modulates empathic neural responses. *The Journal of Neuroscience, 29*(26), 8525–8529.

Yoshida, W., Dolan, R. J., & Friston, K. J. (2008). Game theory of mind. *PLoS Computational Biology, 4*(12), e1000254.

Yoshida, W., Dziobek, I., Kliemann, D., Heekeren, H. R., Friston, K. J., & Dolan, R. J. (2010). Cooperation and heterogeneity of the autistic mind. *The Journal of Neuroscience, 30*(26), 8815–8818.

Zink, C., Tong, Y., Chen, Q., Bassett, D., Stein, J., & Meyer-Lindenberg, A. (2008). Know your place: Neural processing of social hierarchy in humans. *Neuron, 58*(2), 273–283.

Social and Emotional Factors in Decision-Making: Appraisal and Value

Elizabeth A. Phelps[1] *and Peter Sokol-Hessner*[2]

[1]New York University, New York, USA
[2]California Institute for Technology, Pasadena, California, USA

INTRODUCTION

In an effort to understand the basic processes that underlie decisions, scientists have typically exposed individual human participants as well as animals to choice situations in laboratory settings. As this volume and the larger literature demonstrates, this approach has yielded robust and exciting findings delineating the behavioral and neural mechanisms underlying simple choice and complex decisions. In spite of this success, to fully understand the complexity of human decisions, it is necessary to consider the range of factors that may be more prevalent in decisions outside the laboratory. In this chapter we discuss two of these factors: emotion and social interaction.

Neuroscience of Preference and Choice
DOI: 10.1016/B978-0-12-381431-9.00019-X

We consider each of these factors independently, but there is significant overlap in that a primary means to elicit emotions is by introducing social interaction. For instance, it has been shown that members of different social groups elicit distinct emotional responses (Cuddy, Fiske & Glick, 2007; Harris & Fiske, 2007), which can impact subsequent decisions. In addition, as the discipline of social psychology has demonstrated repeatedly over the years, merely the presence of another may alter how we choose to act, one reason (of many) being the emotional discomfort of non-conformity (e.g., Asch, 1956). Importantly, both emotion and social interaction introduce factors that alter the determination of subjective value and, as a result, the decision.

A common mechanism through which this might occur is appraisal. Although economic studies of decision-making often refer to value as if it is a property of the object or choice being evaluated, an equally important component of the value computation is the appraisal by the evaluator. Both individual trait-like factors, such as risk sensitivity, and situational factors, such as satiation when assessing food rewards, can change the subjective value one assigns to a choice. In this chapter, we suggest that both emotion and social interaction can be situational factors that influence decisions, in addition to trait-like emotional dispositions. Furthermore, we suggest that a mechanism by which both emotion and social interaction influence decisions is to alter the appraisal of the choice options, thus influencing the value computation.

Below we review recent studies examining the impact of emotion and social interactions on economic and neuroeconomic decision tasks. We also highlight a few studies examining how these factors interact; for example, how a social stimulus elicits an emotional response that influences the decision. This review is not exhaustive, but rather highlights studies that demonstrate how emotional and social factors can influence the appraisal of the choice options in altering the decision and its underlying neural representation.

EMOTIONAL FACTORS

Emotion is a broad concept that is thought to represent a range of component affective processes (e.g., see Scherer, 2005 for a more detailed discussion). Although a review of component process models of emotion is beyond the scope of this chapter, a few basic definitions are useful in considering the role of emotion in decision-making. The term "emotion" is often used to describe a discrete, synchronized patterned response in reaction to an external or internal event which may include all or some of the following: subjective experience or feeling, bodily responses such as physiological arousal, expression in the face or body, and action tendencies – i.e., the propensity to approach or withdraw. Emotion is often differentiated from mood, which is a more stable, long-lasting state primarily characterized by subjective experience that may or may not be elicited by an internal or external event. Finally in some theories of emotion, preference (and/or attitude) refers to the more enduring properties of people or objects resulting in an affective evaluation and a corresponding action tendency.

Although the relation between emotion and decision-making has long been a topic of philosophical debate (Aristotle, trans 1941), economic theories of judgment and decision-making (Kahneman, 2003) and neuroeconomic investigations (Cohen, 2005), surprisingly few studies of decision-making have explicitly measured or manipulated emotion variables. In our brief review of this literature, we only focus on studies that introduce emotion into the task and assess the impact on choice and/or measure a specific emotional response and link it to choice. We exclude studies that infer an emotional response from patterns of brain activation or brain injury (i.e. reverse inference), unless accompanied by an assessment of emotion.

One of the traditional techniques used in affective science to introduce an affective component into a task is mood induction. In a study designed to assess the impact of mood on a classic economic decision, Lerner, Small & Loewenstein (2004) presented participants with one of three film clips and assessed decisions with an endowment-effect scenario. The first two film clips elicited self-reported feelings of sadness and disgust respectively, and a third, neutral film clip, elicited no emotion in particular. Immediately afterwards, participants were either given a set of highlighters and asked for how much they would sell those highlighters, or they were shown the highlighters and asked several times to choose between different amounts of money and the highlighter set (eliciting a "choosing equivalent," or the amount of money at which point they would be indifferent between the money and the highlighters).

In the neutral condition, participants exhibited a classic endowment effect, in which selling prices for the highlighter set were higher than choosing equivalents (i.e., the prospect of "losing" the highlighters when selling them is worse than the prospect of "gaining" them at the same price). Interestingly, when participants decided on selling and choosing prices after watching the "sad" movie clip, they exhibited a reversal of the endowment effect – choosing prices were higher than selling prices. After the "disgust" clip, choosing and selling prices were both low, and equal.

To explain this striking reversal or elimination of the classic endowment effect with a simple mood manipulation, Lerner and colleagues (2004) suggested that moods can result in an appraisal tendency; that is, a tendency to appraise unrelated events in a manner consistent with that mood. For instance, when sad, the tendency is to move away from the current circumstance to one that is less depressing. In the classic endowment effect, an individual values what he or she already possesses over something new. When sad, however, changing the current circumstance (i.e. what you possess) may be viewed as more valuable than the status quo, resulting in the reversal of the classic endowment effect.

A different approach to induce an emotional response is via a more direct, physiological intervention. Such an approach was used in a study on risk attitudes and stress, in which the researchers immersed participants' hands in near-freezing water before a choice task (Porcelli & Delgado, 2009). This procedure, called the "cold-pressor task," is known to reliably engage an acute stress response, which is capable of affecting behavior (Ishizuka, Hillier & Beversdorf, 2007). The choice task involved a series of simple risky gambles in either the loss domain or the gain domain. The choice was always between less money with higher probability (conservative), and more money with lower probability (risky). The classic finding in these situations is called the reflection effect, in which participants are risk averse in the gain domain (choosing the conservative option more often) and risk seeking in the loss domain (choosing the risky option more often). In this study, that risk attitude profile was exaggerated by acute stress – that is, after the cold-pressor task, participants were more conservative in the gain domain, and more risky in the loss domain. These results demonstrate that directly manipulating participants' levels of stress can contaminate later choices and significantly shift behavior. This result is particularly remarkable for the simplicity of the task, and the presence of feedback. If participants' choices were based on their knowledge and on previous results, then the stress induction should have had no effect – the fact that it did suggests that at least some of the components of the stress response may be integrated directly into the processes behind valuation and decision-making.

Instead of trying to induce a mood state or emotional response, another approach is to introduce stimuli that carry emotional information, such as facial expressions during the choice task. This approach was used in a study in which participants were subliminally shown angry, happy, or neutral facial expressions, before making a number of decisions about drinks (Winkielman, Berridge & Wilbarger, 2005). After completing a task nominally about gender classification (during which a set of one of the types of faces mentioned before was subliminally presented), participants poured, consumed, rated, and priced a drink. The authors found that, preceding the drink, decisions with angry faces reduced the amount poured, the amount consumed, the ratings, and the price participants would be willing to pay for the drink. Subliminal presentation of happy faces had the opposite effect. These results are particularly notable for the subliminal aspect of the experimental manipulation – participants were unaware of the presence of the faces, and the emotional content of those faces was irrelevant to the task at hand. However, the subtle affective signal generated by these cues was sufficient to alter the appraisal of the choice options and change decisions.

By manipulating emotion, the studies described above can determine the causal role of emotion in influencing the computation of value. A second means to examine the role of emotion in decisions is to measure and quantify emotional reactions and relate those measurements to other observable aspects of the decision task. There are as many ways to measure emotion as there are components – one can assess subjective feelings by self-report, action tendencies with decisions and actions, facial expressions, or bodily responses with psychophysiology. Assessing physiological responses has several advantages in that they are quantifiable, objective, continuously graded, and easily measured alongside other variables. Though other components of emotion are no less important, many recent studies of emotion and decision-making have used the physiological arousal response as a measure of emotion, and as a result, we will focus mainly on this physiological assessment of emotion. As will shortly become clear, one reason for this is that physiological arousal is often closely associated with assessments of value – or more specifically, the representation of subjective values over which decisions are actually made.

One commonly used physiological response is the skin conductance response (SCR), an indication of autonomic nervous system arousal. Using SCR, one of the first studies to make the case for arousal as a component of subjective value examined patients with damage to the orbitofrontal cortex (OFC) in the performance of a risky gambling task (Bechara, Damasio, Tranel & Damasio, 1997). The task, called the "Iowa Gambling Task," consisted of repeated choices among four decks of cards. Two decks yielded high rewards quite often, but the occasional very high penalty as well, which in the long-run resulted in a net loss ("bad" decks). The other two decks yielded both smaller rewards and punishments, but led to a net gain over time ("good" decks). Over the course of the task, non-brain damaged controls began to generate anticipatory SCRs to the "bad" decks, shortly after which they began to avoid those decks; the brain damaged patients did neither. Bechara and colleagues suggested that the anticipatory arousal response serves as an important component of the decision process, essentially altering the value of the options and steering control participants away from the "bad" decks. Patients with OFC damage, who failed to generate these anticipatory arousal responses, also failed to incorporate their emotional response into their assessment of value. Though there were numerous methodological concerns with this study (e.g., Dunn, Dalgleish & Lawrence, 2005; Fellows & Farah, 2005; Maia & McClelland, 2004), it remains the first to closely link arousal responses to behavioral decision patterns, and to suggest that arousal might be closely (or even necessarily) linked to assessments of subjective value.

Because of the non-invasive nature of most physiological measurements of arousal, a logical step is to assess arousal 'in the field." One study did exactly this, recording SCR, cardiovascular variables, respiration, and body temperature in professional traders over the course of a normal work day (Lo & Repin, 2002). It was found that SCRs were more frequent and cardiovascular responses were greater during both heightened periods of volatility in the market, as well as discrete market events. Perhaps most interestingly, however, this pattern of autonomic responses was exaggerated for traders with low to moderate experience, and attenuated for those of high experience. While all traders showed significant bodily arousal responses during the course of a normal workday, the connection to experience suggests that physiological arousal was an integral component of these professionals' decisions and reactions. It is possible that the most experienced traders had always had that profile of autonomic responses (and that's why they stuck around long enough to become experienced traders). But it is also possible that, as both their knowledge and skills developed (i.e. they gained experience), their assessment and appraisal of the market decisions changed, and the corresponding bodily responses during the performance of their job was also altered. In fact, one could argue, that shifts in arousal in response to market events is part and parcel of "gaining experience," just as much as growth in explicit knowledge or skills.

The results of Lo and Repin (2002) coincide with a recent study by Sokol-Hessner, Hsu, Curley et al. (2009), examining the impact of perspective shift that influences appraisal on the relation between arousal and choice. In this study, participants were presented with two sets of identical risky monetary choices. For one of the sets, participants were encouraged to "attend" to the individual choices at hand and their potential outcomes, while for the other set they were instructed to "regulate" each choice by considering it in its greater context, as one of many choices in a larger set, or portfolio. An econometric model of valuation and decision-making was used to estimate a number of aspects of participants' behavior including their degree of loss aversion (relative weighting of losses and gains) in that set.

In the "attend" condition, participants were on average more aroused per dollar to losses compared to gains, and this "over-arousal," which could be considered a physiological measure of loss aversion, correlated with the estimated degree of behavioral loss aversion. But perhaps most interestingly, only those participants for whom the portfolio regulation technique significantly affected their choices (in which case they became less loss averse), experienced significantly reduced "over-arousal" to losses relative to gains. The participants in that study (undergraduates) were far from professional traders – yet by taking a perspective similar to that which an experienced trader might take, they showed not only changes in their choices that reflected a greater context, but also changes in SCR not unlike those observed in the experienced traders relative to the inexperienced ones in Lo and Repin (2002). The combination of these results makes a compelling case that at least part of the difference between novices and professionals may be a consequence of how their appraisal of the choice alters both the emotional response to the choice options and its impact on the assessment of value.

Altering the appraisal or interpretation of an event is a primary means of regulating emotional responses (see Ochsner & Gross, 2005). The studies described above demonstrate that influencing the appraisal of a choice through either specific instructions (Sokol-Hessner et al., 2009) or experience (Lo & Repin, 2002) changes arousal response and decisions. More recently, brain imaging studies have demonstrated that neural systems implicated in the representation of value are also influenced by manipulating appraisal. For instance, a study by Delgado, Gillis & Phelps (2008a) conditioned stimuli to be associated with monetary rewards. By introducing a simple emotion regulation instruction in this classical conditioning task, participants were able to reduce their physiological arousal response, as assessed with SCR, to the conditioned stimulus. Simultaneously, they reduced activation in the striatum, a region where the magnitude of the response generally increases with monetary value (Delgado et al., 2008a). The regulation task also led to increased blood oxygenation level dependent (BOLD) responses in dorsolateral prefrontal cortex (DLPFC) consistent with previous studies on the cognitive regulation of emotion (e.g., Ochsner & Gross, 2005), and the ventromedial prefrontal (VMPFC), a region thought to be involved in both emotion regulation (Hartley and Phelps, 2010) and the representation of value (Rangel, Camerer & Montague, 2008).

More recent investigations have demonstrated that introducing an emotion regulation technique to alter appraisal during a decision task has a similar impact on the representation of value in the brain. For example, Martin and Delgado (2011) found that using an imagery-based emotion regulation technique during a gambling task, reduced striatal activation during the decision, along with the tendency towards risky choices. Similarly, Sokol-Hessner and colleagues (submitted) examined the impact of introducing the portfolio perspective during a decision task. Replicating their behavioral results described above (Sokol-Hessner et al., 2009), they found the portfolio regulation technique diminished loss aversion. In addition, implementing the regulation technique reduced amygdala activation to losses, and led to an overall increase in BOLD responses in the striatum, VMPFC and DLPFC.

As the studies described above demonstrate, the components of emotion have undoubtedly complex relationships with the many processes that contribute to valuation and decision-making. Our understanding of those relationships can only be aided by increased specificity and measurement of emotion. The characterization of the precise impact of emotion on decision-making will vary depending on range factors, including the specific affective manipulation and corresponding state change, individual factors, the decision task and additional task demands, and the means of assessing the emotional response. However, across tasks and manipulations, it is clear that a primary impact of emotion on decisions is to temporarily shift the appraisal of the choice options, thus influencing the assessment of value.

SOCIAL FACTORS AND EMOTIONAL INFLUENCES

Given the complex social nature of everyday human life, it is not surprising that there are a range of means by which social information can influence decisions. In fact, the simple presence of social stimuli can be rewarding or punishing, thus altering choices. For example, in a clever series of study examining the rewarding properties of social interaction, Platt and colleagues showed that monkeys will "pay" (i.e. forego juice) to view socially relevant images of other monkeys, and that these social rewards engage the same reward circuitry as non-social decision-making tasks (Deaner, Khera & Platt, 2005; Klein, Deaner & Platt, 2008).

Economic decisions are, by their very nature, social transactions. However, in spite of the social dependence of economic decisions, literature examining how specific social features, such as the social quality of the interaction or the characteristics of the social partner, influence economic choices has only recently emerged. Although it is possible to study economic decision-making isolated from the larger societal context that imbues value in monetary reinforcement, there are some classic behavioral economic games that are completely dependent on social interaction. Below we highlight recent studies examining the influence of social factors on three such games: experimental economic auctions, the trust game and the ultimatum game.

One of the anomalies of experimental economics is the tendency for participants to overbid, or pay "too much" in auctions. In this case, just the mere presence of competition with another person alters decisions to pay. It has been proposed that one of the factors mediating overbidding is the "joy of winning" over a social partner (Goeree, Holt & Palfrey, 2003). In an effort to obtain support for this hypothesis by examining BOLD responses in the reward circuitry, Delgado, Schotter, Ozbay & Phelps (2008b) scanned participants while playing an auction game vs. a lottery game. Winning or losing the lottery resulted in a predictable striatal response (Delgado, Locke, Stegner & Fiez, 2003) – that is an increase in BOLD signal to a monetary gain, but no difference from baseline when losing the lottery since there was no monetary loss. In contrast, the auction game showed the same predictable increased BOLD response to monetary gains, but a decrease in BOLD signal when losing the auction, even though there was no monetary loss. In contrast to the "joy of winning" hypothesis, it appeared the social loss experienced when losing the auction mirrored a monetary loss in the pattern of striatal response. Furthermore, the magnitude of the decrease in the striatal BOLD response to losing the auction correlated with amount of overbidding. There was no relationship between the striatal response to monetary gains and bids chosen. These results suggest that it was the anticipation or fear of losing the social competition that drove overbidding. To provide further evidence for this hypothesis, an additional behavioral economics experiment was conducted in which the choices were framed to emphasize losses, gains or neither. Emphasizing loss resulted in greater overbidding. There results indicate that just the presence of another person in an economic exchange can change subjective value and its underlying neural representation. Furthermore, these results demonstrate how an investigation of the impact of social factors on decisions can begin to isolate how specific factors, such as social loss, elicited by the social interaction may uniquely impact decisions, such as decisions bid.

This relationship between the presence of social others and decisions has also been investigated with the trust game. This game involves deciding whether to trust a social partner to maximize reward. In a typical version of this task, the investor is endowed with a sum of money that she or he can either keep or choose to share with a partner, the trustee. If the investor decides to share, the sum is multiplied so that the trustee receives, for instance, 3 or 4 times the sum invested. The trustee then has a choice, to either share the larger sum with the investor, in which case both investor and trustee profit, or keep the entire sum, in which case the investor's trust is violated resulting in a monetary loss, along with a relatively larger gain for the trustee.

In the first neuroeconomic study to examine the impact of social interaction on the neural systems mediating the trust game, King-Casas, Tomlin, Anen et al. (2005) simultaneously scanned two partners playing repeated rounds of this game. In the early rounds, they observed the same pattern of BOLD responses to monetary reinforcement in the striatum one might expect in a non-social task (e.g., Delgado et al., 2003) – that is, an increase in BOLD signal when the investor is rewarded and receives a profit and a decrease in BOLD signal when the trust is violated and there is a monetary loss. However, as the partners played repeated rounds with each other, this pattern shifted. Once a "reputation" was acquired, there was no longer a striatal response to monetary outcome, instead the striatal response was now shifted to the presentation of the partner whose past actions may have led to monetary gain or loss. King-Casas and colleagues (2005) suggest that this pattern is similar to that observed with learning the value of non-social cues through prediction errors, in which reward responses in the striatum serve to update the value of a predictive cue while learning is ongoing, but once the cue value is acquired the striatal response shifts to the cue and striatal activation to the reward outcome is diminished (McClure, Berns & Montague, 2003).

It is not surprising that interacting repeatedly with a social partner might alter the appraisal of the value of that interaction. As King-Casas et al. (2005) suggest, in this case the pattern of brain response indicates that we learn about the predictive nature of social cues much like we learn about the predictive nature of non-social cues. However, social reputations are not only acquired by a history of repeated interactions. Social reputations can be linked to a number of factors, including knowledge of previous, unrelated social interactions and social group membership. Two studies examining how other types of social information can influence trust decisions explored the impact of moral character (Delgado, Frank & Phelps, 2005) and race (Stanley, Sokol-Hessner, Banaji & Phelps, 2011) of the trustee.

In a study by Delgado and colleagues (2005), moral character of the trustee partners was manipulated by introducing short vignettes about their life path and previous actions suggesting "good," "bad," or "neutral" moral character. Importantly, none of these vignettes mentioned previous economic transactions. After this introduction, participants played several rounds of the trust game with each partner. Each trustee partner was equally likely to reward the participant investor (i.e. share profits) across rounds. In spite of this equal pattern of reinforcement, the participant investor was more likely to choose to trust the "good" partner, than the "bad" or "neutral" partner, even after several rounds of the game when the participant could verbally report that the likelihood of each partner sharing profits was equivalent. In other words, the outcomes of previous trust decisions did not seem to update future trust decisions to the same degree when interacting with partners of "good" or "bad" moral character. An examination of the neural systems mediating the influence of moral character on trust decisions provides insight into why actions may not be updated based on previous outcomes as one might expect. Specifically, an examination of BOLD responses in the striatum to the outcome of trust decisions revealed diminished responses overall when interacting with the "good" partner relative to the "neutral" partner. A weaker, but similar pattern was observed with the "bad" partner. As mentioned above, the striatal response to outcomes is proposed to update knowledge about the predictive nature of the cue (McClure et al., 2003). If this striatal signal is diminished, one might expect that previous outcomes may not influence future decisions to the same degree. In other words, the moral character of the partner may have taken this feedback learning mechanism "offline," resulting in choices driven by social factors as much as, or more than, previous interactions and reward history.

Finally, a recent study explored the interaction of social factors and emotion in decision-making by assessing whether the implicit affective response to members of different race groups can be linked to decisions to trust (Stanley et al., 2011). The automatic affective response to race groups was assessed with the implicit association test (IAT), which is a Stroop-like task that measures differences in reaction time when pairing affective judgments (e.g, good or bad) with categorization judgments of race (e.g., Black or White). Unlike explicit measures of race attitudes, which are thought to assess the cognitive component corresponding to beliefs, this version of IAT is thought to assess the affective component of attitudes. For many stimuli, such as consumer goods, implicit and explicit assessments of attitudes align. However, when assessing attitudes towards stimuli where intention and beliefs may be at odds with affective responses, such as race groups in the United States, implicit (IAT) and explicit assessments of attitudes often do not correspond (Greenwald, Poehlman, Uhlmann & Banaji, 2009). This pattern was also observed in the decisions to trust. Stanley et al. (2011) found that the relative amount the participant investor chose to invest in White or Black trustee partners was correlated with IAT scores, but not explicit measures of race attitudes. In other words, a participant investor whose IAT score indicated a pro-White bias invested relatively more with White trustees than Black trustees and vice versa. Importantly, there was no overall difference in the average amount invested with Black and White trustees, but rather individual variability in implicit race attitudes correlated with variability in decisions to invest with Black or White trustee partners.

The examination of the influence of social factors in the trust game demonstrates that we can update our value representation of social others much like we learn the value of other environmental cues, and that social factors which one might expect to be independent of assessments of economic value can nevertheless be incorporated into the value computation and influence decisions. Another behavioral economic game that critically depends on social interaction, and has been shown to be modulated by specific social factors, is the "ultimatum game." In this game there is a sum of money to be divided between two partners. The proposer has the option to divide the money however she or he sees fit. The proposer can keep the entire sum, give all of it to the partner, or anything in between. However, the partner also has a choice. The responder can reject the offer. If this happens, both the proposer and the responder receive nothing. One of the puzzling aspects of the ultimatum game from an economic perspective is that the responder will often reject offers they deem to be unfair. For example, if the proposer chooses to offer the responder $20 out of $100 and keep $80, this offer has about a 50% chance of being rejected (Roth, 1995; Thaler, 1988) in spite of the fact the responder will lose $20 in this transaction.

In the first published neuroeconomic study of the ultimatum game, Sanfey, Rilling, Aronson et al. (2003), showed that the social aspect of this transaction is critical. Participants (playing the responder role) were much more likely to reject a low or unfair offer (i.e. an 8/2 split of $10), if the proposer was another person as opposed to a computer. In other words, the appraisal of the value of $2 shifted dramatically depending on the social nature of the task. Furthermore, Sanfey and colleagues showed that greater BOLD responses in the anterior insular cortex were correlated with a higher rate of rejection for unfair offers from social others. The insular cortex is a region implicated in a broad range of mental processes, including affective responses in a social context. These results led the authors to conclude that emotion might play a role in the rejection of unfair offers. Furthermore, the authors suggested that the relative BOLD response in the insular cortex and DLPFC, a region thought to be involved in the control of emotion, might be linked to whether an unfair offer is accepted or rejected.

This modulation of the insular cortex by unfair offers in the Sanfey et al. (2003) study, and the involvement of the DLPFC, has been replicated and extended in other studies examining the relationship between social factors and decisions in the ultimatum game. For example, a recent study by Güroğlu, van den Bos, Rombouts & Crone (2010) manipulated whether the proposer had a choice in offering a fair option. If the proposer had no choice to offer anything other than an unfair split, the responder was much more likely to accept the offer. The insular cortex response was modulated by this constrained choice set, and the perceived intention of the proposer, however activation of the DLPFC primarily reflected acceptance or rejection of the unfair offers regardless of constraints on the proposer.

Another recent study examined how the social context of the offer might alter the perception of fairness and decisions to reject. Wright, Symmonds, Fleming & Dolan (2011) presented participants with a range of ultimatum game offers within the larger social context of a group of proposers. One of the groups proposed a standard range of potential offers. The other two groups proposed the same range of offers, but these were interleaved with offers that were either "more fair" or "less fair." This allowed the researcher to compare responses to the same offers, but in varying social contexts. The results suggest that the social context of the group mattered in determinations of fairness. The same offer was more likely to be accepted when presented in the context of the "less fair" group, relative to the standard group, and less likely to be accepted in context of the "more fair" group. They also observed that the modulation of assessments of fairness or inequality by the social context was integrated into the insular cortex response. In addition, increased DLPFC activation reflected the rejections of offers perceived as more fair, which was modulated by the social context.

Across these neuroeconomic studies of the ultimatum game, there is a suggestion that the right DLPFC plays a role in the decision to accept or reject unfair offers, even when this judgment incorporates social context. In an effort to determine if the right DLPFC plays a causal role on decisions to accept or reject, Knoch, Pascual-Leone, Meyer et al. (2006) used transcranial magnetic stimulation (TMS) to temporarily disrupt neural processing in this region. They found that when TMS was applied to the right DLPFC, participants were more likely to accept unfair offers. Interestingly, the participants still rated the offers as unfair, suggesting that the DLPFC plays a role in implementing the rejection of offers.

As mentioned in the discussion of emotion above, the DLPFC is a region that has been linked to the control of emotional responses through altering the appraisal of the emotional event. Studies on the neural basis of the ultimatum game suggest that unfair offers elicit an emotional response (as represented in the insular cortex) that requires regulatory control by DLPFC, which in turn modulates decisions to accept or reject. Although the involvement of this neural circuitry implies a role for emotion in the ultimatum game, these studies do not assess or manipulate emotion. To address this issue, Sanfey and colleagues (2003) have conducted a series of studies exploring the relationship between emotion and decisions in the ultimatum game.

In the first study in this series van't Wout, Kahn, Sanfey and Aleman (2006) assessed SCR while participants played the ultimatum game (in the role of the responder) with either another person or a computer. They found that SCRs were greatest to unfair offers relative to fair offers, but only when playing against a human being – no difference was observed when the opponent was a computer. Furthermore, the size of the SCR was positively correlated, across participants, with rejection rate, suggesting a parallel to Bechara et al. (1997). Rather than "bad" decks, SCRs indicated "bad" offers, but in both cases, the action of rejection was the same. Importantly, the human/computer dimension of van't Wout et al. (2006) argues that value (and arousal) in their study was not simply a reflection of money, but also a reflection of the appraisal of the social value inherent in such a task.

In a second study, Harlé and Sanfey (2007) examined the impact of a mood induction procedure on performance in the ultimatum game. Participants viewed movie clips that elicited sadness, amusement or neutral mood. After viewing the movie clip, they played a series of ultimatum games with different partners. The induction of amusement had no effect on decisions, but participants who responded particularly intensely to the sad movie clip rejected even more of the unfair offers. Much like the Lerner et al. (2004) study described earlier, the underlying mood state altered the appraisal of the decision options, resulting in a shift in the pattern of choices. Finally, in a recent study, van't Wout, Chang and Sanfey (2010) explicitly instructed participants to use an emotion regulation technique to alter the appraisal of the emotional meaning of the offers. When utilizing this technique, participants were more likely to accept unfair offers. These findings explicitly assessing, manipulating and regulating emotion during the social interaction of the ultimatum game demonstrate a strong link between the influence of social stimuli and emotion on decision-making.

Given the necessity for social interaction in many economic decisions, whether it is the interaction with an institution, a social group, or another individual, it is surprising how little is known about the impact of specific social factors on decisions. The studies above represent a subset of an emerging literature that is beginning to explore this complex relationship. This research suggests the range of social factors that alter decisions is immense. By delineating the impact of specific factors, such as race, social group context, or simply the presence of another, on specific components of the decision process, we are starting to uncover the commonalities and differences among social and emotional influences on choice behavior.

CONCLUSIONS

The emerging research on the psychological and neural processes underlying decision-making has provided a foundation for understanding decisions in an emotional and social context. What is abundantly apparent in the current literature is that transient emotional and social factors can significantly alter judgments of subjective value, and as a result, choices. Less apparent is how, precisely, this occurs.

In this chapter we have highlighted the overlap and commonalities in the impact of emotional and social factors on decisions. One overlap is the fact that social stimuli may influence decisions by virtue of the emotional responses they elicit. Research on emotion has demonstrated a variety of means by which emotions can alter choice. For instance, some studies have shown a link between physiological arousal and decisions. Manipulations that regulate arousal responses by altering appraisal also alter choices. Interestingly, some judgments of social decisions mirror this pattern demonstrating that arousal, and its control, may also be driving social influences on choice behavior. The overlap in the neural circuitry mediating emotional and social influences on decision-making provides further support for this link.

It is also clear that there are unique contributions of emotional and social factors to the decision process. Non-social stimuli can elicit emotional responses, and social stimuli engage distinct processes such as mentalizing about others and judgments of intent. In addition, as the review of the impact of the specific emotional and social factors highlights, they each may contribute distinctly to different decision variables. However, a common mechanism across the range of social and emotional factors that change decisions is that they alter the interpretation or appraisal of the significance of the choice. We suggest that appraisal is an important, and perhaps underappreciated, aspect of the determination of subjective value. The social and emotional context in which the choice options are encountered is likely to be expressed in the appraisal of the options, which may be a critical component of the value computation.

References

Aristotle (1941). In R. McKeon (Ed.), *The basic works of aristotle*. New York: Random House.

Asch, S. (1956). Studies of independence and conformity: 1. A minority of one against a unanimous majority. *Psychological Monographs, 70*, 1–70.

Bechara, A., Damasio, H., Tranel, D., & Damasio, A. R. (1997). Deciding advantageously before knowing the advantageous strategy. *Science, 275*(5304), 1293–1295.

Cohen, J. D. (2005). The vulcanization of the human brain: A neural perspective on interactions between cognition and emotion. *Journal of Economic Perspectives, 19*, 13–24.

Cuddy, A. J., Fiske, S. T., & Glick, P. (2007). The BIAS map: Behaviors from intergroup affect and stereotypes. *Journal of Personality and Social Psychology, 92*(4), 631–648.

Deaner, R. O., Khera, A. V., & Platt, M. L. (2005). Monkeys pay per view: Adaptive valuation of social images by rhesus macaques. *Current Biology, 15*(6), 543–548.

Delgado, M. R., Locke, H. M., Stegner, V. A., & Fiez, J. A. (2003). Dorsal striatum responses to reward and punishment: Effects of valence and magnitude manipulations. *Cognitive Affective and Behavioral Neuroscience, 3*(1), 27–38.

Delgado, M. R., Frank, R. H., & Phelps, E. A. (2005). Perceptions of moral character modulate the neural systems of reward during the trust game. *Nature Neuroscience, 8*(11), 1611–1618.

Delgado, M. R., Gillis, M., & Phelps, E. A. (2008). Regulating the expectation of reward via cognitive strategies. *Nature Neuroscience, 11*(8), 880–881.

Delgado, M. R., Schotter, A., Ozbay, E. Y., & Phelps, E. A. (2008). Understanding overbidding: Using the neural circuitry of reward to design economic auctions. *Science, 321*(5897), 1849–1852.

Fellows, L. K., & Farah, M. J. (2005). Different underlying impairments in decision-making following ventromedial and dorsolateral frontal lobe damage in humans. *Cerebral Cortex, 15*, 58–63.

Greenwald, A. G., Poehlman, T. A., Uhlmann, E. L., & Banaji, M. R. (2009). Understanding and using the Implicit Association Test: III. Meta-analysis of predictive validity. *Journal of Personality and Social Psychology, 97*(1), 17–41.

Güroğlu, B., van den Bos, W., Rombouts, S. A., & Crone, E. A. (2010). Unfair? It depends: Neural correlates of fairness in social context. *Social Cognitive and Affective Neuroscience*, 5(4), 414–423.

Harlé, K., & Sanfey, A. G. (2007). Incidental sadness biases social economic decisions in the Ultimatum Game. *Emotion*, 7(4), 876–881.

Harris, L. T., & Fiske, S. T. (2007). Social groups that elicit disgust are differentially processed in mPFC. *Social Cognitive and Affective Neuroscience*, 2(1), 45–51.

Hartley, C. A., & Phelps, E. A. (2010). Changing fear: The neurocircuity of emotion regulation. *Neuropsychopharmacology*, 35(1), 136–146.

Ishizuka, K., Hillier, A., & Beversdorf, D. Q. (2007). Effect of the cold pressor test on memory and cognitive flexibility. *Neurocase*, 13, 154–157.

Kahneman, D. (2003). A perspective on judgment and choice: Mapping bounded rationality. *American Psychologist*, 58, 697–720.

King-Cass, B., Tomlin, D., Anen, C., Camerer, C. F., Quartz, S. R., & Montague, P. R. (2005). Getting to know you: Reputation and trust in a two-person economic exchange. *Science*, 308(5718), 78–83.

Klein, J. T., Deaner, R. O., & Platt, M. L. (2008). Neural correlates of social target value in macaque parietal cortex. *Current Biology*, 18(6), 419–424.

Knoch, D., Pascual-Leone, A., Meyer, K., Treyer, V., & Fehr, E. (2006). Diminishing reciprocal fairness by disrupting the right prefrontal cortex. *Science*, 314(5800), 829–832.

Lerner, J. S., Small, D. A., & Loewenstein, G. F. (2004). Heart strings and purse strings: Carryover effects of emotions on economic decisions. *Psychological Science*, 15(5), 337–341.

Lo, A. W., & Repin, D. V. (2002). The Psychophysiology of Real-Time Financial Risk Processing. *JOCN*, 14, 323–339.

Maia, T. V., & McClelland, J. L. (2004). A reexamination of the evidence for the somatic marker hypothesis: What participants really know in the Iowa gambling task. *Proceedings of the National Academy of Sciences*, 101, 16709–16710.

Martin, L. N., & Delgado, M. R. (2011, January 21). The influence of emotion regulation on decision making under risk. *Journal of Cognitive Neuroscience* (Epub ahead of print).

McClure, S. M., Berns, G. S., & Montague, P. R. (2003). Temporal prediction errors in a passive learning task activate human striatum. *Neuron*, 38(2), 339–346.

Ochsner, K. N., & Gross, J. J. (2005). The cognitive control of emotion. *Trends in Cognitive Sciences*, 9(5), 242–249.

Porcelli, A. J., & Delgado, M. R. (2009). Acute stress modulates risk taking in financial decision making. *Psychological Science*, 20(3), 278–283.

Rangel, A., Camerer, C., & Montague, P. R. (2008). A framework for studying the neurobiology of value-based decision making. *Nature Reviews Neuroscience*, 9(7), 545–556.

Roth, A. E. (1995). Bargaining experiments. In J. H. Kagel & A. E. Roth (Eds.), *Handbook of experimental economic* (pp. 253–348). Princeton, NJ: Princeton University Press.

Sanfey, A. G., Rilling, J. K., Aronson, J. A., Nystrom, L. E., & Cohen, J. D. (2003). The neural basis of economic decision-making in the ultimatum game. *Science*, 300, 1755–1757.

Scherer, K. R. (2005). What are emotions? And how can they be measured? *Social Science Information*, 44, 695–729.

Sokol-Hessner, P., Hsu, M., Curley, N. G., Delgado, M. R., Camerer, C. F., & Phelps, E. A. (2009). Thinking like a trader selectively reduces individuals' loss aversion. *Proceedings of the National Academy of Sciences*, 106(13), 5035–5040.

Stanley, D. A., Sokol-Hessner, P., Banaji, M. R., & Phelps, E. A. (2011). Implicit race attitudes predict trustworthiness judgments and economic trust decisions. *Proceedings of the National Academy of Sciences*, 108(19), 7710–7715.

Thaler, R. H. (1988). Anomolies: The ultimatum game. *Journal of Economic Perspectives*, 2, 195–206.

van't Wout, M., Kahn, R., Sanfey, A. G., & Aleman, A. (2006). Affective state and decision-making in the Ultimatum Game. *Exprimental Brain Research, 169*(4), 564–568.

van't Wout, M., Chang, L. J., & Sanfey, A. G. (2010). The influence of emotion regulation on social interactive decision-making. *Emotion, 10*(6), 815–821.

Winkielman, P., Berridge, K. C., & Wilbarger, J. L. (2005). Unconscious affective reactions to masked happy versus angry faces influence consumption behavior and judgments of value. *Personality and Social Psychology Bulletin, 31*(1), 121–135.

Wright, N. D., Symmonds, M., Fleming, S. M., & Dolan, R. J. (2011). Neural segregation of objective and contextual aspects of fairness. *Journal of Neuroscience, 31*(14), 5244–5252.

PART IV

PERCEPTUAL
FACTORS

Auditory Preferences and Aesthetics: Music, Voices, and Everyday Sounds

Josh H. McDermott

Center for Neural Science, New York University, USA

INTRODUCTION

Some sounds are preferable to others. We enjoy the hypnotic roar of the ocean, or the voice of a favorite radio host, to the point that they can relax us in the midst of an otherwise stressful day. However, sounds can also drive us crazy, be it a baby crying next to us on a plane or the high-pitched whine of a dentist's drill. Hedonic and aversive responses to sound figure prominently in our lives. Unpleasant sounds warn us of air raids, fires, and approaching police cars, and are even used as coercive tools during interrogation procedures. Pleasant sounds are fundamental to music, our main sound-driven form of art, and the pleasantness of voices is important in evaluating members of the opposite sex.

This chapter will present an exploration of sounds that evoke hedonic and aversive responses in humans. Our central interest is in what determines our auditory preferences, and why. *A priori* we can imagine that many different factors might come into play, including acoustic properties, learned associations between sounds and emotional situations, the surrounding context, input from other senses, and the listener's personality and mood. As we shall see, all of these factors can at times be important. We will consider the aesthetic response to isolated sounds, annoying and pleasant, as well as to the complex sound sequences produced by music.

ANNOYING SOUNDS

At their worst, sounds can be flat out cringe-worthy. The most commonly cited example is the sound of fingernails on a chalkboard, the mere thought of which is enough to make many people grimace. What makes such sounds so awful? Apart from their intrinsic interest, much of the motivation for studying annoying sounds comes from industry. Manufacturers have long had an interest in understanding what makes sounds aversive so as to avoid these properties in products that emit noise (electric saws, refrigerators, cars, trains, etc.), and studies on this topic date back many decades.

When people are asked to rate the annoyingness of large sets of real-world sounds, considerable agreement is usually observed across listeners (Cardozo & van Lieshout, 1981; Terhardt & Stoll, 1981), at least within the groups of Westerners typically studied. A fair bit of the variance in pleasantness across sounds can be explained by a few simple acoustic properties.

In addition to overall loudness, two properties that have substantial influence are "sharpness" and "roughness" (Fastl, 1997; Terhardt & Stoll, 1981). Sharpness describes the proportion of energy at high frequencies, with sharper sounds (those with more high-frequency energy) generally found to be less pleasant. Frequencies in the range of 2–4 kHz contribute the most to annoyingness (Kumar, Forster, Bailey & Griffiths, 2008). This range is high in absolute terms, but well below the upper limit of what is audible to a human listener with normal hearing. Screech-like sounds (Figure 10.1), much like fingernails on a blackboard, lose some of their

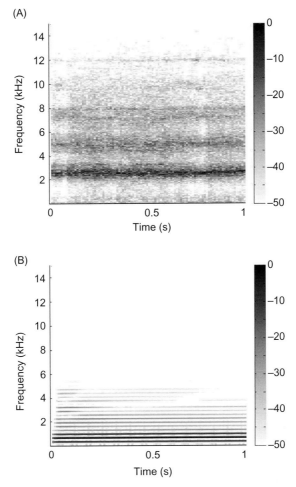

FIGURE 10.1   **Spectrograms of (A) a screech (produced by scraping a metal garden tool down a glass window) and (B) a note played on a saxophone.** Gray level displays sound amplitude in dB. Note that the screech has a concentration of energy between 2–4 kHz. The saxophone, by contrast, has harmonics that are more closely spaced, as it has a lower pitch, with most of the energy concentrated below 2 kHz.

aversiveness when frequencies in the 2–4 kHz range are filtered out, but not when frequencies above this range are removed (Halpern, Blake & Hillenbrand, 1986).

Roughness is the perceptual correlate of fluctuations in energy (intensity) that occur over time, analogous to the fluctuations in surface depth that determine the roughness of an object to the touch. Fluctuations at rates between ~20–200 Hz are those that determine roughness (Terhardt,

1974a); any lower, and the fluctuations can be heard individually rather than contributing to a sound's timbre. In general, the rougher a sound, the less pleasant it tends to be. For instance, studies of automobile interior noise, which manufacturers aim to make as pleasant as possible, indicate that roughness is a major determinant of unpleasantness (Takao, Hashimoto & Hatano, 1993). Roughness is also a characteristic of many scraping and screeching sounds, including that of fingernails on a blackboard. The amplitude fluctuations in these sounds result from an object (e.g. a fingernail) rapidly catching and then releasing on the surface being scraped, producing many brief bursts of sound that cause amplitude fluctuations.

Why are these sound properties unpleasant? The annoying effects of sharpness may be rooted in the frequency sensitivity of the ear, which peaks in the range of 2–4 kHz. The ear canal has a resonance in this range, boosting sound levels of these frequencies by as much as 30 dB (Henoch & Chesky, 1999). Exposure to noise in this frequency range is thus most likely to damage the ear. The aversive response to these frequencies could simply be because they sound the loudest, and are potentially the most dangerous to listen to. Notably, most highly unpleasant sounds are much less aversive at low volume.

It is less clear why roughness is unpleasant, as it does not obviously pose any danger to the auditory system. It is also unclear at present whether the reaction to roughness and other sound properties is universal and obligatory. In Western music, as discussed below, rough sounds are thought to be unpleasant (Helmholtz, 1863; McDermott, Lehr & Oxenham, 2010; Plomp and Levelt, 1965). In some cultures, however, roughness is a staple of musical expression, and its aesthetic interpretation may be different than in the Western world (Vassilakis, 2005). Even in Western music, rough sounds have become common in some subgenres since the introduction of distortion in rock music in the 1950s and 1960s, and in this context are enjoyed by listeners. It thus remains possible that the aversion to roughness is partially context-dependent.

Associations between sounds and the events in the world that cause them also clearly play some role in whether we experience a sound as pleasant or unpleasant. In a large internet-based experiment, the sound rated most awful out of a large set was that of someone vomiting (Cox, 2008a). Though not particularly sharp or rough, the associations most of us have with vomiting no doubt contributed to its status as the most annoying sound. The same study found that seeing the image of fingernails on a chalkboard, or a dentist, while listening to the corresponding sound yielded a worse rating for the sound (Cox, 2008b). Visual input can also render sounds less annoying – white noise is deemed less objectionable when accompanied by a picture of a waterfall that suggests a natural sound source (Abe, Ozawa, Suzuki & Sone, 1999).

PLEASANT ENVIRONMENTAL SOUNDS

Fortunately, not all sounds are annoying. Indeed, many people find natural environmental soundscapes (ocean waves, rainfall, etc.) to be relaxing and pleasant, to the point that recordings of such sounds are marketed to aid sleep and relaxation. Why are these sounds enjoyable? Very little research has addressed the pleasantness of environmental sounds, but emotional associations with relaxing circumstances surely play some role. Such sounds also typically lack the acoustic properties described above that are found in many annoying sounds. The sounds of oceans, rain, wind, etc. usually have more energy at low frequencies than at high (Voss & Clarke, 1975), and feature slow temporal modulations (Attias & Schreiner, 1997; Singh & Theunissen, 2003), rather than prominent modulations in the roughness range. It is also interesting that the perception of many natural sound "textures," such as water sounds, can be explained in relatively simple terms with generic statistics of the early auditory system (McDermott, Oxenham & Simoncelli, 2009; McDermott & Simoncelli, 2011), though a relationship between this sort of simplicity and pleasantness remains conjecture at present.

VOICES

The attractiveness of voices is of particular interest because the voice is believed to provide important signals of the fitness of potential partners. Humans historically tended to engage in sexual activity at night, when little light was available to judge visual characteristics. The voice may thus have been critical for judgments of mate quality, with vocal characteristics that signal fitness coming to be perceived as attractive. Empirically, voices vary significantly in their attractiveness, and listeners largely agree on what sounds attractive (Zuckerman & Driver, 1989).

Consistent with the notion that it functions to signal mate quality, vocal attractiveness co-varies with body symmetry (e.g. between the widths of the two hands), a known marker of fitness (Hughes, Harrison & Gallup, 2002), and co-varies with other sexually dimorphic traits – males with larger shoulder-to-hip ratios, and females with larger hip-to-waist ratios, have more attractive voices (Hughes, Dispenza & Gallup, 2004). Individuals with more attractive voices also have more sexual partners, and become sexually active earlier in life, than those with unattractive voices. In females, voice attractiveness is actually a better predictor of promiscuity than the visual signal provided by hip-to-waist ratio (Hughes, Dispenza & Gallup, 2004), and varies across the menstrual cycle, providing an "honest" signal of fertility (Pipitone & Gallup, 2008). The influence of voice attractiveness in most life situations is at least

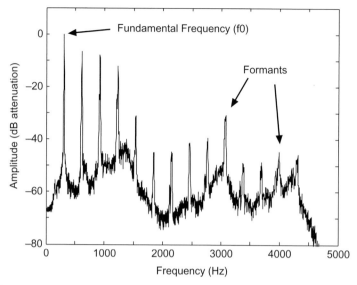

FIGURE 10.2 **Power spectrum of a vowel.** Note that the frequencies in the vowel are integer multiples of the fundamental frequency (harmonics), and are thus equally spaced along the spectrum.

partly confounded with other aspects of attractiveness (which are correlated with attractive voice characteristics), but when provided only with vocal samples, listeners judge themselves more likely to affiliate with people with attractive voices (Miyake & Zuckerman, 1993).

Acoustically, voices are characterized in part by their characteristic pitch and formant frequencies, each of which correspond to a component of the sound production process. The sound signal that leaves a speaker's mouth can be thought of as sound from a source (the vocal folds) that is then passed through a filter (the vocal tract) that alters its frequency characteristics. Sound is created when the vocal folds open and close as air passes through them. The opening and closing happens at a regular rate and generates a sound waveform that repeats at this rate. The pitch corresponds to the rate of repetition, i.e. the fundamental frequency (f0) of the voice (the other frequencies are integer multiples of the f0, i.e. harmonics of it; Figure 10.2). The f0 can be controlled to some extent by the musculature surrounding the vocal folds, but the central tendency is set by the thickness and size of the folds (Fant, 1960).

Formants, in contrast, are global peaks in the spectrum that result from the filtering effects of the vocal tract, which amplifies some frequencies and dampens others (Figure 10.2). They can also be varied by changing the shape of the throat and mouth, as when articulating different vowels, but their central tendency and spacing are determined by the vocal tract's intrinsic shape, most notably its length. Formants thus provide a cue to vocal tract length, which in turn is tightly linked to body size. In this respect they are distinct from pitch, which depends instead on the size of the vocal folds, a trait that is largely independent of overall body size (Fitch, 1997).

Both pitch and formant frequencies are sexually dimorphic. Testosterone causes the vocal folds to increase in size, lowering voice pitch, as happens in males at puberty. Pitch is high in children and decreases during development in both sexes, but at a faster rate in males, with the decrease accelerated during puberty. The result is that adult male f0s are on average about half that of adult females (Bachorowski and Owren, 1999). Individual variation within each sex is high, however, and likely provides a signal of hormone levels that relate to reproductive potential. Formant frequencies and their spacing similarly differ on average between sexes, but also vary considerably across individuals of a particular sex.

Low voice pitch in men is more attractive to female listeners, who associate it with other attractive male physical features and signals of sexual maturity (Collins, 2000). Moreover, artificially lowering the pitch and formants of a male voice increases its attractiveness to female listeners, an effect that is enhanced during the fertile phase of the menstrual cycle (Feinberg, Jones, Law Smith et al., 2006). Low voice pitch is also predictive of reproductive success in hunter-gatherers, for whom the absence of birth control methods allows this success to be readily measured (Apicella, Feinberg & Marlow, 2007).

In female voices, the reverse trend occurs: they are judged as more attractive by men if they are higher in pitch and have more widely spaced formants (Collins and Missing, 2003), and artificially raising female voice pitch increases attractiveness (Feinberg, DeBruine, Jones & Perrett, 2008). Like male voice pitch, female vocal pitch appears to signal fitness. Women with feminine facial features tend to have high pitched voices (Collins and Missing, 2003), suggesting that both provide cues to femininity and reproductive fitness. Voice pitch in women is also correlated (negatively) with health risk factors like overall weight, body mass index, and body fat percentage (Vukovic, Feinberg, DeBruine et al., 2010), suggesting that it signals properties indirectly related to reproductive success as well. Moreover, when communicating with men that they find attractive, women speak with higher pitch, indicating implicit knowledge of the fitness-related signal provided by the pitch of their voice (Fraccaro, Jones, Vukovic et al., 2011).

In addition to gender-specific effects, there is a general tendency to prefer voices that have clearer (less noisy) pitch. Voice pitch clarity, indicated quantitatively by a high "harmonic-to-noise" ratio, is another signal of fitness – it decreases in the elderly (Ferrand, 2002), and is lower in voices that are hoarse (Yumoto, Gould & Baer, 1982), as when suffering from a cold or other illness. Preferences for clear pitch are evident in recent studies of voice averaging. Averaging recordings of vowels produced by a large set of different speakers (using voice morphing software) tends to yield a voice that is almost as attractive as the most attractive individual voice in the set (Bruckert, Bestelmeyer, Latinus et al., 2010). Some of this effect is due to the smoothing inherent to averaging, which enhances the harmonic frequencies of the pitch and averages out the aperiodic noisy components (imperfections in the pitch), producing a cleaner pitch.

The effect of averaging multiple voices together is similar in some respects to the effect of reverberation, which also tends to make voices sound better, as most of us have experienced when singing in the shower. The hard walls of a shower reflect the sound that comes out of our mouths, such that the ears receive sound indirectly from each of the reflections in addition to the sound that comes directly from the mouth. Because the path from the source to our ears is longer for the reflected versions, they reach our ears a small fraction of a second later than the direct sound. The ear thus receives a sum of many copies of the original sound, each with a different delay, and each filtered to some extent by the reflective surface (which generally absorbs some frequencies more than others). The effect is somewhat like taking an average of different voices in that the noisy components tend to get averaged out, leaving a pitch that sounds more pure than it would without the reverberation. Recording engineers in fact typically incorporate reverberation into the music recording process for its aesthetic effect, either by choosing a room with pleasant reverberation in which to record, or by adding reverberation as a digital effect (Gardner, 1998).

It has also been argued that the effect of voice averaging is not limited to simply smoothing out imperfections, and that voices whose pitch and formants are closer to the average values for that sex are more attractive (Bruckert et al., 2010), as is the case for faces (Langlois & Roggman, 1990) and other stimuli (Halberstadt & Rhodes, 2000). One piece of evidence is that "moving" an individual's voice towards the average, by altering the f0 and formant frequencies in the direction of the average voice, tends to make the voice more attractive (Bruckert et al., 2010). This would seem inconsistent with the many findings, cited earlier, that lowering male pitch, or raising female pitch, increases attractiveness irrespective of where the voice lies relative to the average, but it is possible the effect is driven mainly by the formants rather than the pitch. As with faces (Perrett, May & Yoshikawa, 1994), however, it seems likely that at least some highly attractive voices are non-average (consider the deep baritone of Barry White, or the gravelly character of Louis Armstrong's voice, for instance), though it remains to be tested.

MUSIC

Music is the domain in which our aesthetic response to sound is most obvious and striking. For the typical human listener, music is a highly rewarding stimulus. Listening to our favorite music activates the same reward pathways that are stimulated by good food, cocaine, and sex (Blood & Zatorre, 2001; Menon & Levitin, 2005; Salimpoor, Benovoy, Larcher et al., 2011). The reward of listening to music motivates us to consume a startling amount of it, expending considerable resources as a result. Before the availability of free online music caused the industry to nosedive, music sales were in tens of billions of dollars (Geter and Streisand, 1995), and major record labels were viewed by Wall Street as lucrative investment opportunities (Knopper, 2010). When randomly probed via their cell phones, British adults were recently found to be in the presence of music 39% of the time (North, Hargreaves & Hargreaves, 2004). Music is ubiquitous in restaurants and department stores (Bruner, 1990), and has been shown to improve sales (North, Shilcock & Hargreaves, 2003), presumably because of its positive influence on mood.

There are many open questions concerning the nature of our response to music, including where music is processed in the brain (Peretz & Zatorre, 2003), whether it interacts with other cognitive abilities and resources (Patel, 2008), why we experience emotion when listening to it (Juslin & Sloboda, 2010; Zentner, Grandjean & Scherer, 2008), and why we have it to begin with (Huron, 2001; McDermott, 2008; Wallin, Merker & Brown, 2001). In this chapter our interest is specifically in why people like the music that they do. Ultimately we will discuss the preferences that a listener has for one piece of music over another, but it is first worth considering the aesthetic response to the simpler sound elements that music is made of.

Instrument Sounds

Music often is played on instruments, the sound of which is obviously important to the aesthetic value of the end product. Not every violin, or every guitar, sounds the same. Different instruments are distinguished by having different "timbre" – aspects of their sound that are not captured by pitch or loudness, that vary both across and within an instrument category. Some instrument brands are especially sought-after for their aesthetically pleasing timbre, the most famous example being the Stradivarius violin (Beamen, 2000). However, the sound of the music that enters our ears, and that determines our aesthetic response, is not just determined by the instrument and manner of playing. The experience of live music, for instance, depends crucially on the concert hall, the sound of which is the product of painstakingly adjusted reflection patterns that add reverberation while ensuring that clarity is preserved throughout the listening space (Lokki, Patynen, Tervo et al., 2011). Other factors influence music that is recorded for later mass consumption, including the brand of microphone used to record the instrument, the placement of the microphone(s) around the instrument and room, the room where the instrument is recorded, and the filters and other effects applied post-recording (Milner, 2009). Many music producers and engineers develop an idiosyncratic and elusive mixture of these factors that gives their recordings of drums, horns, etc., a distinctive and characteristic sound.

The sound of a great recording can be remarkably difficult to replicate. One place where this is evident is in the modern-day practice of "sampling", in which a contemporary music maker excerpts a brief segment of another recording (often from an earlier era) to obtain a sound with the desired timbre that would otherwise be difficult to recreate in a modern recording studio. In the golden age of hip-hop, for instance, drum "breaks" were often excerpted from older recordings and looped (often combined with other samples) to create rhythm tracks. Certain drum breaks were used over and over, in some cases on literally hundreds of different recordings, as producers valued particular drum sounds that had been recorded decades earlier (Crate Kings, 2007). Some of these classic samples were bits of records by well-known artists like James Brown or Kool and the Gang, but many were taken from otherwise obscure recordings, by artists such as the Honeydrippers, the Winstons, and the Incredible Bongo Band. They may not have produced hit records at the time, but they achieved a sound that contemporary listeners value even if they are unaware of its origins.

Although instrument sounds matter greatly to music listeners, with much time and effort devoted to achieving the right sound during the recording process, drawing scientific conclusions about why people like particular instrument sounds is a challenge. It is likely that preferences vary across genre and culture, and that individual differences are substantial. Instrument sounds may also matter most in how they combine with other sounds to create a piece of music (Schloss, 2004).

Consonance and Dissonance of Chords

One domain in which preferences have been more rigorously measured and modeled is that of musical chords – combinations of notes played at the same time. It has been known for thousands of years that some combinations of notes are more pleasing than others (to Western listeners, at least). Figure 10.3a shows an example data set of pleasantness ratings given by American undergraduates to various two- and three-note chords (McDermott et al., 2010). It is apparent that some chords are rated higher than others, and that the general pattern is

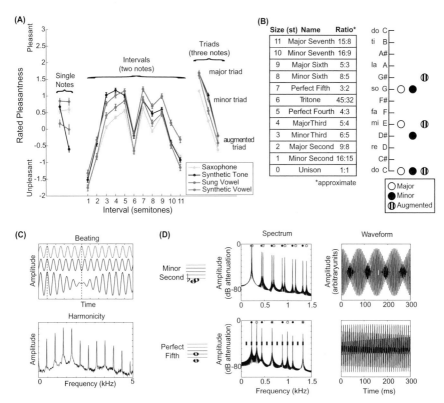

FIGURE 10.3 **Consonance preferences and their possible acoustic basis.** (A) Average pleasantness ratings of individual notes and chords, for a large group of American undergraduates. The two single-note conditions differed in pitch (lower pitch on left). Error bars denote standard errors. (B) Intervals and chords from A, with diatonic scale (on left) as reference. Ratios in stimuli approximated those listed in table, due to use of the equal-tempered scale. (C) Beating and harmonicity. Top – two sinusoids of different frequencies are plotted in red and blue; their superposition (in black) contains amplitude modulation known as "beating." Bottom – amplitude spectrum for the note A440 played on an oboe. The frequencies in the note are all integer multiples of the fundamental frequency of 440 Hz, and as a result are regularly spaced along the frequency axis. (D) Spectra and waveforms for the minor second and perfect fifth, in which beating and harmonicity are apparent. The intervals were generated by combining two synthetic complex tones with different fundamental frequencies. Red (open) and blue (closed) circles denote the frequencies belonging to each note. The frequencies of the fifth are approximately harmonically related (black lines denote harmonic series). Amplitude modulation (from beating) is evident in the waveform of the minor second, but not the fifth (Modified from McDermott et al., 2010).

largely independent of what instrument is used to play the notes. The highly-rated chords are conventionally called consonant, and the low-rated chords are termed dissonant.

The distinction between consonance and dissonance is central to Western music. Computational analysis of scores and scales indicates that many classical composers made choices to maximize consonance in their compositions (Huron, 1991), and that the structure of the Western diatonic scale itself may have resulted from an attempt to maximize the number of possible consonant note combinations (Huron, 1994). However, when dissonance is used, it plays an important role, being routinely employed to create tension in music, as is often apparent in movie or television soundtracks.

Why is it that only some combinations of notes sound consonant? Debates over the basis of consonance date back to the Greeks, who famously believed that aesthetics derived from ratios, and noted that consonant intervals are produced by strings whose lengths form simple integer ratios (Figure 10.3b). Modern-day theories have instead tried to explain consonance in terms of acoustic properties – roughness, mentioned above in the context of aversive noises, or harmonicity, an important property of the frequency spectra of many natural sounds.

The proposal of a role for roughness in consonance is generally credited to Helmholtz, who noticed that dissonant chords tend to have a preponderance of a phenomenon known as beating (Helmholtz, 1863). Beating occurs whenever similar frequencies are present simultaneously (Figure 10.3c, top). Over time, two frequencies shift in and out of phase, causing them to constructively and then destructively interfere. This cyclical pattern results in a sound (shown in black in the figure) that waxes and wanes in amplitude. Beating is one way to produce a sound that is rough, which tends to be heard as unpleasant, as discussed earlier (Terhardt, 1974a). Helmholtz noted that dissonant chords tend to produce many pairs of frequencies that are close but not identical, and that thus beat. One example, the minor second, is shown in Figure 10.3c (middle row, left column). Each pair of nearby frequencies beat, producing a rough waveform (Figure 10.3c, bottom left). Consonant intervals, in contrast (for example, the perfect fifth, shown in Figure 10.3c, middle row, right column), have frequencies that are either identical or widely spaced, and that produce little beating as a result. This difference in roughness has been widely proposed to underlie the differences in pleasantness of different musical chords (Hutchinson & Knopoff, 1978; Kameoka & Kuriyagawa, 1969; Plomp & Levelt, 1965; Sethares, 1999; Vassilakis, 2005).

Although roughness has arguably been the standard explanation of consonance since the 1960s, an alternative explanation in terms of "harmonicity" has also retained proponents. In the earlier section on voice, we discussed how sounds that have a pitch contain frequencies that are

harmonically related – the frequencies are integer multiples of a fundamental frequency, producing regular peaks in the spectrum (an example is
shown in Figure 10.3c, bottom). Vocal and instrument sounds tend to have
a pitch, and to be harmonic, and thus when a chord is played in music,
each of the component notes is generally a harmonic tone. However, it
turns out that for consonant chords, the combined frequencies of all the
notes together are also harmonically related. Every frequency in the spectrum of the perfect fifth, for instance (Figure 10.3c, middle right), corresponds to an element of the harmonic series (indicated by the thick
black line segments superimposed on the spectrum), although generally not every harmonic is present. Dissonant chords, in contrast, produce sets of frequencies that are inharmonic. Harmonic frequencies have
been supposed to be preferred over inharmonic frequencies due to their
resemblance to single tones, thus potentially explaining the preference
for consonance over dissonance. The harmonicity theory of consonance
has a long history (Bidelman & Krishnan, 2009; Ebeling, 2008; Stumpf,
1890; Terhardt, 1974b; Tramo, Cariani, Delgutte & Braida, 2001), but has
in recent decades been disregarded in favor of roughness. The theories
proved difficult to definitively distinguish because they make many of the
same predictions (Mathews & Pierce, 1980; McDermott & Oxenham, 2008).

   In recent work, my colleagues and I were able to disentangle these
factors using individual differences (McDermott et al., 2010). The logic
of our approach was that the strength of the preference for consonance
over dissonance ought to vary somewhat across individuals, as should
the aversion to either inharmonicity or beating, the two acoustic factors
thought to possibly contribute (negatively) to consonance. If one of the
acoustic factors is causally related to preferences for consonance, then
the strength of the aversion for that factor ought to correlate with the
strength of consonance preferences. We measured the aversions to beating and to inharmonicity in a large set of subjects, then measured preferences for isolated musical chords, and examined correlations between the
different preferences. To assess the aversion to the two candidate acoustic
factors, we measured the preference for stimuli that lacked beating over
stimuli that produced beating, and for harmonic tones over inharmonic
tones (in which the frequencies of a harmonic tone were perturbed such
that they were no longer multiples of a common fundamental frequency).

   The results were surprisingly decisive: only harmonicity preferences correlated significantly with consonance preferences. Our results
indicate that roughness (caused by beating in this case) is orthogonal
to dissonance rather than covarying with it – rough sounds are clearly
unpleasant in many contexts, but the difference between consonant
and dissonant chords seems to be due to another acoustic variable, that
of harmonicity. This may be because in practice, consonant chords are
in fact not consistently rougher than dissonant chords, or because the

overall amount of beating varies considerably with instrument timbre, such that it is not diagnostic of the note combinations in a chord. In any case, roughness does not seem to be causally related to consonance. We concluded that much of the pleasantness of musical chords derives from whether their frequencies are harmonically related or not.

Concluding that harmonicity underlies consonance leaves open the question of why we prefer harmonic sounds, and thus consonance. In particular, there is longstanding interest in whether the response to consonance is learned from exposure to music (Cazden, 1945; Lundin, 1947), which tends to have more consonant than dissonant chords. It would be useful to know to what extent consonance preferences are present in foreign cultures, some of which have music that departs considerably from Western music in scales and harmony, but regrettably few such studies have addressed this issue thus far (Butler & Daston, 1968; Fritz, Jentschke, Gosselin et al., 2009). Consonance has been studied more extensively in developmental psychology. Several investigators have found that young infants seem to prefer consonance to dissonance (Trainor & Heinmiller, 1998; Zentner & Kagan, 1998; Trainor, Tsang & Cheung, 2002), suggesting that preferences emerge without much exposure to music. On the other hand, in our individual differences work, we found that preferences for consonance, as well as for harmonicity, were correlated with the number of years undergraduate subjects had spent playing an instrument, suggesting that the preference is at least modified considerably by musical experience (McDermott et al., 2010). Further work is needed to definitively address the universality and/or innate nature of this basic aspect of music perception.

Musical Pieces and Genres

We will now explore the basis of preferences for more complex and extended pieces of music. Everyone has experience listening to music, and many of us have intuitions about why we like what we like. Much of the research in this area has thus far confirmed with controlled experiments phenomena that have a good deal of intuitive plausibility. Many factors come into play. Social influences loom large, as people use music to project an identity (North, Hargreaves & O'Neill, 2000), and are strongly influenced by what others around them listen to when making their own listening choices (Salganik, Dodds & Watts, 2006). There are also factors involving something like intrinsic aesthetic quality, at least within a culture, as well as factors idiosyncratic to particular listeners, such as their past experience and personality.

Exposure and Familiarity

One of the largest influences on music preferences is prior exposure – we are inclined to like things we have heard before, and to dislike those we have not. People tend to prefer to listen to the music of their culture, even as young infants (Soley & Hannon, 2010), and often find the music of foreign cultures to be uninteresting or unpleasant by comparison (Fung, 1993). In modern times, foreign music has in fact been used in warfare and interrogation. During the Iraq War, the BBC reported that uncooperative prisoners were exposed for prolonged periods of time to heavy metal music and American children's songs (e.g. the Sesame Street theme) in order to coerce them into talking to US interrogators, a practice that is apparently standard operating procedure in the US "war on terror" (Cusick, 2008). The stress from exposure to foreign music is evidently considerable. A related much-publicized incident occurred during the US invasion of Panama. Manuel Noriega, the Panamanian dictator, took refuge in the Vatican embassy, which US forces could not enter without violating international law. To induce him to surrender, US troops supposedly set up loudspeakers outside the embassy, from which they played Van Halen, Guns and Roses, and other hard rock music around the clock. Noriega surrendered within a week.

Even within a familiar culture and genre, many people have had the experience of finding a piece of music relatively unrewarding upon first listen, but coming to love it with repeated listens. Any DJ can tell you that the single most important factor predicting whether people will dance to a song is whether or not they are familiar with it; even highly danceable music is unlikely to evoke enthusiasm on the dancefloor the first time it is played. Familiarity effects are also well-documented experimentally. Across genres, familiar musical pieces are generally liked more than unfamiliar pieces (Hargreaves, Messerschmidt & Rubert, 1980; Hargreaves, 1987). This by itself could be explained by the possibly higher quality of familiar pieces (in that better songs would be more likely to become hits). However, repeated exposure also increases liking in unfamiliar music pieces, whether in music from a familiar idiom (Gilliland & Moore, 1924; Ali & Peynircioglu, 2010) or an unfamiliar or foreign style – modern classical (Mull, 1957), or Pakistani music heard by Americans (Heingartner & Hall, 1974). Effects are typically observed over the course of a few exposures. The effect of "mere exposure" is not unique to music. Stimuli across the board are liked more with repeated exposure, at least up to a point (Zajonc, 1968), often plateauing after about 10 exposures (Bornstein, 1989). It often feels subjectively that the effects of repetition on liking are more pronounced for music than for other stimuli, but I know of no empirical evidence to support that notion at present.

Many people also have the sense that there is a special relationship between music and the memories induced by prior exposure. We tend to have a fondness for the music that surrounded us during childhood and adolescence, and often feel we like particular pieces in part because they remind us of good times from the past. However, at least as has been tested thus far, memory for music does not seem to be any more accurate than memory for other kinds of stimuli to which we have comparable exposure. For instance, when recollection of songs from people's high school years were compared to that for faces from their high school yearbooks, no memory advantage was observed (Schulkind, 2009). The sense we have of a privileged link between music and memory may be an illusion like that associated with memory for emotional life events. People often believe that their memories of emotionally traumatic events (e.g. the Kennedy assassination, or the September 11 terrorist attacks) are more accurate than memories of more mundane events, but when tested in the lab, accuracy is in fact no different (Talarico & Rubin, 2003). What *is* different is the sense of vividness with which people recollect emotional events (Talarico & Rubin, 2003), believed to be because the emotion centers of the brain alter the experience of remembering (Sharot, Delgado & Phelps, 2004). Memory for music may involve a similar phenomenon – the emotional content of music may cause the experience of remembering it to be enhanced even if the memory itself is no more precise or robust to decay.

Complexity

The effect of familiarity on our aesthetic response to music is substantial, but it also seems obvious that it cannot be the only factor influencing what we like – some pieces are clearly much easier to like than others. One widely discussed idea in experimental aesthetics is that the aesthetic response is related to complexity. The notion is that stimuli that are too simple or too complex are not aesthetically pleasing, but that somewhere in the middle lies an optimum. Complexity and aesthetic value are thus proposed to be related via an inverted U-shaped function (Berlyne, 1971); Figure 10.4. Exactly how complexity should be measured is unclear. Some authors have argued for information theoretic measures; others define it for music in terms of the degree of conformity to the rules of the dominant musical idiom.

Intuitively, the idea of an inverted U-shaped curve relating complexity and pleasingness is at least partly consistent with what we know from experience: something that is too repetitive is boring, while something that is completely random has no structure, and thus cannot be related to things we've previously heard. The idea is also consistent with observations that the complexity of typical musical melodies is moderate, with note-to-note changes being dominated by small intervals (Voss & Clarke, 1975; Vos & Troost, 1989).

Numerous experimental findings support a role for complexity in musical preferences. Moderately complex pieces tend to be preferred over pieces of lesser or greater complexity, be they piano solos whose complexity is varied by changing the number of chords and degree of

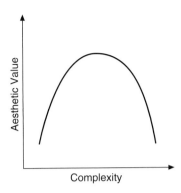

FIGURE 10.4 The inverted U-curve proposed to related subjective complexity to aesthetic value (Berlyne, 1971).

syncopation (Heyduk, 1975), instrumental classical music (Radocy, 1982), random tone sequences whose temporal correlation is varied (Voss & Clarke, 1975), contemporary pop music (North & Hargreaves, 1995), or ambient electronic music (North & Hargreaves, 1996) whose complexity is assessed by listener ratings. One problematic issue is that if the range of complexity that is sampled by an experiment is not centered near the optimal level of complexity, or is not broad enough, one would expect to see either a monotonic dependence of liking on complexity, or none at all, both of which are sometimes reported (Martindale & Moore, 1989; Smith & Melara, 1990; Orr & Ohlsson, 2001). That said, many investigators have found an inverted U-shape.

Importantly, what is purported to matter is subjective rather than objective complexity. This distinction is critical because the subjective complexity of a piece of art or music is thought to decrease with repeated exposure, as the observer internalizes its structure. The preference discussed earlier for the music of one's own culture may derive from this principle. As we develop in a musical culture, we internalize its rules and characteristic forms. We become more adept at encoding idiomatic pitch and rhythm structures, and at recognizing the emotions they represent, relative to those of other cultures (Dalla Bella, Peretz, Rousseau & Gosselin, 2001; Hannon & Trehub, 2005; Lynch, Eilers, Oller & Urbano, 1990; Thompson and Balkwill, 2010; Trehub, Schellenberg & Kamenetsky, 1999). Because our expertise is culture-specific, the subjective complexity of typical foreign music may be prohibitively high, preventing us from enjoying it.

Repeated exposure is expected to shift the location of a particular piece leftwards on the complexity axis of the inverted U-curve, one interesting consequence of which is that the effect of exposure on aesthetic evaluation should depend on the starting location. A piece that initially has a level of complexity that is close to optimal is thus predicted to be liked less with repetition, as it falls off the peak, whereas a piece that is initially too complex is expected to improve with repeated exposure, at least up to a point. These predictions have been confirmed in multiple studies. Smith and Cuddy found that repetition decreased pleasingness for harmonically simple melodies, but increased it for more complex melodies, with the biggest increases occurring for the most complex melodies in the stimulus set (Smith & Cuddy, 1986). Schellenberg and colleagues found that repetition increased the liking of moderately complex piano pieces if listeners were doing another task while listening to the music, but that focused listening caused liking to increase and then decrease with repetition, as though listeners had passed over the peak of the curve (Schellenberg, Peretz & Vieillard, 2008; Szpunar, Schellenberg & Pliner, 2004). Similar results have been obtained by other groups (Tan, Spackman & Peaslee, 2006; Verveer, Barry & Bousfield, 1933).

The proposed interaction of repetition and complexity fits qualitatively with aspects of everyday musical experience. A jazz recording, with chords and scales that are not fully familiar, might initially be too complex to be maximally rewarding to an untrained listener, but with repeated listens becomes increasingly comprehensible until it reaches the peak of the curve. A catchy pop tune, in contrast, may be instantly appealing due to its simple, familiar structure, but with repeated listens becomes annoying rather than enjoyable – it starts out at the peak of the curve, and repetition makes it more predictable than is optimal. An inverted U-shaped trajectory is also consistent with the finite shelf-life of hit songs. We do not listen to the same songs forever, but rather consume them for a time, eventually moving on, at least for the moment.

Introspectively, we know that a song we've tired of can become pleasing again following a delay, and experiments confirm this in some cases (Verveer, Barry & Bousfield, 1933). How this recovery relates to memory remains unclear, as we often have the impression of remembering the piece perfectly even following the delay. It may be that the sensory trace has decayed, rather than the abstract memory of the musical structure, allowing us to enjoy the piece even though we know it by heart. There are also cases in which music may be too complex for an unschooled listener to show much of an improvement in liking with repetition, as with highly avant-garde jazz (Hargreaves, 1984). Much remains to be studied about the representational changes that underlie these effects of exposure on preference.

The effect of complexity on preferences also interacts with the musical expertise of the listener. Experts (e.g. music graduate students) sometimes have a higher optimal complexity than novices (e.g. undergraduate non-music majors), at least when complexity is measured by the degree of deviation from the conventions of a musical idiom (Smith & Melara, 1990). This fits with the expectation that expert listeners have internalized musical conventions to a greater degree, such that the prototypical structure that is optimal for the untrained listener may strike them as overly simplistic. Other evidence indicates that expertise reduces the influence of complexity on preferences (Orr & Ohlsson, 2005), suggesting that other aesthetic factors become more important in people who have engaged in unusual amounts of focused listening and/or production. Such findings may relate to the differences that frequently exist between the music assessments of professional critics and lay listeners (North & Hargreaves, 1998).

Emotion

It has long been acknowledged that the emotional effects of music are central to its aesthetic value. Listeners report that the emotional content of music is one of the main reasons they listen to it, and they can typically identify the emotion that a piece of music was intended to convey (Fritz et al., 2009; Hevner, 1936). However, enjoyment of music is determined not by what emotion people judge it to be conveying, but rather by what they themselves experience when they listen to it. Enjoyment is maximal at moments of peak emotional arousal (Salimpoor, Benovoy, Longo et al., 2009), and listeners prefer pieces that induce emotion over those that do not (Schubert, 2010). Listeners in fact give low liking ratings to musical works when there is a large gap between what they deem it to be intending to convey and what they actually experience emotionally when they listen to it (Schubert, 2007). This likely corresponds to the common experience of hearing a piece of music that seems to be trying too hard to impart a particular feeling, and that comes across as inauthentic or "cheesy" as a result.

People often report enjoying happy music more than sad, other things being equal (Thompson, Schellenberg & Husain, 2001). Happiness in Western music is typically conveyed by fast tempos and major keys, which probably explains the general preference for fast tempos across age groups (LeBlanc, Colman, McCrary et al., 1988). That said, it is well known that people enjoy listening to sad music, especially if it is made familiar through repeated exposure (Ali & Peynircioglu, 2010; Schellenberg et al., 2008). Moreover, peak emotional experiences in music are more often produced by sad music than happy, as discussed below. The enjoyment of sad music is often viewed as paradoxical, though it is perhaps no more so than the fact that people also enjoy sad films.

The emotional effects of music give it important functions in our lives. For instance, listeners frequently report using music for mood regulation (Thayer, Newman & McClain, 1994; Phillips, 1999). In some contexts, listeners who are put in a bad mood beforehand are more prone to listen to energetic and joyful music than listeners who are put in a positive emotional state (Knobloch & Zillman, 2002). However, there is also evidence that people placed in a sad mood (e.g. by watching a documentary about the last letters written home by soldiers killed in battle) are initially drawn to sad music, particularly if they describe themselves as prone to ruminating on negative emotions (Chen, Zhou & Bryant, 2007). It seems likely that mood regulation is a domain with substantial individual differences, reflected in how people interact with music.

One striking feature of popular music is the preponderance of love-related lyrical themes, often of a mournful variety. The intuitive notion that the romantically rejected are drawn to music describing unrequited or lost love is in fact born out experimentally. When people are given the option of freely sampling different pieces of music, those who describe themselves as romantically discontented prefer to listen to mournful, love-lamenting music sung by members of the same gender, whereas people in satisfying relationships prefer to listen to music that celebrates love (Knobloch, Weisbach & Zillman, 2004; Knobloch & Zillman, 2003). This is thus another instance where people seek music that is congruent with their emotional state, rather than using music to alter it.

Personality

Music preferences also depend on and convey personality. Standard personality trait assessments correlate with the music a person likes: people ranking high in "sensation seeking" prefer intense and arousing music (rock, punk, rap, etc.) over less arousing music such as soundtracks (Little & Zuckerman, 1986; McNamara & Ballard, 1999), extroverts prefer music with enhanced bass (McCown, Keiser, Mulhearn, & Williamson, 1997) and that is energetic and rhythmic (Rentfrow & Gosling, 2003), and people who rank high in "openness" tend to like music labeled as reflective and complex (Rentfrow & Gosling, 2003). There are also gender differences: bass enhancement is more popular with men than women (McCown et al., 1997). Stereotypes about what types of people like different types of music thus have some empirical validity (Rentfrow & Gosling, 2007). They also influence our interactions. Undergraduates believe that their music preferences reveal as much about themselves as their hobbies (Rentfrow & Gosling, 2003). They talk about music more than other topics when getting to know another person, and can use a person's music preferences to form accurate impressions of their personality (Rentfrow & Gosling, 2006).

Pleasurable Moments in Music

Even for preferred recordings of music, the pleasure we derive from listening varies considerably over the course of the piece, and can be assessed with continuously obtained ratings during listening. These rating trajectories are highly reliable for individual listeners, with the temporal pattern for a particular piece replicating across multiple presentations (Madsen, Britten & Capperella-Sheldon, 1993). Consistency is sometimes seen across listeners as well, though it is also common for a piece to evoke peak pleasure responses in one listener while not in another. The variations in pleasure appear to be partly due to variations in felt emotion. Although the emotions associated with great pleasure can be muted and relaxed (Gabrielsson, 2001), peak pleasure often occurs at moments of peak emotional arousal (Salimpoor et al., 2009), and these have received the most study thus far.

The moments at which listeners experience a peak aesthetic and emotional response are typically brief, lasting on the order of a few seconds. Many listeners report experiencing "chills" during such moments of pleasure – palpable physical sensations of arousal, such as goose bumps (Goldstein, 1980; Panksepp, 1995; Sloboda, 1991). Although chills evoked in other contexts are not always pleasant (they can be produced by fear, for instance), in musical contexts, the arousal they signal is typically pleasurable. They are not a rare phenomenon, but not everyone experiences them. Studies with random samples of non-musicians indicate that perhaps half of the general population experiences chills (Goldstein, 1980; Grewe, Nagel, Kopiez, & Altenmüller, 2007), with the experience being more common among people closely involved with music, such as music degree students.

Chills have been of particular interest because they provide a time-stamp for an emotional and aesthetic crescendo. They also provide an objective measure of the aesthetic experience, because of the associated physiological response – chills co-occur with peaks in measures of physiological arousal, such as galvanic skin response, heart rate, and respiration rate (Grewe, Kopiez & Altenmüller, 2009; Guhn, Hamm & Zentner, 2007; Rickard, 2004; Salimpoor et al., 2009). Consistent with a relation to musical enjoyment, chill-evoking music tends to be rated overall as aesthetically pleasing, and when people listen to such music, their reward pathways are activated (Blood & Zatorre, 2001). The response in the striatum (a key part of this pathway) covaries with the pleasure experienced during listening, and peaks during the experience of a chill, which is specifically accompanied by striatal dopamine release (Salimpoor et al., 2011).

What happens in music to cause these aesthetic peaks? There is consensus that most of the acoustical and musical correlates involve unexpected changes, be they to harmony, texture, pitch range, loudness, or the number of instruments or voices (Grewe et al., 2007; Guhn et al., 2007; Panksepp, 1995; Sloboda, 1991). Some such changes require implicit knowledge of musical "syntax" acquired through enculturation (e.g. to recognize that a chord is unexpected) but others involve basic acoustic variables like overall intensity, that could be detected at the earliest stages of the auditory system. Chills are more commonly evoked by sad or nostalgic music than happy (Panksepp, 1995), and by slow movements than fast, but the moments when they are induced are distinct from those that elicit tears, which are presumably less related to high arousal (Sloboda, 1991).

Although they often occur at moments where something unpredictable happens, the incidence of chills increases with exposure to a piece of music, at least for moderate amounts of exposure, as though learning the musical structure helps the listener recognize the critical chill-evoking deviations (Sloboda, 1991). These findings underscore the importance and paradoxical nature of expectation in our experience of music (Huron, 2006; Meyer, 1961; Narmour, 1990). Our aesthetic response seems to hinge on violations of the expectations induced by our knowledge of musical rules, yet repeated exposure enhances the response. It is as though the aesthetic response is driven by something that lacks direct access to our explicit memory (because the response to the expectation violation is enhanced even though consciously we know in advance what will happen), but that benefits from the enhanced structural representations attained from repeated exposure.

Open Issues in Music Aesthetics

The literature reviewed here reveals that many influences on our musical preferences can be verified and studied scientifically. However, science has yet to broach a number of the most intriguing aspects of musical aesthetics. For instance, we still know little about what sets a great recording apart from one that is merely good or passably competent. Beauty is to some extent in the eye of the beholder, and individuals certainly differ in their tastes. But within a culture and genre, there is often considerable agreement on what is great, embodied in the observation that nearly everyone likes "Kind of Blue," "Abbey Road," or "Songs in the Key of Life." Professional critics are not as correlated in their ratings of music as are individuals rating vocal (Zuckerman & Driver, 1989) or facial (Cunningham, Roberts, Barbee et al., 1995) attractiveness, but the correlations are still substantial (Lundy, 2010): ~0.5 for music versus ~0.9 for faces/voices.

Moreover, although music consumption, and critical assessments, are clearly affected by social influence (observations of what others choose to listen to), they are also constrained by the extremes of musical quality. When download choices in an artificial online music market were monitored in the presence of social influence, the very best songs (as measured via download choices in the absence of social influence) usually did well, and the very worst songs usually did poorly, even though the popularity of most other recordings was unpredictable (due to the instability that seems inherent to complex marketplace interactions) (Salganik, Dodds & Watts, 2006). Quality, at least as reflected in what people like, is thus measurable, and matters.

What, then, underlies the quality of exceptional works of music liked or loved by nearly everyone within a cultural group? Some of the factors reviewed earlier, namely complexity and unpredictability, are surely part of intrinsic aesthetic quality. But can they explain what differentiates Mozart or Bob Dylan from their peers? It is often said that great pieces of music strike a balance between originality and conformity to the rules of a genre, which sounds a lot like the peak of Berlyne's inverted U-shaped curve. But for this to be more than a vague intuition, we have to understand the relevant measure of complexity or predictability, and at present we lack such a formulation for realistic musical structure.

Apart from songwriting and composition, there are elusive variables that determine whether a particular recording of a composition turns out great rather than terrible. To realize how critical such factors can be, one has only to endure Britney Spears' cover of the Rolling Stones "Satisfaction," or to compare Otis Redding's competent but not quite spectacular version of "Respect" to Aretha Franklin's definitive rendition. These recordings are based on the same score, but the choices made in the recording studio, from the instrumental and vocal arrangement to the levels in the mix, make a vast difference in whether we are enchanted or unmoved by the result. Powerful perceptual and cognitive principles are at work, and represent important targets for future research.

CONCLUSIONS

Sounds can make us sigh in contentment, spend our time and money, or cringe in pain. We have reviewed some of the factors that cause us to react to sounds with pleasure or disgust. Some aspects of our auditory preferences can be explained by relatively simple acoustic properties, such as sharpness, roughness, or harmonicity. These preferences in some cases were likely shaped by evolution to help us avoid danger or select quality mates. Others may simply be flukes of the auditory system. Context matters, as does experience – we like things we have heard before. Music preferences additionally involve the interaction of personality traits and emotional content, aesthetic principles such as optimal complexity, and physiologically realized episodes of peak emotional arousal. Important aspects of musical aesthetics remain for the moment impenetrable, but they represent powerful effects that scientists will hopefully attempt to understand as experimental aesthetics proceeds in the coming decades.

Acknowledgments

The author thanks Marion Cousineau, Tali Sharot, and Gaye Soley for comments on an earlier draft of this chapter.

References

Abe, K., Ozawa, K., Suzuki, Y., & Sone, T. (1999). The effects of visual information on the impression of environmental sounds. *Inter-Noise, 99*, 1177–1182.

Ali, S. O., & Peynircioglu, Z. F. (2010). Intensity of emotion conveyed and elicited by familiar and unfamiliar music. *Music Perception, 27*(3), 177–182.

Apicella, C. L., Feinberg, D. R., & Marlowe, F. W. (2007). Voice pitch predicts reproductive success in male hunter-gatherers. *Biology Letters, 3*, 682–684.

Attias, H., & Schreiner, C. E. (1997). Temporal low-order statistics of natural sounds. In M. Mozer, M. Jordan, & T. Petsche (Eds.), *Advances in neural information processing* (Vol. 9). Cambridge: MIT Press.

Bachorowski, J., & Owren, M. J. (1999). Acoustic correlates of talker sex and individual talker identity are present in a short vowel segment produced in running speech. *Journal of the Acoustical Society of America, 106*(2), 1054–1063.

Beamen, J. (2000). *The violin explained: Components, mechanism, and sound*. Oxford: Oxford University Press.

Berlyne, D. E. (1971). *Aesthetics and psychobiology*. New York: Appleton.

Bidelman, G. M., & Krishnan, A. (2009). Neural correlates of consonance, dissonance, and the hierarchy of musical pitch in the human brainstem. *Journal of Neuroscience, 29*, 13165–13171.

Blood, A. J., & Zatorre, R. J. (2001). Intensely pleasurable responses to music correlate with activity in brain regions implicated in reward and emotion. *Proceedings of the National Academy of Sciences of the United States of America, 98*(20), 11818–11823.

Bornstein, R. F. (1989). Exposure and affect: Overview and meta-analysis of research, 1968–1987. *Psychological Bulletin, 106*(2), 265–289.

Bruckert, L., Bestelmeyer, P., Latinus, M., Rouger, J., Charest, I., & Rousselet, G. A., et al. (2010). Vocal attractiveness increases by averaging. *Current Biology, 20*(2), 116–120.

Bruner, G. C. (1990). Music, mood, and marketing. *J Market, 54*(4), 94–104.

Butler, J. W., & Daston, P. G. (1968). Musical consonance as musical preference: A cross-cultural study. *Journal of General Psychology, 79*, 129–142.

Cardozo, B. L., & van Lieshout, R. A. J. M. (1981). Estimates of annoyance of sounds of different character. *Applied Acoustics, 14*, 323–329.

Cazden, N. (1945). Musical consonance and dissonance: A cultural criterion. *Journal of Aesthetics and Art Criticism, 4*(1), 3–11.

Chen, L., Zhou, S., & Bryant, J. (2007). Temporal changes in mood repair through music consumption: Effects of mood, mood salience, and individual differences. *Media Psychology, 9*, 695–713.

Collins, S. A. (2000). Mens voices and women's choices. *Animal Behaviour, 60*, 773–780.

Collins, S. A., & Missing, C. (2003). Vocal and visual attractiveness are related in women. *Animal Behaviour, 65*, 997–1004.

Cox, T. J. (2008a). Scraping sounds and disgusting noises. *Applied Acoustics, 69*(12), 1195–1204.

Cox, T. J. (2008b). The effect of visual stimuli on the horribleness of awful sounds. *Applied Acoustics, 69*(8), 691–703.

Crate Kings (2007). *The 30 greatest hip hop drum breaks & samples of all Time!* Hip-Hop Samples, Beats, Drums, and Production News.

Cunningham, M. R., Roberts, R., Barbee, A. P., Druen, P., & Wu, C. (1995). Their ideas of beauty are, on the whole, the same as ours: Consistency and variability in the cross-cultural perception of female physical attractiveness. *Journal of Personality and Social Psychology, 68*, 261–279.

Cusick, S. G. (2008). "You are in a place that is out of the world...": Music in the detention camps of the "Global War on Terror". *Journal of the Society for American Music, 2*(1), 1–26.

Dalla Bella, S., Peretz, I., Rousseau, L., & Gosselin, N. (2001). A developmental study of the affective value of tempo and mode. *Cognition, 80*, B1–B10.

Ebeling, M. (2008). Neuronal periodicity detection as a basis for the perception of consonance: A mathematical model of tonal fusion. *Journal of the Acoustical Society of America, 124*(4), 2320–2329.

Fant, G. (1960). *Acoustic theory of speech production*. The Hague, Netherlands: Mouton.

Fastl, H. (1997). The psychoacoustics of sound-quality evaluation. *Acustica, 83*, 754–764.

Feinberg, D. R., DeBruine, L. M., Jones, B. C., & Perrett, D. I. (2008). The role of femininity and averageness of voice pitch in aesthetic judgments of women's voices. *Perception, 37*, 615–623.

Feinberg, D. R., Jones, B. C., Law Smith, M. J., Moore, F. R., DeBruine, R. E., & Cornwell, R. E., et al. (2006). Menstrual cycle, trait estrogen level, and masculinity preferences in the human voice. *Horticultural Research, 49*, 215–222.

Ferrand, C. T. (2002). Harmonics-to-noise ratio: An index of vocal aging. *Journal of Voice, 16*(4), 480–487.

Fitch, W. T. (1997). Vocal tract length and formant frequency dispersion correlate with body size in rhesus macaques. *Journal of the Acoustical Society of America, 102*(2), 1213–1222.

Fraccaro, P. J., Jones, B. C., Vukovic, J., Smith, G. F., Watkins, C. D., & Feinberg, D. R., et al. (2011). Experimental evidence that women speak in a higher voice pitch to men they find attractive. *Journal of Evolutionary Psychology, 9*, 57–67.

Fritz, T., Jentschke, S., Gosselin, N., Sammler, D., Peretz, I., & Turner, R., et al. (2009). Universal recognition of three basic emotions in music. *Current Biology, 19*, 1–4.

Fung, C. V. (1993). A review of studies on non-Western music preference. *Update: Applications of Research in Music Education, 12*, 26–32.

Gabrielsson, A. (2001). Emotions in strong experiences with music. In P. N. Juslin & J. A. Sloboda (Eds.), *Music and emotion: Theory and research* (pp. 431–449). Oxford: Oxford University Press.

Gardner, W. G. (1998). Reverberation algorithms. In M. Kahrs & K. Brandenburg (Eds.), *Applications of digital signal processing to audio and acoustics*. Norwell, MA: Kluwer Academic Publishers.

Geter, T., & Streisand, B. (1995). Recording sound sales: The music industry rocks and rolls to the newest financial rhythms. *US New and World Report, 70*, 67–68.

Gilliland, A., & Moore, H. (1924). The immediate and long-time effects of classical and popular phonograph selections. *Journal of Applied Psychology, 8*, 309–323.

Goldstein, A. (1980). Thrills in response to music and other stimuli. *Physiological Psychology, 8*(1), 126–129.

Grewe, O., Kopiez, R., & Altenmüller, E. (2009). The chill parameter: Goose bumps and shivers as promising measures in emotion research. *Music Perception, 27*(1), 61–74.

Grewe, O., Nagel, F., Kopiez, R., & Altenmüller, E. (2007). Listening to music as a re-creative process: Physiological, psychological, and psychoacoustical correlates of chills and strong emotions. *Music Perception, 24*(3), 297–314.

Guhn, M., Hamm, A., & Zentner, M. (2007). Physiological and musico-acoustic correlates of the chill response. *Music Perception, 24*(5), 473–483.

Halberstadt, J., & Rhodes, G. (2000). The attractiveness of nonface averages: Implications for an evolutionary explanation of the attractiveness of average faces. *Psychological Science, 11*(4), 285–289.

Halpern, D. L., Blake, R., & Hillenbrand, B. (1986). Psychoacoustics of a chilling sound. *Perception and Psychophysics, 39*(2), 77–80.

Hannon, E. E., & Trehub, S. E. (2005). Metrical categories in infancy and adulthood. *Psychological Science, 16*(1), 48–55.

Hargreaves, D. J. (1984). The effects of repetition on liking for music. *Journal of Research in Music Education, 32*(1), 35–47.

Hargreaves, D. J. (1987). Verbal and behavioral responses to familiar and unfamiliar music. *Current Psychological Research and Reviews, 6*(4), 323–330.

Hargreaves, D. J., Messerschmidt, P., & Rubert, C. (1980). Musical preference and evaluation. *Psychology of Music, 8*, 13–18.

Heingartner, A., & Hall, J. V. (1974). Affective consequences in adults and children of repeated exposure to auditory stimuli. *Journal of Personality and Social Psychology, 29*(6), 719–723.

Helmholtz, H. v. (1863). *Die lehre von den tonempfindungen als physiologische grundlage fur die theorie der musik*. Braunschweig, F. Vieweg und Sohn.

Henoch, M. A., & Chesky, K. (1999). Ear canal resonance as a risk factor in music-induced hearing loss. *Medical Problems of Performing Artists, 14*, 103–106.

Hevner, K. (1936). Experimental studies of the elements of expression in music. *American Journal of Psychology, 48*, 246–268.

Heyduk, R. G. (1975). Rated preference for musical compositions as it relates to complexity and exposure frequency. *Perception and Psychophysics, 17*(1), 84–91.

Hughes, S. M., Dispenza, F., & Gallup, G. G., Jr. (2004). Ratings of voice attractiveness predict sexual behavior and body configuration. *Evolution and Human Behavior, 25*, 295–304.

Hughes, S. M., Harrison, M. A., & Gallup, G. G., Jr. (2002). The sound of symmetry: Voice as a marker of developmental instability. *Evolution and Human Behavior, 23*, 173–180.

Huron, D. (1991). Tonal consonance versus tonal fusion in polyphonic sonorities. *Music Perception, 9*(2), 135–154.

Huron, D. (1994). Interval-class content in equally tempered pitch-class sets: Common scales exhibit optimum tonal consonance. *Music Perception, 11*(3), 289–305.

Huron, D. (2001). Is music an evolutionary adaptation?. *Annals New York Academy Sciences, 930*, 43–61.

Huron, D. (2006). *Sweet anticipation: Music and the psychology or expectation*. Cambridge, MA: MIT Press.

Hutchinson, W., & Knopoff, L. (1978). The acoustical component of western consonance. *Interface, 7*, 1–29.

Juslin, P. N., & Sloboda, J. A. (Eds.). (2010). *Handbook of music and emotion: Theory, research, applications*. Oxford: Oxford University Press.

Kameoka, A., & Kuriyagawa, M. (1969). Consonance theory. *Journal of the Acoustical Society of America, 45*, 1451–1469.

Knobloch, S., Weisbach, K., & Zillman, D. (2004). Love lamentation in pop songs: Music for unhappy lovers? *Zeitschrift fur Medienpsychologie, 16*(3), 116–124.

Knobloch, S., & Zillman, D. (2002). Mood management via the digital jukebox. *Journal of Communication, 52*, 351–366.

Knobloch, S., & Zillman, D. (2003). Appeal of love themes in popular music. *Psychological Reports, 93*, 653–658.

Knopper, S. (2010). *Appetite for self-destruction: The spectacular crash of the record industry in the digital age*. New York: Soft Skull Press.

Kumar, S., Forster, H. M., Bailey, P. J., & Griffiths, T. D. (2008). Mapping unpleasantness of sounds to their auditory representation. *Journal of the Acoustical Society of America, 124*(6), 3810–3817.

Langlois, J. H., & Roggman, L. A. (1990). Attractive faces are only average. *Psychological Science, 1*, 115–121.

LeBlanc, A., Colman, J., McCrary, J., Sherrill, C., & Malin, S. (1988). Tempo preferences of different age music listeners. *Journal of Research in Music Education, 36*(3), 156–168.

Little, P., & Zuckerman, M. (1986). Sensation seeking and music preferences. *Personality and Individual Differences, 7*(4), 575–578.

Lokki, T., Patynen, J., Tervo, S., Siltanen, S., & Savioja, L. (2011). Engaging concert hall acoustics is made up of temporal envelope preserving reflections. *Journal of the Acoustical Society of America, 129*(6), EL223–EL228.

Lundin, R. W. (1947). Toward a cultural theory of consonance. *Journal of Psychology, 23*, 45–49.

Lundy, D. E. (2010). A test of consensus in aesthetic evaluation among professional critics of modern music. *Empirical Studies of the Art, 28*(2), 243–258.

Lynch, M. P., Eilers, R. E., Oller, D., & Urbano, R. C. (1990). Innateness, experience, and music perception. *Psychological Science, 1*(4), 272–276.

Madsen, C. K., Britten, R. V., & Capperella-Sheldon, D. A. (1993). An empirical method for measuring the aesthetic experience to music. *Journal of Research in Music Education, 41*(1), 57–69.

Martindale, C., & Moore, K. (1989). Relationship of musical preference to collative, ecological, and psychophysical variables. *Music Perception, 6*(4), 431–446.

Mathews, M. V., & Pierce, J. R. (1980). Harmony and nonharmonic partials. *Journal of the Acoustical Society of America, 68*(5), 1252–1257.

McCown, W., Keiser, R., Mulhearn, S., & Williamson, D. (1997). The role of personality and gender in preference for exaggerated bass in music. *Personality and Individual Differences, 23*(4), 543–547.

McDermott, J. (2008). The evolution of music. *Nature, 453*, 287–288.

McDermott, J. H., & Oxenham, A. J. (2008). Music perception, pitch, and the auditory system. *Current Opinion in Neurobiology, 18*, 452–463.

McDermott, J. H., Oxenham, A. J., & Simoncelli, E. P. (2009). Sound texture synthesis via filter statistics. In *IEEE Workshop on applications of signal processing to audio and acoustics* (pp. 297–300). New Paltz, New York.

McDermott, J. H., Lehr, A. J., & Oxenham, A. J. (2010). Individual differences reveal the basis of consonance. *Current Biology, 20*, 1035–1041.

McDermott, J. H., & Simoncelli, E. P. (2011). Sound texture perception via statistics of the auditory periphery: Evidence from sound synthesis. *Neuron 71*, 926–940.

McNamara, L., & Ballard, M. E. (1999). Resting arousal, sensation seeking, and music preference. *Genetic Social and General Psychology Monographs, 125*(3), 229–250.

Menon, V., & Levitin, D. J. (2005). The rewards of music listening: Response and physiological connectivity of the mesolimbic system. *Neuroimage, 28*, 175–184.

Meyer, L. (1961). *Emotion and meaning in music.* Chicago: University of Chicago Press.

Milner, G. (2009). *Perfecting sound forever: An aural history of recorded music.* New York: Faber & Faber.

Miyake, K., & Zuckerman, M. (1993). Beyond personality impressions: Effects of physical and vocal attractiveness on false consensus, social comparison, affiliation, and assumed and perceived similarity. *Journal of Personality, 61*(3), 411–437.

Mull, H. (1957). The effect of repetition on enjoyment of modern music. *Journal of Psychology, 43*, 155–162.

Narmour, E. (1990). *The analysis and cognition of basic melodic structures: The implication-realization model.* Chicago: University of Chicago Press.

North, A. C., & Hargreaves, D. J. (1995). Subjective complexity, familiarity, and liking for popular music. *Psychomusicology, 14*, 77–93.

North, A. C., & Hargreaves, D. J. (1996). Responses to music in aerobic exercise and yogic relaxation classes. *British Journal of Psychology, 87*, 535–547.

North, A. C., & Hargreaves, D. J. (1998). Affective and evaluative responses to pop music. *Current Psychology, 17*(1), 102–108.

North, A. C., Hargreaves, D. J., & Hargreaves, J. J. (2004). Uses of music in everyday life. *Music Perception, 22*(1), 41–77.

North, A. C., Hargreaves, D. J., & O'Neill, S. A. (2000). The importance of music to adolescents. *British Journal of Educational Psychology, 70*, 255–272.

North, A. C., Shilcock, A., & Hargreaves, D. J. (2003). The effect of musical style on restaurant customers' spending. *Environment and Behavior, 35*, 712–718.

Orr, M. G., & Ohlsson, S. (2001). The relationship between musical complexity and liking in jazz and bluegrass. *Psychology of Music, 29*, 108–127.

Orr, M. G., & Ohlsson, S. (2005). Relationship between complexity and liking as a function of expertise. *Music Perception, 22*(4), 583–611.

Panksepp, J. (1995). The emotional sources of chills induced by music. *Music Perception, 13*(2), 171–207.

Patel, A. D. (2008). *Music, language, and the brain.* Oxford: Oxford University Press.

Peretz, I., & Zatorre, R. J. (Eds.). (2003). *The cognitive neuroscience of music.* New York: Oxford University Press.

Perrett, D. I., May, K. A., & Yoshikawa, S. (1994). Facial shape and judgements of female attractiveness. *Nature, 368*, 239–242.

Phillips, D. D. (1999). Adolescent uses of audio: Personal music and radio music. *Journal of Radio Studies, 6*(2), 222–235.

Pipitone, R. N., & Gallup, G. G. (2008). Women's voice attractiveness varies across the menstrual cycle. *Evolution and Human Behavior, 29*(4), 268–274.

Plomp, R., & Levelt, W. J. M. (1965). Tonal consonance and critical bandwidth. *Journal of the Acoustical Society of America, 38*, 548–560.

Radocy, R. E. (1982). Preference for classical music: A test for the hedgehog. In Psychology of music special issue: Proceedings of the ninth international seminar on research in music education (pp. 91–95).

Rentfrow, P. J., & Gosling, S. D. (2003). The do re mi's of everyday life: The structure and personality correlates of music preferences. *Journal of Personality and Social Psychology, 84*(6), 1236–1256.

Rentfrow, P. J., & Gosling, S. D. (2006). Message in a ballad: The role of music preferences in interpersonal perception. *Psychological Science, 17*(3), 236–242.

Rentfrow, P. J., & Gosling, S. D. (2007). The content and validity of music-genre stereotypes among college students. *Psychology of Music, 35*(2), 306–326.

Rickard, N. S. (2004). Intense emotional responses to music: A test of the physiological arousal hypothesis. *Psychology of Music, 32*(4), 371–388.

Salganik, M. J., Dodds, P. S., & Watts, D. J. (2006). Experimental study of inequality and unpredictability in an artificial cultural market. *Science, 311*, 854–856.

Salimpoor, V. N., Benovoy, M., Larcher, K., Dagher, A., & Zatorre, R. J. (2011). Anatomically distinct dopamine release during anticipation and experience of peak emotion to music. *Nature Neuroscience, 14*(2), 257–262.

Salimpoor, V. N., Benovoy, M., Longo, G., Cooperstock, J. R., & Zatorre, R. J. (2009). The rewarding aspects of music listening are related to degree of emotional arousal. *PLoS ONE, 4*(10), e7487.

Schellenberg, E. G., Peretz, I., & Vieillard, S. (2008). Liking for happy- and sad-sounding music: Effects of exposure. *Cognition and Emotion, 22*(2), 218–237.

Schloss, J. G. (2004). *Making beats: The art of sample-based hip-hop.* Hanover, CT: Wesleyan University Press.

Schubert, E. (2007). The influence of emotion, locus of emotion and familiarity upon preference in music. *Psychology of Music, 35*(3), 499–515.

Schubert, E. (2010). Affective, evaluative, and collative responses to hated and loved music. *Psychology of Aesthetics, Creativity, and the Arts, 4*(1), 36–46.

Schulkind, M. D. (2009). Is memory for music special? *Annals the New York Academy of Sciences, 1169*, 216–224.

Sethares, W. A. (1999). *Tuning, timbre, spectrum, scale.* Berlin: Springer.

Sharot, T., Delgado, M. R., & Phelps, E. A. (2004). How emotion enhances the feeling of remembering. *Nature Neuroscience, 7*(12), 1376–1380.

Singh, N. C., & Theunissen, F. E. (2003). Modulation spectra of natural sounds and ethological theories of auditory processing. *Journal of the Acoustical Society of America, 114*(6), 3394–3411.

Sloboda, J. A. (1991). Music structure and emotional response: Some empirical findings. *Psychology of Music, 19*, 110–120.

Smith, J. D., & Melara, R. J. (1990). Aesthetic prefernce and syntactic prototypicality in music: 'Tis the gift to be simple. *Cognition, 34*, 279–298.

Smith, K. C., & Cuddy, L. L. (1986). The pleasingness of melodic sequences: Contrasting effects of repetition and rule-familiarity. *Psychology of Music, 14*, 17–32.

Soley, G., & Hannon, E. E. (2010). Infants prefer the musical meter of their own culture: A cross-cultural comparison. *Developmental Psychology, 46*(1), 286–292.

Stumpf, C. (1890). *Tonpsychologie.* Leipzig: Verlag S. Hirzel.

Szpunar, K. K., Schellenberg, E. G., & Pliner, P. (2004). Liking and memory for musical stimuli as a function of exposure. *Journal of Experimental Psychology: Learning, Memory Cognition, 30*(2), 370–381.

Takao, H., Hashimoto, T., & Hatano, S. (1993). Quantification of subjective unpleasantness using roughness level. In *Noise and vibration conference & exposition.* Traverse City, MI, USA.

Talarico, J. M., & Rubin, D. C. (2003). Confidence, not consistency, characterizes flashbulb memories. *Psychological Science, 14*, 455–461.

Tan, S., Spackman, M. P., & Peaslee, C. L. (2006). The effects of repeated exposure on liking and judgments of musical unity of intact and patchwork compositions. *Music Perception, 23*(5), 407–421.

Terhardt, E. (1974a). On the perception of periodic sound fluctuations (roughness). *Acustica, 30*(4), 201–213.

Terhardt, E. (1974b). Pitch, consonance, and harmony. *Journal of the Acoustical Society of America, 55*, 1061–1069.

Terhardt, E., & Stoll, G. (1981). Skalierung des Wohlklangs von 17 Umweltschallen und Untersuchung der beteiligten Hörparameter. *Acustica, 48*, 247–253.

Thayer, R. E., Newman, J. R., & McClain, T. M. (1994). Self-regulation of mood: Strategies for changing a bad mood, raising energy, and reducing tension. *Journal of Personality and Social Psychology, 67*(5), 910–925.

Thompson, W. F., Schellenberg, E. G., & Husain, G. (2001). Arousal, mood and the Mozart effect. *Psychological Science, 12*(3), 248–251.

Thompson, W. F., & Balkwill, L. L. (2010). In P. N. Juslin & J. A. Sloboda (Eds.), *Cross-cultural similarities and differences. Music and Emotion* (pp. 755–788). Oxford: Oxford University Press.

Trainor, L. J., & Heinmiller, B. M. (1998). The development of evaluative responses to music: Infants prefer to listen to consonance over dissonance. *Infant Mental Health Journal, 21*(1), 77–88.

Trainor, L. J., Tsang, C. D., & Cheung, V. H. W. (2002). Preference for sensory consonance in 2- and 4-month-old infants. *Music Perception, 20*(2), 187–194.

Tramo, M. J., Cariani, P. A., Delgutte, B., & Braida, L. D. (2001). Neurobiological foundations for the theory of harmony in Western tonal music. *Annals the New York Academy of Sciences, 930*, 92–116.

Trehub, S. E., Schellenberg, E. G., & Kamenetsky, S. B. (1999). Infants' and adults' perception of scale structure. *Journal of Experimental Psychology Human Perception and Performance, 25*(4), 965–975.

Vassilakis, P. (2005). Auditory roughness as a means of musical expression. *Selected Reports in Ethnomusicology, 12*, 119–144.

Verveer, E. M., Barry, H., & Bousfield, W. A. (1933). Change in affectivity with repetition. *American Journal of Psychology, 45*(1), 130–134.

Vos, P., & Troost., J. (1989). Ascending and descending melodic intervals: Statistical findings and their perceptual relevance. *Music Perception, 6*, 383–396.

Voss, R. F., & Clarke, J. (1975). "1/f noise" in music and speech. *Nature, 258*, 317–318.

Vukovic, J., Feinberg, D. R., DeBruine, L., Smith, F. G., & Jones, B. C. (2010). Women's voice pitch is negatively correlated with health risk factors. *Journal of Evolutionary Psychology, 8*, 217–225.

Wallin, N. L., Merker, B., & Brown, S. (Eds.), (2001). *The origins of music*. Cambridge: MIT Press.

Yumoto, E., Gould, W. J., & Baer, T. (1982). Harmonics-to-noise ratio as an index of the degree of hoarseness. *Journal of the Acoustical Society of America, 71*(6), 1544–1550.

Zajonc, R. B. (1968). Attitudinal effects of mere exposure. *Journal of Personality and Social Psychology, 9*, 1–27.

Zentner, M. R., Grandjean, D., & Scherer, K. (2008). Emotions evoked by the sound of music: Characterization, classification, and measurement. *Emotion, 8*(4), 494–521.

Zentner, M. R., & Kagan, J. (1998). Infants' perception of consonance and dissonance in music. *Infant Mental Health Journal, 21*(3), 483–492.

Zuckerman, M., & Driver, R. E. (1989). What sounds beautiful is good: The vocal attractiveness stereotype. *Journal of Nonverbal Behavior, 13*(2), 67–82.

11

The Flexibility of Chemosensory Preferences

Géraldine Coppin[*] and David Sander

Swiss Center for Affective Sciences, and Laboratory for the Study of
Emotion Elicitation and Expression, University of Geneva

[*]Corresponding author.

INTRODUCTION

Immutability Versus Flexibility of Chemosensory Preferences

The title of this chapter – *"The flexibility of chemosensory preferences"* is constructed of three key terms that we need to define before moving forward: chemosensory, preferences and flexibility. *Chemosensory* preferences refer to preferences regarding odors, flavors and tastes (*a definition of these terms is provided in the glossary at the end of the chapter*). There are numerous definitions of the term *preferences* in the literature. Here, we will consider a preference towards X (X being a smell or a flavor) as an *"indicator of the subjective expected value of engaging in goal-directed behavior towards object X"* (Changizi & Shimojo, 2008). This definition includes both a component of valuation (liking X more than Y) and a behavioral tendency (in particular, approaching/withdrawing and choosing/rejecting). Finally, *flexibility* underlines the focus of this chapter on the changeable character of chemosensory preferences.

The central topic of this chapter will be how and to what extent chemosensory preferences can be modulated. However, before tackling this question, we will consider evidence that adopts the opposite perspective, i.e., in what way chemosensory preferences can be argued to be, at least partially, innate and hard wired (Steiner, 1979), and as such, rather inflexible. To do so, we will briefly review two main lines of evidence supporting this idea.

First, newborns display characteristic facial patterns as a function of hedonic variations of odors and tastes (Steiner, 1974, 1979). For example, in three-day-old infants, butyric acid, which is a rather unpleasant smell, elicits significantly more facial markers of disgust than vanillin, which is on average considered a pleasant smell (Soussignan, Schaal, Marlier & Jiang, 1997). These facial patterns are strongly homologous to those of other primate species, especially the species that are phylogenetically our closest relatives (Steiner, Glaser, Hawilo & Berridge, 2001). Chemosensory preferences could be shared across species and be partially genetically determined. For example, the lifelong taste preference for sweetness might have a genetically coded component (Keskitalo, Knaapila, Kallela et al., 2007). This suggests a genetically based and predetermined component of olfactory and gustatory preferences (see also the work of Mandairon, Poncelet, Bensafi & Didier, 2009, on olfactory preferences shared across mice and humans).

Second, some physicochemical properties of odorant molecules can predict how humans will perceive their hedonic character (Khan, Luk, Flinker et al., 2007). Moreover, such properties can allow an artificial nose to categorize the odors according to their pleasantness with high accuracy (Haddad, Medhanie, Roth et al., 2010). These results suggest that there is a predictable link between odorant structures and stimulus pleasantness.

Chemosensory preferences develop very early, as early as when a child is still in the womb, influenced by the mother's diet (Beauchamp & Mennella, 2009; Schaal, Marlier & Soussignan, 2000; Schaal, Soussignan & Marlier, 2002). Thus, early and regular exposure at the mother's breast to a given smell has an impact on a child's preference for the smell itself. There is also evidence that such exposure has an impact on a child's behavior at 7 and 21 months of age toward objects with the same smell (Delaunay-El Allam, Soussignan, Patris et al., 2010). Moreover, early exposure to a smell (Poncelet, Rinck, Bourgeat et al., 2010), such as first associations between an odor and an object, may have a *"privileged brain representation,"* in particular in terms of hippocampus activity (see Yeshurun, Lapid, Dudai & Sobel, 2009).

Despite this evidence, some authors have insisted that the olfactory system allows flexibility in dealing with the environment and high plasticity in responsiveness to odors (e.g., Engen, 1979, 1988). A significant body of evidence substantiates the claim of flexibility. For instance, cultural background and experience have been shown to matter more than genetics for preferences regarding sweet tastes (Mennella, Pepino & Reed, 2005). An appreciation of bitter tastes can be acquired across life, despite a newborn's reaction of disgust to these tastes (Mennella et al., 2005). The extent to which chemosensory preferences are flexible has been increasingly investigated not only at the behavioral level, but also at the cerebral level. Our perspective on the corpus of data on this research topic will be presented in greater detail below.

Aims and Structure of this Chapter

Here, we will review empirical results regarding the flexibility of chemosensory preferences, focusing on two aspects: (1) the nature of the influences that could modulate chemosensory preferences, and their neural underpinnings, as far as they are understood today; and (2) the generality of the mechanisms underlying the flexibility of chemosensory preferences, i.e., the degree to which the same mechanisms extend to other types of sensory preferences.

To do so, we will present the key factors that modulate chemosensory preferences. The valuation of any sensory stimulus depends on a number of factors, some of them shared across sensory modalities, others more tightly linked to specific sensory systems. We will restrict our attention to the factors that are most strongly linked with chemosensory processing. In particular, we will discuss modulatory factors as they relate to the three basic functions of the chemosensory system. These functions can be classified into three main categories: ingestion (e.g., appetite regulation); avoidance of environmental hazards (e.g., detection of microbial threats); and social communication (e.g., emotional contagion) (Stevenson, 2010).

This chapter is structured as follows. First, we briefly present how needs, goals and values relate to chemosensory preferences. Second, we emphasize the role of learning and exposure in the flexibility of chemosensory preferences. Third, we discuss the importance of other sensory (e.g., visual inputs), decision-making (i.e., choices) and cognitive (e.g., verbal labels) information in the flexibility of chemosensory preferences. Pertinent data on the neural underpinnings of the described phenomena are presented throughout the chapter, where applicable.

NEEDS, GOALS AND VALUES IN THE FLEXIBILITY OF CHEMOSENSORY PREFERENCES

As mentioned earlier, the chemosensory system is intimately linked to ingestion, avoidance of environmental hazards and social communication. As such, it provides the organism with assistance for answering questions such as: *"Shall I eat the rest of this dish that has been sitting in my fridge for two days, or should I cook something new?" "Shall I approach or stay away from this person?" "What self-image do I want to communicate by wearing a perfume during this romantic dinner?"*

The aim of the following part of this chapter is to provide insight into how needs, goals and values are able to modify preferences regarding smells and flavors. The role of the amygdala and the orbitofrontal cortex (OBC) in such modulation is also presented.

Impact of Needs, Goals and Values on Chemosensory Preferences

Needs and Chemosensory Preferences

Needs refer to a psychological entity assumed to arouse actions towards goals that a person would pursue for their satisfaction (Gendolla, 2009). For instance, the need for nourishment is intimately connected with the goal of food intake. Preferences regarding such a goal depend notably on both the quality and quantity of previously ingested food. Regarding the quantity of already ingested food, one study showed that even if chocolate was rated as pleasant at the beginning of an experiment, the more chocolate that participants consumed, the less the chocolate was rated as pleasant (Small, Zatorre, Dagher et al., 2001).

There is evidence that such preferences related to food intake may be coded by the activity of the OBC cortex. Its activity is related to the representation of the affective value of smells (Rolls, Grabenhorst & Parris, 2010; Small, Bender, Veldhuizen et al., 2007) and tastes (Rolls, Critchley,

Verhagen & Kadoshisa, 2010) and more generally to the affective value of stimuli, independently from their sensory modalities. Even more generally, there is evidence that OBC activity is related to social and monetary stimuli (Grabenhorst & Rolls, 2011; Rolls & Grabenhorst, 2008). Crucially for our point here, appetite modulates OBC cortex activity, the activity of which is decreased after consumption to satiety of chocolate (Small et al., 2001), bananas (O'Doherty, Rolls, Francis et al., 2000), tomato juice or chocolate milk (Kringelbach, O'Doherty, Rolls & Andrews, 2003). Regarding the quality of already ingested food, a phenomenon called *sensory-specific satiety* elicits modulation of food pleasantness. Thus, the decrease in pleasantness of the sensory properties of a food eaten to satiety (Rolls, Rolls, Rowe & Sweeney, 1981) is larger than the corresponding decrease for foods that have not been eaten. This phenomenon also applies to foods that share sensory properties of the eaten food (Rolls, Vanduijvenvoorde & Rolls, 1984).

Thus, the pleasantness of food items can be modulated by how much and which type of food has already been ingested. As appetite and food intake are by definition in constant change, preferences regarding food items appear to have a highly modulated character.

Goals and Chemosensory Preferences

Goals are another important factor for food intake preferences. You might, for example, eat tofu rather than meat if your current goal is to keep your vegetarian co-workers happy, even if you like meat more than tofu. The impact of diet on food consumption preferences has been experimentally studied, both at the behavioral and at the neural level. Results have shown that participants trying to control themselves chose unhealthy but tasty food items less frequently than did participants who were not trying to control themselves (Hare, Camerer & Rangel, 2009). Moreover, in a food consumption context, the decisions of dieters, as well as non-dieters, were related to the activity of the ventromedial prefrontal cortex; this activity correlated with the expected reward that was associated with the consumption of a given food. But in contrast to the decisions of non-dieters, dieters' decisions were also related to dorsolateral prefrontal cortex activity, which plays a role in self-control (Hare et al., 2009). The pursuit of a goal such as losing weight consequently modified food consumption preferences in dieters.

This impact of goals on chemosensory preferences is far from being an exception. Goals have been shown to impact preferences in many domains (see Warren, McGraw & Van Boven, 2011, for a review).

Values and Chemosensory Preferences

Values are defined as broad motivational constructs that determine what we consider important and which goals we choose to pursue (Rohan, 2000). For instance, the value of self-interest has been shown to impact the amount of money given during a charitable task, as well as the activity of the reward system while engaging in this activity (Brosch, Coppin, Scherer et al., 2011). But does the influence of values also extend to the domain of chemosensory preferences? The answer is yes: when there is a match between one's most important values and the value symbolized by a product, the product tastes better (Allen, Gupta & Monnier, 2008). To give a concrete example, we first need to point out that the consumption of red meat is correlated with the value of social power, while vegetables and fruits symbolize the rejection of social power (Allen & Ng, 2003). This correlation has been shown in three studies that measured human values together with attitudes towards different types of foods by means of questionnaires (Allen & Ng, 2003). Experimental evidence showed that participants who reject social power evaluated a vegetarian alternative to a sausage roll as more tasty and had a higher purchase intention. This effect was independent of the food they actually tried (Allen et al., 2008).

Here again, the impact of values on chemosensory preference is similar to results reported in other areas of research regarding preferences. For example, values have been shown to at least partially influence preferences for a relationship partner (Goodwin & Tinker, 2002).

Amygdala: A Central Structure in Needs-, Goals- and Values-Related Relevance

According to Gottfried (2010), "the function of sensory systems is optimized to detect and encode behaviorally relevant events (objects) that are encountered in the real world" (p. 637). Needs, goals and values all contribute to the relevance of a given chemosensory stimulus. In other words, a smell or a flavor can have a different significance across different individuals and different contexts, depending on current needs, goals and values. This differential importance of a smell or a flavor across individuals and contexts may lead to changeable preferences towards it, as discussed earlier.

In terms of neural underpinnings, the amygdala is known to be activated by odors (de Araujo, Rolls, Velazco et al., 2005; Gottfried, 2008). The amygdala has also been proposed to act as a "relevance detector" (Coppin & Sander, in press; Pessoa, 2010; Sander, Grafman & Zalla, 2003; Sander, 2009; Sander, in press). In the olfactory domain, highly aversive olfactory (Zald & Pardo, 1997) and gustatory (Zald, Lee, Fluegel & Pardo, 1998) stimulation can be considered very relevant – and has been shown to elicit amygdala activity. However, amygdala activity is not just related to aversive chemosensory stimulations. Winston, Gottfried, Kilner and Dolan (2005) have shown that the amygdala responds to a combination of valence and intensity, which could reflect the overall relevance of a given smell. In the gustatory domain, Small, Velduizen, Felsted et al.

(2008) suggested that the amygdala, together with the thalamus, may predict the meaningfulness or biological relevance of tastes and flavors.

What experimental evidence exists for this hypothesis? Amygdala activity was higher when participants were hungry compared with when they were not hungry during the visual presentation of food items (LaBar, Gitelman, Parrish et al., 2001). A similar pattern was found when participants were told to imagine being in a restaurant and choosing their favorite food from the menu (Hinton, Parkinson, Holland et al., 2004). The activation of the amygdala was also more pronounced when participants were reading food names that they particularly liked, compared with reading names of more neutral food (Arana, Parkinson, Hinton et al., 2003). By recording single neuron activity from the amygdala while participants were making purchase decisions about food items, Jenison, Rangel, Oya et al. (2011) showed that the amygdala response was linearly related to the value assigned to a given food. This representation seems to flexibly depend on the level of hunger of the participants. Both the amygdala and the OBC cortex responses to a predictive cue of the presentation of food-related smells were shown to decrease as a function of satiety (Gottfried, O'Doherty & Dolan, 2003).

Taken together, the findings suggest that amygdala activity is related to the importance of smells and flavors in a given context, and more generally, to the relevance of a stimulus or a situation (Sander et al., 2003). Note that relevance is highly flexible across time and space and can be acquired and modified. This may explain why some authors such as Köster (2002) think that it is absurd to ask people the question *"Why do you like this food?"* Why we like smelling or eating something at a particular point can depend on numerous and variable factors that we are not necessarily aware of.

In the next part of this chapter, we will discuss the role of learning and exposure in modulating, and even creating, preferences for smells and flavors.

THE ROLE OF LEARNING AND EXPOSURE FACTORS IN THE FLEXIBILITY OF CHEMOSENSORY PREFERENCES

The Role of Aversive and Appetitive Conditioning

Olfactory learning is crucial for allowing flexibility and adaptation in a given olfactory environment (Hudson & Distel, 2002). One way to achieve learning is by means of conditioning, which has been extensively studied in the context of preference creation and modulation (e.g., De Houwer, Thomas & Baeyens, 2001). Conditioning can allow the discrimination, both perceptually and cortically, of previously non-discriminable odor enantiomers, which are mirror-image molecules (Li, Howard, Parrish & Gottfried, 2008).

Aversive and appetitive conditioning also plays an important role in preference learning and modulation, both for olfactory and gustatory stimuli. Regarding aversive conditioning, the smell of eugenol, for example, is rated as unpleasant by participants who fear going to the dentist, probably by association with potentially painful dental treatment (e.g., Robin, Alaoui-Ismaïli, Dittmar & Vernet-Maury, 1999). In the gustatory domain, the power of aversive conditioning is even more impressive: a conditioned taste aversion requires only single-trial learning and works despite a long delay between a given taste and an illness (Bernstein, 1991; Garcia, Hankins & Rusiniak, 1974). Regarding appetitive conditioning, food preference reinforcement can have two sources: so-called *flavor–flavor* learning and *flavor–nutrient* learning (Ackroff, 2008). Flavor–flavor learning consists of the increased evaluation of flavors by association with already preferred flavors (such as a sweet flavor; e.g., Fanselow & Birk, 1982). Flavor–nutrient learning represents the increased hedonic character of flavors whose ingestion is followed by the pleasant effects of the nutrients. The cerebral structures underlying this second type of preference learning have recently been investigated in humans. The cerebral areas involved are the striatum, the amygdala and the medial OBC cortex (Fobbs, Veldhuizen, Douglas et al., 2011).

Learning includes – but is not restricted to – aversive and appetitive conditioning. The mere repetitive exposure of a smell or a flavor can be another important type of learning, as it leads to the creation or the modulation of its hedonic evaluations.

The Role of Exposure

The Creation of Chemosensory Perception

Androstenone is a particularly interesting molecule for investigating the creation of olfactory perception and preference. A large part of the population cannot smell this odorant (approximately 40% of individuals cannot perceive an odor when presented with androstenone; Boyle, Lundström, Knecht et al., 2006). When not perceived, "smelling" androstenone is rated as neither unpleasant nor pleasant. It is, however, possible to acquire the capacity to perceive androstenone after repeated exposure (e.g., Wysocki, Dorries & Beauchamp, 1989). Once this ability has developed, androstenone is on average perceived as unpleasant.

The Mere Exposure Effect with Olfactory Stimuli

The mere exposure effect (Zajonc, 1968) has been reported to affect preferences in a variety of settings. Its effect on smells is of particular interest, because the effect of repeated exposure to smells may be somewhat complementary to the survival relevance of quick and power-ful taste aversions (Delplanque, Grandjean, Chrea et al., 2008).

As in other sensory domains, a high correlation between familiarity and pleasantness has been reported in the olfactory domain (e.g., Cain & Johnson, 1978). If this is a causal relationship, then repeated exposure to a smell increases one's preference towards it. However, this relation-ship would not be true for all types of smells: an increased preference occurs only for initially neutral to positively evaluated odors, but not for negatively evaluated odors (Delplanque et al., 2008). This non-extension of the mere exposure effect to negative odors may be understood regard-ing the high relevance of malodors for survival. Given the importance of olfaction for ingestion behaviors (Stevenson, 2010), it would not be adap-tive to start liking, and possibly to have an appetite for or to approach, something that is potentially lethal just because of repeated exposures.

This selectiveness of the mere exposure effect has also been demon-strated in the case of interpersonal evaluations, depending on the reward versus punishment associated with seeing a particular person. The more a given person was seen, the more the person was liked, but only when he or she was associated with the delivery of a reward and not when associated with a punishment (Swap, 1977).

The Role of Culture and Social Appraisal

The frequency of exposure to different kinds of smells and foods is highly dependent on one's culture. Although common folk psychology suggests that *"there is no accounting for taste,"* it is in practice not uncom-mon for people to criticize one another's culinary tastes when these are perceived as bizarre – particularly across cultural divides. For example, anecdotally, the British humorously call the French "frogs" – because frog's legs are considered a delicacy in France, but not typically in Britain. Similarly, baby mouse wine (a bottle of rice wine containing the bodies of baby mice) may not sound as appealing to foreigners as it does to its consumers in China and Korea, where it is available.

In scientific research, culture has been demonstrated to be a pow-erful force in olfactory (Ayabe-Kanamura, Schicker, Laska et al., 1998; Ferdenzi, Schirmer, Roberts et al., in press; Moncrieff, 1966) and gusta-tory (Bourdieu, 1984; Wright, Nancarrow & Kwok, 2001) preferences. The smell of durian, a common fruit in Asia with a very powerful and characteristic aroma, evoked feelings of disgust in Geneva or Liverpool, while it was evaluated as mainly pleasant in Singapore (Ferdenzi et al., in press). Directly related to culture, identity might also be an impor-tant factor in the perception of identity-relevant smells. Thus, the olfac-tory perception of chocolate, for which Switzerland is world famous, is modulated by accessibility to the Swiss identity in Swiss participants (Coppin, Delplanque, Cayeux et al., 2011a).

The concept of social appraisal probably also plays a particularly important role here. Social appraisal (Manstead & Fischer, 2001) globally refers to social influence on the appraisal of a given stimulus or person. It has been shown to notably influence women's preferences for men. Thus, a man being looked at by a smiling woman is going to be perceived as more attractive than is a man being looked at by a woman with a neutral facial expression (Jones, DeBruine, Little et al., 2007). Similarly, in the chemosensory domain, seeing another person eating meat with a neutral or happy facial expression versus a disgusted facial expression affects the desire to eat such food (Rousset, Schlich, Chatonnier et al., 2008).

More generally, social context can impact food preferences. For example, the desire to eat decreased when an obese person was observed, independently of his or her facial expression. In contrast, the desire to eat increased when a person of normal weight was seen with a facial expression of pleasure (Barthomeuf, Rousset & Droit-Volet, 2010). This effect was not observed in children (Barthomeuf, Droit-Volet & Rousset, 2011). In children, the desire to eat seems to be more influenced by the eater's emotional facial expression than by his or her weight. Such results suggest flexibility in the relevance of social context factors for food preferences, as well as flexibility in food preferences.

The extent to which social factors modulate preferences generally is discussed by Campbell-Meilkejohn and Frith (Chapter 8 of this book).

In summary, learning, such as through conditioning or mere exposure, if broadly embedded in a given social context and culture, constitutes a very powerful way to modulate preferences for smells and flavors. The intrinsic emergence of chemosensory perception in an information-rich environment invites further consideration of the role of other sensory inputs and/or cognitive information in smell and food preferences. This is the topic of the following section of this chapter.

THE ROLE OF OTHER SENSORY INPUTS AND COGNITIVE FACTORS IN THE FLEXIBILITY OF CHEMOSENSORY PREFERENCES

Information from Other Sensory Modalities and Decision-Making Influences

Impact of Inputs from Other Sensory Modalities

Chemosensory perception occurs in a world full of simultaneous visual, auditory and tactile sensory inputs, which influence it. Olfactory detection is more rapid and more accurate when smells are presented with congruent visual cues (e.g., Gottfried & Dolan, 2003). In addition to mere perception, the interaction between the different sensory systems, such as the perceived match between a color and a smell, alters chemosensory pleasantness. For example, the smell of strawberry-flavored drinks is more pleasant when the drink is red than when it is green. The activity of caudal regions of the OBC, as well as those of the insular cortex, increases with the perceived congruency between a given color and an odor (Österbauer, Matthews, Jenkinson et al., 2005). Smells can also be associated with some abstract, visually presented symbols, and the congruency between a given smell and a given symbol can modify its pleasantness (Seo, Arshamian, Schemmer et al., 2010). In addition to visual information, the presentation of smells and foods is often associated with sounds. Auditory information may also play a role in the perception of olfactory pleasantness. A smell is perceived as being more pleasant when evaluated while listening to a congruent sound rather than an incongruent sound. Moreover, hearing a pleasant sound right before the presentation of a smell will increase the smell's perceived pleasantness (Seo & Hummel, 2011).

Impact of Decision-Making Processes

In the plethora of information available, that related to decision-making is particularly relevant for the discussion of the flexibility of preferences. The influence of choice on subsequent preferences (Brehm, 1956) is discussed in other chapters of this book (Johansson, Hall & Chater, Chapter 6; Sharot, Chapter 3) and will consequently not be developed here. It is, however, worth noting that post-choice preference modulation has also been reported for smells (Coppin, Delplanque, Cayeux et al., 2010) and tastes (Hall, Johansson, Tärning et al., 2010). After a choice between two similarly liked smells or tastes, the chosen one is rated as "more pleasant," and the rejected one as "less pleasant," in comparison to the ratings made before the choice. This modulation of olfactory preferences by choice may be implicit (Coppin et al., 2010) and remains stable even a week later (Coppin, Delplanque, Cayeux et al., 2011b).

Verbal Labeling and Expectations

Verbal Labels and Smells

The impact of verbal labels on olfactory-perceived pleasantness has been shown to be quite dramatic: the same odor is perceived as more pleasant when presented with a positive rather than a negative verbal label (Djordjevic, Lundstrom, Clément et al., 2008; Herz & von Clef, 2001; Herz, 2003). For example, when presented without a label, isovaleric acid is typically evaluated as highly unpleasant. Labeling this smell as "cheddar cheese" leads to more pleasant evaluations than does labeling it as "body odor" (de Araujo et al., 2005). Moreover, correlated with these pleasantness ratings, the rostral anterior cingulate cortex (ACC) and the medial OBC were more activated during the "cheddar cheese" label than during the "body odor" label, although the odor was identical (de Araujo et al., 2005). The activation of the ACC can be related to its supposed role in coding the subjective pleasantness of many types of stimuli and to link such a representation to goal-directed actions. These results suggest that the medial OBC may respond to pleasant smells, even when pleasantness is modulated by cognitive information. Grabenhorst, Rolls and Bilderbeck (2008) similarly found that a flavor labeled "boiled vegetables water" led to less pleasant evaluations than the same flavor labeled "rich and delicious flavor." The labels modulated the activity of the OBC, as well as the pregenual cingulate cortex, in response to flavors.

Expectations and Flavors

Expectations can be driven by many factors, but a very common one in everyday choices is price. Studies in this area have notably been conducted using wines. Wine is an interesting beverage because it is considered much more than simply a source of nutrition, being more related to culture, values and social status (Colman, 2008). While pleasantness between different wines was not significantly different when they were presented with no price information, pleasantness was correlated with price when the wines were presented with made-up prices. When the wines were presented with prices, pleasantness ratings were correlated with medial OBC activity (Plassman, O'Doherty, Shiv & Rangel, 2008). Expectations may therefore modulate the hedonic value of a wine via the activity of the OBC, whose role in hedonic representation was discussed earlier. Interestingly, during blind tastings, the correlation between price and the overall rating of a wine was small and negative. Such a result means that, on average, people enjoyed drinking the more expensive wine used in the study slightly *less*, when they did not know its price (Goldstein, Almenberg, Dreber et al., 2008).

Similarly, receiving positive or negative information from wine experts about a wine that is about to be tasted also influences its hedonic evaluation and the willingness to pay for a bottle of this wine (Siegrist & Cousin, 2009). It is important to point out that expectations seem to modulate preference by influencing the *tasting experience itself* (see Lee, Frederick & Ariely, 2006). This conclusion is drawn from the observation that hedonic evaluation and willingness to pay for a bottle of wine are not affected if the wine expert's information is given *after* the tasting.

The Influence of Brands on Beverage Preferences

The role played by brands has been an important topic in understanding the dynamics of chemosensory preferences. Nevid (1981) used carbonated water beverages of two different statuses to investigate how an advertisement could lead to a particular preference. He used a high-status (Perrier brand) and a low-status (Old Fashioned brand) beverage. His results suggest that the quality was evaluated as being better when the beverage was Perrier in comparison to Old Fashioned. Such a preference towards Perrier was not found when the brands were not presented.

More recently, McClure, Li, Tomlin et al. (2004) have focused on two very famous soda brands – Coke and Pepsi – and demonstrated their impact on chemosensory preferences. These two drinks are almost the same in terms of their chemical composition, but most people display a strong preference for one rather than the other. Behaviorally, results were very similar to what Nevid found with carbonated water beverages: when the two beverages were tasted with no information about the brand in a double-blind taste test, participants' preferences were split equally. In terms of neural underpinnings, two different systems seemed to be involved in preferences. When no brand information was available, the activity of the ventromedial prefrontal cortex was correlated with participants' preferences for the drinks. However, when brand information was available, the dorsolateral prefrontal cortex, hippocampus and midbrain also showed activation that correlated with participants' preferences for the drinks. These results raise the possibility that hedonic perception is modulated by prior affective experience. Consistent results have been obtained for car brands, where culturally familiar car logos have been shown to activate the medial prefrontal cortex (Schaefer, Berens, Heinze & Rotte, 2006). Thus, brands may lead to strong preferences towards one item rather than another, despite the absence of important differences in the product attributes.

CONCLUSION

In terms of neural underpinnings, several cerebral areas are known as generally important in chemosensory processing, such as the amygdala, the piriform cortex or the rostral entorhinal cortex (see Gottfried, 2010, for a review). Regarding the flexibility of chemosensory preferences more specifically, two cerebral regions seem to be particularly involved: the amygdala and the OBC cortex. The amygdala appears to act as a relevance detector (Sander et al., 2003) for stimuli that are particularly important in a given context, notably smells and flavors. The evidence further suggests that OBC cortex activity encodes the current hedonic value of a smell. Other structures such as the ACC also appear to be involved in the coding of olfactory and gustatory preferences, notably in directing goal-directed actions, such as approach or withdrawal.

The extent to which chemosensory preferences are fixed is highly debated in the literature. While some authors have argued that chemosensory pleasantness perception is to some extent predetermined (e.g., Khan et al., 2007), other authors have insisted that the intrinsic ambiguity of olfactory perception makes it more likely to be modulated by non-olfactory factors (e.g., Gottfried, 2008). According to the latter view, chemosensory preferences are related to physicochemical properties, but can be strongly modulated by factors such as those addressed in this chapter, in particular learning, exposure, needs, goals and values.

Glossary

Flavor Flavor perception is the result of the combination of different sensorial inputs: tastes, smells and oral somatosensory sensations (McBurney, 1986; Small, 2008b) and possibly of visual and auditory cues (Auvray & Spence, 2008).

Odor Odors have been defined as the *"perceived smells that emanate from an odorant or mixture of odorants,"* an odorant being *"a chemical stimulus that is capable of evoking a smell"* (Gottfried, 2010). As such, odors refer to subjective constructs and can consequently be modulated by many influences (Hudson & Distel, 2002).

Taste The term "taste" can be considered as a more commonly used word to refer to the concept of "flavor" (Small, 2008a).

References

Ackroff, K. (2008). Learned flavor preferences: The variable potency of post-oral nutrient reinforcers. *Appetite, 51*, 743–746. (doi: 10.1016/j.appet.2008.05.059).

Allen, M. W., Gupta, R., & Monnier, A. (2008). The interactive effect of cultural symbols and human values on taste evaluation. *Journal of Consumer Research, 35*, 294–308. (doi: 10.1086/590319).

Allen, M. W., & Ng, S. H. (2003). Human values, utilitarian benefits and identification: The case of meat. *European Journal of Social Psychology, 33*, 37–56. (doi: 10.1002/ejsp.128).

Arana, F. S., Parkinson, J. A., Hinton, E., Holland, A. J., Owen, A. M., & Roberts, A. C. (2003). Dissociable contributions of the human amygdala and orbitofrontal cortex to incentive motivation and goal selection. *Journal of Neuroscience, 23*, 9632–9638.

Auvray, M., & Spence, C. (2008). The multisensory perception of flavor. *Consciousness and Cognition, 17*, 1016–1031. (doi: 10.1016/j.concog.2007.06.005).

Ayabe-Kanamura, S., Schicker, I., Laska, M., Hudson, R., Distel, H., & Kobayakawa, T., et al. (1998). Differences in the perception of everyday odors – A Japanese-German cross-cultural study. *Chemical Senses, 23*, 31–38.

Barthomeuf, L., Rousset, S., & Droit-Volet, S. (2010). The desire to eat in the presence of obese or normal-weight Eaters as a function of their emotional facial expression. *Obesity, 18*, 719–724. (doi: 10.1038/oby.2009.357).

Barthomeuf, L., Droit-Volet, S., & Rousset, S. (2011). Differences in the desire to eat in children and adults in the presence of an obese eater. *Obesity, 19*, 939–945. (doi: 10.1038/oby.2011.26).

Beauchamp, G. K., & Mennella, J. A. (2009). Early flavor learning and its impact on later feeding behavior. *Journal of Pediatrics Gastroenterology and Nutrition, 48*, 25–30. (doi: 10.1097/MPG.0b013e31819774a5).

Bernstein, I. L. (1991). Flavor aversion. In T. V. Getchell, R. L. Doty, L. M. Bartoshuk & J. B. Snow (Eds.), *Smell and taste in health and disease* (pp. 417–428). New York: Raven Press.

Bourdieu, P. (1984). *Distinction: A social critique of the judgment of taste* (R. Nice, Trans.). London, UK: Routledge.

Boyle, J. A., Lundström, J., Knecht, M., Jones-Gotman, M., Schaal, B., & Hummel, T. (2006). On the trigeminal percept of androstenone and its implications on the rate of specific anosmia. *Journal of Neurobiology, 66*, 1501–1510. (doi: 10.1002/neu.20294).

Brehm, J. W. (1956). Post-decision changes in desirability of choice alternatives. *Journal of Abnormal Social Psychology, 52*, 384–389. (doi:10.1037/h0041006).

Brosch, T., Coppin, G., Scherer, K. R., Schwartz, S., & Sander, D. (2011). Generating value(s): Psychological value hierarchies reflect context-dependent sensitivity of the reward system. *Social Neuroscience, 6*, 198–208. (doi: 10.1080/17470919.2010.506754).

Cain, W. S., & Johnson, F. (1978). Lability of odor pleasantness: Influence of mere exposure. *Perception, 7*, 459–465. (doi: 10.1068/p070459).

Changizi, M. A., & Shimojo, S. (2008). A functional explanation for the effects of visual exposure on preference. *Perception, 37*, 1510–1519. (doi: 10.1068/p6012).

Colman, T. (2008). Wine politics: How governments, environmentalists: *Mobsters, and critics influence the wines we drink*. California: University of California Press.

Coppin, G., Delplanque, S., Cayeux, I., Porcherot, C., & Sander, D. (2010). I'm no longer torn after choice: How explicit choices can implicitly shape preferences for odors. *Psychological Science, 21*, 489–493. (doi:10.1177/0956797610364115).

Coppin, G., Delplanque, S., Cayeux, I., Porcherot, C., Margot, C., Velazco, M. I., et al. (2011a). Swiss identity smells like chocolate: Accessible social identities shape olfactory experience (in preparation).

Coppin, G., Delplanque, S., Cayeux, I., Porcherot, C., & Sander, D. (2011b). When flexibility is stable: Implicit long-term shaping of olfactory preferences (in preparation).

Coppin, G., & Sander, D. (in press). Neuropsychologie affective et Olfaction: Etudier la sensibilité de l'amygdale aux odeurs pour tester les théories de l'émotion [Affective neuropsychology and olfaction: Investigate amygdala sensitivity to smells to test theories of emotion]. In B. Schaal, Ferdenzi, C. & O. Wathelet (Eds.), *Odeurs et émotions [Smells and emotions]*. Dijon: Presses Universitaires de Dijon.

de Araujo, I. E., Rolls, E. T., Velazco, M. I., Margot, C., & Cayeux, I. (2005). Cognitive modulation of olfactory processing. *Neuron, 46*, 671–679. (doi: 10.1016/j.neuron.2005.04.021).

De Houwer, J., Thomas, S., & Baeyens, F. (2001). Associative learning of likes and dislikes: A review of 25 years of research on human evaluative conditioning. *Psychological Bulletin, 127*, 853–859. (doi: 10.1037//0033-2909.127.6.853).

Delaunay-El Allam, M., Soussignan, R., Patris, B., Marlier, L., & Schaal, B. (2010). Long-lasting memory for an odor acquired at the mother's breast. *Developmental science, 13*, 849–863. (doi: 10.1111/j.1467-7687.2009.00941.x).

Delplanque, S., Grandjean, D., Chrea, C., Aymard, L., Cayeux, I., & Le Calvé, B., et al. (2008). Emotional processing of odours: Evidence for a non-linear relation between pleasantness and familiarity evaluations. *Chemical Senses, 33*, 469–479. (doi:10.1093/chemse/bjn014).

Djordjevic, J., Lundstrom, J. N., Clément, F., Boyle, J. A., Pouliot, S., & Jones-Gotman, M. (2008). A rose by another name: Would it smell as sweet? *Journal of Neurophysiology, 99*, 386–393. (doi: 10.1152/jn.00896.2007).

Engen, T. (1979). The origin of preferences in taste and smell. In J. H. A. Kroeze (Ed.), *Preference behaviour and chemoreception* (pp. 263–273). London: Information Retrieval.

Engen, T. (1988). The acquisition of odour hedonics. In S. Van Toller & G. H. Dodd (Eds.), *Perfumery: The psychology and biology of fragrance* (pp. 79–90). London, UK: Chapman and Hall.

Fanselow, M. S., & Birk, J. (1982). Flavor-flavor associations induce hedonic shifts in taste preference. *Learning and Behavior: A Psychonomic Society Publication, 10*, 223–228. (doi: 10.3758/BF03212273).

Ferdenzi, C., Schirmer, A., Roberts, S. C., Delplanque, S., Porcherot, C., Cayeux, I., et al. (in press). Affective dimensions of odor perception: A comparison between swiss, british and singaporean populations,Emotion PMID, 21534667. (epub ahead of print).

Fobbs, W., Veldhuizen, M., Douglas, D. M., Lin, T., Yeomans, M., Flammer, L., et al. (2011). Neural correlates of flavor-nutrient conditioning in humans.

Garcia, J., Hankins, W. G., & Rusiniak, K. W. (1974). Behavioral regulation of the milieu interne in man and rat. *Science, 185*, 824–831. (doi: 10.1126/science.185.4154.824).

Gendolla, G. H. E. (2009). Needs. In D. Sander & K. R. Scherer (Eds.), *The Oxford companion to emotion and the affective sciences* (pp. 273–274). New York and Oxford: Oxford University Press.

Goldstein, R., Almenberg, J., Dreber, A., Emerson, J. W., Herschkowitsch, A., & Katz, J. (2008). Do more expensive wines taste better? Evidence from a large sample of blind tastings. *Journal of Wine Economics, 3*, 1–9.

Goodwin, R., & Tinker, M. (2002). Value priorities and preferences for a relationship partner. *Pers Ind Differ, 32*, 1339–1349. (doi: 10.1016/S0191-8869(01)99122-2).

Gottfried, J. A. (2008). Perceptual and neural plasticity of odor quality coding in the human brain. *Chemosensory Perception, 1*, 127–135. (doi: 10.1007/s12078-008-9017-1).

Gottfried, J. A. (2010). Central mechanisms of odour object perception. *Nature Reviews Neuroscience, 11*, 628–641. (doi: 10.1038/nrn2883).

Gottfried, J. A., & Dolan, R. J. (2003). The nose smells what the eye sees: crossmodal visual facilitation of human olfactory perception. *Neuron, 17*, 375–386. (doi: 10.1016/S0896-6273(03)00392-1).

Gottfried, J. A., O'Doherty, J., & Dolan, R. J. (2003). Encoding predictive reward value in human amygdala and orbitofrontal cortex. *Science, 301*, 1104–1107.

Grabenhorst, F., Rolls, E. T., & Bilderbeck, A. (2008). How cognition modulates affective responses to taste and flavor: Top-down influences on the orbitofrontal and pregenual cingulate cortices. *Cerebral Cortex, 18*, 1549–1559. (doi: 10.1093/cercor/bhm185).

Grabenhorst, F., & Rolls, E. T. (2011). Value, pleasure and choice in the ventral prefrontal cortex. *Trends in Cognitive Sciences, 15*, 56–67. (doi: 10.1016/j.tics.2010.12.004).

Haddad, R., Medhanie, A., Roth, Y., Harel, D., & Sobel, N. (2010). Predicting odor pleasantness with an electronic nose. *PLoS Computational Biology, 6*, e1000740. (doi: 10.1371/journal.pcbi.1000740).

Hall, L., Johansson, P., Tärning, B., Sikström, S., & Deutgen, T. (2010). Magic at the marketplace: Choice blindness for the taste of jam and the smell of tea. *Cognition, 117*, 54–61. (Elsevier B.V. doi: 10.1016/j.cognition.2010.06.010).

Hare, T. A., Camerer, C. F., & Rangel, A. (2009). Self-control in decision-making involves modulation of the vmPFC valuation system. *Science, 324*, 646–648. (doi: 10.1126/science.1168450).

Herz, R. S. (2003). The effect of verbal context on olfactory perception. *Journal of Experimental Psychology: General, 132*, 595–606. (doi: 10.1037/0096-3445.132.4.595).

Herz, R. S., & von Clef, J. (2001). The influence of verbal labeling on the perception of odors: Evidence for olfactory illusions? *Perception, 30*, 381–391. (doi: 10.1068/p3179).

Hinton, E. C., Parkinson, J. A., Holland, A. J., Arana, F. S., Roberts, A. C., & Owen, A. M. (2004). Neural contribution to the motivational control of appetite in humans. *European Journal of Neuroscience, 20*, 1411–1418. (doi: 10.1111/j.1460-9568.2004.03589.x).

Hudson, R., & Distel, H. (2002). The individuality of odor perception. In C. Rouby, B. Schaal, D. Dubois, R. Gervais & A. Holley (Eds.), *Olfaction, taste, and cognition* (pp. 408–420). Cambridge: Cambridge University Press.

Jenison, R. L., Rangel, A., Oya, H., Kawasaki, H., & Howard, A. (2011). Value encoding in single neurons in the human amygdala during decision making. *Journal of Neuroscience, 31*, 331–338. (doi: 10.1523/JNEUROSCI.4461-10.2011).

Jones, B. C., DeBruine, L. M., Little, A. C., Burriss, R. P., & Feinberg, D. R. (2007). Social transmission of face preferences among humans. *Proceedings Biological Sciences/The Royal Society, 274*, 899–903. (doi: 10.1098/rspb.2006.0205).

Keskitalo, K., Knaapila, A., Kallela, M., Palotie, A., Wessman, M., & Sammalisto, S., et al. (2007). Sweet taste preferences are partly genetically determined: identification of a trait locus on chromosome 161-3. *American Journal of Clinical Nutrition, 86*, 55–63.

Khan, R. M., Luk, C. H., Flinker, A., Aggarwal, A., Lapid, H., & Haddad, R., et al. (2007). Predicting odour pleasantness from odorant structure: Pleasantness as a reflection of the physical world. *Journal of Neuroscience, 27,* 10015–10023. (doi: 10.1523/JNEUROSCI.1158-07.2007).

Köster, E. G. (2002). The specific characteristics of the sense of smell. In C. Rouby, B. Schaal, D. Dubois, R. Gervais & A. Holley (Eds.), *Olfaction, taste, and cognition* (pp. 27–43). Cambridge: Cambridge University Press.

Kringelbach, M. L., O'Doherty, J., Rolls, E. T., & Andrews, C. (2003). Activation of the human orbitofrontal cortex to a liquid food stimulus is correlated with subjective pleasantness. *Cerebral Cortex, 13,* 1064–1071. (doi: 10.1093/cercor/13.10.1064).

LaBar, K. S., Gitelman, D. R., Parrish, T. B., Kim, Y. H., Nobre, A., & Mesulam, M. M. (2001). Hunger selectivity modulates corticolimbic activation to food stimuli in humans. *Behavioral Neuroscience, 115,* 493–500. (doi:10.1037//0735-7044.115.2.493).

Lee, L., Frederick, S., & Ariely, D. (2006). Try it, you'll like it. The influence of expectation, consumption, and revelation on preferences for beer. *Psychological Science, 17,* 1054–1058. (doi: 10.111/j.1467-9280.2006.01829.x).

Li, W., Howard, J. D., Parrish, T., & Gottfried, J. A. (2008). Aversive learning enhances perceptual and cortical discrimination of initially indiscriminable odor cues. *Science, 319,* 1842–1845. (doi: 10.1126/science.1152837).

Mandairon, N., Poncelet, J., Bensafi, M., & Didier, A. (2009). Humans and mice express similar olfactory preferences. *Plos One, 4,* e4209. (doi: 10.1371/journal.pone.0004209).

Manstead, A. S. R., & Fischer, A. H. (2001). Social appraisal: The social world as object and influence on appraisal processes. In K. R. Scherer, A. Schorr & T. Johnstone (Eds.), *Appraisal processes in emotion: Theory, method, research* (pp. 221–232). New York, NY: Oxford University Press.

McBurney, D. H. (1986). Taste, smell and flavor terminology: taking the confusion out of confusion. In H. L. Meiselman & R. S. Riykin (Eds.), *Clinical measurement of taste and smell* (pp. 117–124). New York: Macmillan.

McClure, S. M., Li, J., Tomlin, D., Cypert, K. S., Montague, L. M., & Montague, P. R. (2004). Neural correlates of behavioral preference for culturally familiar drinks. *Neuron, 44,* 379–387. (doi: 10.1016/j.neuron.2004.09.019).

Mennella, J. A., Pepino, M. Y., & Reed, D. R. (2005). Genetic and environmental determinants of bitter perception and sweet preferences. *Pediatrics, 115,* e216–222. (doi: 10.1542/peds.2004-1582).

Moncrieff, R. W. (1966). *Odour preferences.* New York: Wiley.

Nevid, J. S. (1981). Effects of brand labeling on ratings of product quality. *Perceptual and Motor Skills, 53,* 407–410.

O'Doherty, J., Rolls, E. T., Francis, S., Bowtell, R., McGlone, F., & Kobal, G., et al. (2000). Sensory-specific satiety-related olfactory activation of the human orbitofrontal cortex. *Neuro Report, 11,* 893–897.

Österbauer, R. A., Matthews, P. M., Jenkinson, M., Beckmann, C. F., Hansen, P. C., & Calvert, G. A. (2005). Color of scents: Chromatic stimuli modulate odor responses in the human brain. *Journal of Neurophysiology, 93,* 3434–3441. (doi: 10.1152/jn.00555.2004).

Pessoa, L. (2010). Emotion and cognition and the amygdala: From "what is it?" to "what's to be done?". *Neuropsychologia, 48,* 3416–3429. (doi: 10.1016/j.neuropsychologia.2010.06.038).

Plassmann, H., O'Doherty, J., Shiv, B., & Rangel, A. (2008). Marketing actions can modulate neural representations of experienced pleasantness. *Proceedings of the National Academy of Sciences of the United States of America, 105,* 1050–1054. (doi: 10.1073/pnas.0706929105).

Poncelet, J., Rinck, F., Bourgeat, F., Schaal, B., Rouby, C., & Bensafi, M., et al. (2010). The effect of early experience on odor perception in humans: psychological and physiological correlates. *Behavioural Brain Research, 208,* 458–465. (doi: 10.1016/j.bbr.2009.12.011).

Robin, O., Alaoui-Ismaïli, O., Dittmar, A., & Vernet-Maury, E. (1999). Basic emotions evoked by eugenol odor differ according to the dental experience. A neurovegetative analysis. *Chemical Senses, 24*, 327–335. (doi: 10.1093/chemse/24.3.327).

Rohan, M. J. (2000). A rose by any name? The values construct. *Personality and Social Psychology Review, 4*, 255–277. (doi: 10.1207/S15327957PSPR0403_4).

Rolls, E. T., Critchley, H. D., Verhagen, J. V., & Kadohisa, M. (2010). The representation of information about taste and odor in the orbitofrontal cortex. *Chemosensory Perception, 3*, 16–33. (doi: 10.1007/s12078-009-9054-4).

Rolls, E. T., & Grabenhorst, F. (2008). The orbitofrontal cortex and beyond: from affect to decision-making. *Progress in Neurobiology, 86*, 216–244. (doi: 101.1016/j.pneurobio.2008.09.001).

Rolls, E. T., Grabenhorst, F., & Parris, B. (2010). Neural systems underlying decisions about affective odors. *Journal of Cognitive Neuroscience, 22*, 1069–1082. (doi: 10.1162/jocn.2009.21231).

Rolls, B. J., Rolls, E. T., Rowe, E. A., & Sweeney, K. (1981). Sensory specific satiety in man. *Physiology and Behavior, 27*, 137–142. (doi: 10.1016/0031-9384(81)90310-3).

Rolls, E. T., Vanduijvenvoorde, P. M., & Rolls, E. T. (1984). Pleasantness changes and food-intake in a varied 4-course meal. *Appetite, 5*, 337–348.

Rousset, S., Schlich, P., Chatonnier, A., Barthomeuf, L., & Droit-Volet, S. (2008). Is the desire to eat familiar and unfamiliar meat products influenced by the emotions expressed on eater's faces? *Appetite, 50*, 110–119. (doi: 10.1016/j.appet.200.06.005).

Sander, D. (2009). The amygdala. In D. Sander & K. R. Scherer (Eds.), *The Oxford companion to emotion and the affective sciences* (pp. 28–32). New York and Oxford: Oxford University Press.

Sander, D. (in press). An affective neuroscience approach to models of emotion. In J. L. Armony, & P. Vuilleumier (Eds.), *Handbook of human affective neuroscience*. Cambridge, UK: Cambridge University Press.

Sander, D., Grafman, J., & Zalla, T. (2003). The human amygdala: an evolved system for relevance detection. *Reviews in the Neurosciences, 14*, 303–316.

Schaal, B., Marlier, L., & Soussignan, R. (2000). Human foetuses learn odours from their pregnant mother's diet. *Chemical Senses, 25*, 729–737. (doi: 10.1093/chemse/25.6.729).

Schaal, B., Soussignan, R., & Marlier, L. (2002). Olfactory cognition at the start of life: the perinatal shaping of selective odor responsiveness. In C. Rouby, B. Schaal, D. Dubois, R. Gervais & A. Holley (Eds.), *Olfaction, taste, and cognition* (pp. 421–440). Cambridge: Cambridge University Press.

Schaefer, M., Berens, H., Heinze, H. J., & Rotte, M. (2006). Neural correlates of culturally familiar brands of car manufacturers. *NeuroImage, 31*, 861–865. (doi: 10.1016/j.neuroimage.2005.12.047).

Seo, H. S., Arshamian, A., Schemmer, K., Scheer, I., Sander, T., & Ritter, G., et al. (2010). Cross-modal integration between odors and abstract symbols. *Neuroscience Letters, 478*, 175–178. (doi: 10.1016/j.neulet.2010.05.011).

Seo, H. S., & Hummel, T. (2011). Auditory-olfactory integration: Congruent or pleasant sounds amplify odor pleasantness. *Chemical Senses, 36*, 301–309. (doi: 10.1093/chemse/bjq129).

Siegrist, M., & Cousin, M. E. (2009). Expectations influence sensory experience in a wine tasting. *Appetite, 52*, 762–765. (doi: 10.1016/j.appet.2009.02.002).

Small, D. M. (2008). How does food's appearance or smell influence the way it tastes? *Scientific American, 299*, 100. (doi :10.1038/scientificamerican0708-100).

Small, D. M. (2008). Flavor and the formation of category-specific processing in olfaction. *Chemosensory Perception, 1*, 136–146. (doi: 10.1007/s12078-008-9015-3).

Small, D. M., Bender, G., Veldhuizen, M. G., Rudenga, K., Nachtigal, D., & Felsted, J. (2007). The role of the human orbitofrontal cortex in taste and flavor processing. *Annals New York Academy Sciences, 1121*, 136–151. (doi: 10.1196/annals.1401.002).

Small, D. M., Veldhuizen, M. G., Felsted, J., Mak, Y. E., & McGlone, F. (2008). Separable substrates for anticipatory and consummatory food chemosensation. *Neuron, 57*, 786–797. (doi: 10.1016/j.neuron.2008.01.021).

Small, D. M., Zatorre, R. J., Dagher, A., Evans, A. C., & Jones-Gotman, M. (2001). Changes in brain activity related to eating chocolate: from pleasure to aversion. *Brain, 124,* 1720–1733. (doi: 10.1093/brain/124.9.1720).

Soussignan, R., Schaal, B., Marlier, L., & Jiang, T. (1997). Facial and autonomic responses to biological and artificial olfactory stimuli in human neonates: Re-examining early hedonic discrimination of odours. *Physiology and Behavior, 62,* 745–758. (doi: 10.1016/S0031-9384(97)00187-X).

Steiner, J. E. (1974). Innate discriminative human facial expression to taste and smell stimulation. *Annals New York Academy Sciences, 237,* 229–233. (doi: 10.1111/j.1749-6632.1974.tb49858.x).

Steiner, J. E. (1979). Human facial expressions in response to taste and smell stimulation. *Advances in Child Development and Behavior, 13,* 257–295. (doi: 10.1016/S0065-2407(08)60349-3).

Steiner, J. E., Glaser, D., Hawilo, M. E., & Berridge, K. C. (2001). Comparative expression of hedonic impact: affective reactions to taste by human infants and other primates. *Neuroscience and Biobehavioral Reviews, 25,* 53–74. (doi: 10.1016/S0149-7634(00)00051-8).

Stevenson, R. J. (2010). An initial evaluation of the functions of human olfaction. *Chemical Senses, 35,* 3–20. (doi: 10.1093/chemse/bjp083).

Swap, W. C. (1977). Interpersonal attractions and repeated exposure to rewarders and punishers. *Personality And Social Psychology Bulletin, 3,* 248–251. (doi: 10.1177/014616727700300219).

Warren, C., McGraw, A. P., & Van Boven, L. (2011). Values and preferences: defining preference construction. *Wiley Interdiscip Rev Cogn Sci, 21,* 1438–1445.

Winston, J. S., Gottfried, J. A., Kilner, J. M., & Dolan, R. J. (2005). Integrated neural representations of odor intensity and affective valence in human amygdala. *Journal of Neuroscience, 25,* 8903–8907. (doi: 10.1523/JNEUROSCI.1569-05-2005).

Wright, L. T., Nancarrow, C., & Kwok, P. M. (2001). Food taste preferences and cultural influences on consumption. *British Food Journal, 103,* 348–357. (doi: 10.1108/00070700110396321).

Wysocki, C. J., Dorries, K. M., & Beauchamp, G. K. (1989). Ability to perceive androstenone can be acquired by ostensibly anosmic people. *Proceedings of the National Academy of Sciences of the United States of America, 86,* 7976–7978. (doi: 10.1073/pnas.86.20.7976).

Yeshurun, Y., Lapid, H., Dudai, Y., & Sobel, N. (2009). The privileged brain representation of first olfactory associations. *Current Biology, 19,* 1869–1874. (doi: 10.1016/j.cub.2009.09.066).

Zajonc, R. B. (1968). Attitudinal effects of mere exposure. *Journal of Personality and Social Psychology, 9,* 1–27. (doi:10.1037/h0025848).

Zald, D. H., Lee, J. T., Fluegel, K. W., & Pardo, J. V. (1998). Aversive gustatory stimulation activates limbic circuits in humans. *Brain, 121,* 1143–1154. (doi: 10.1093/brain/121.61143).

Zald, D. H., & Pardo, J. V. (1997). Emotion, olfaction, and the human amygdala: Amygdala activation during aversive olfactory stimulation. *Proceedings of the National Academy of Sciences of the United States of America, 94,* 4119–4124. (doi: 10.1073/pnas.94.8.4119).

12

Dynamic Preference Formation via Gaze and Memory

Hsin-I Liao[1] and Shinsuke Shimojo[2]

[1]Department of Psychology, National Taiwan University, Taipei, Taiwan
[2]Division of Biology, Computation and Neural Systems, California Institute of Technology, Pasadena, CA, USA

GAZE AND PREFERENCE

Eye gaze has distinctive functions in a variety of species in the wilderness. In humans, even the newborn, it has an intrinsic link to social preference decisions (Emery, 2000; Fantz, 1961).

Gaze shifting is just one example of a bodily orienting mechanism. Other examples may include head turning, body turning, turning of toes, finger pointing, and weight shifting. Orienting mechanisms serve significantly for communication (e.g., joint attention, Laughlin, 1968), not only to achieve cognitive goals, but also to share emotions. It starts early in development; for example, infants/toddlers/children show their intention and preference by pointing at things. In the laboratory, infant studies employ the preferential-looking paradigm to examine infants' preferences for various visual objects by observing their head and eye

Neuroscience of Preference and Choice
DOI: 10.1016/B978-0-12-381431-9.00022-X

movements (e.g., Fantz, 1961, 1963; Teller, 1979). The approaching behavior is considered to reflect the underlying mechanism for preference.

Shimojo, Simion, Shimojo and Scheier (2003) demonstrated what they call the "gaze cascade effect," in which the observer's gaze is biased towards the to-be-chosen stimulus prior to the preference decision.

In a typical experiment, two face images (taken from the Ekman and AR face databases) were presented side-by-side on the screen, and participants were required to do a two-alternative-forced-choice (2AFC) preference task. Participants were allowed to freely view the faces until they reached their decision and responded by pressing one of the keys. The faces were kept on the screen until the subject had made a response. Attractiveness was matched to remove any potential effects on preference. Their eye movements were recorded with an eye tracker during the inspection period, to examine whether gaze could predict preference choices.

The eye tracking data revealed a common pattern which is time-locked to the final decision. Starting at approximately 800 ms ahead of the decision response, their gaze became biased towards the face that would be chosen later (Figure 12.1). In contrast, when the participants were asked to perform an objective judgment task (to judge the roundness of faces), or a task to chose the less-attractive one, the gaze bias was not as strong as for the preference choice task. This and other studies have also explored different types of objects (other than faces); and the gaze cascade effect has turned out to be very robust and common across a variety of objects, including geometric figures, commercial products, artwork, etc. The

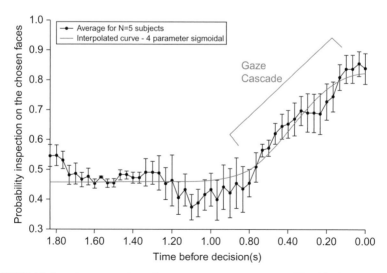

FIGURE 12.1   "Gaze cascade" effect in the preference task.   The likelihood of gaze on the chosen face is plotted against the time to the button pressing moment (by courtesy of Clavdiu Simion).

overall results suggest that gaze contributes to preference choice, which cannot be solely accounted for by the response bias in general.

Orienting not only reflects but also affects preference. To test the hypothesis that eye orienting is intrinsically involved in the preference-generating mechanism, the presentation durations of the faces were manipulated to direct participants' gaze (Shimojo et al., 2003). They were 900 ms for one face and 300 ms for another, and repeated either twice, 6 or 12 times. The participants were instructed to naturally follow the faces with their gaze. Afterwards, the face pair was briefly presented together and the participants asked to make a 2AFC preference response. The result was consistent with the gaze cascade effect in that the faces with the longer duration, thus gazed at longer, were preferred significantly more. A couple of control experiments were designed to isolate the gaze cascade from the "mere exposure" effect (Zajonc, 1968). One of the experiments, which duplicated the main experiment, including the spatio-temporal relationship of the two face images, added a fixation point in the middle, on which the participant was asked to keep fixating. In the second experiment, the participant was again to fixate at the central fixation point, while the two face images were sequentially superimposed with the 900/300 ms durations in the fovea. No bias in the preference choice was observed in either of the control experiments. Thus, a spontaneous engagement with gaze shift seemed to be necessary for the gaze cascade to occur. Shimojo and colleagues proposed an interactive positive loop between the orienting and preference.

Follow-up studies (Simion and Shimojo, 2006, 2007) further suggested that it is the dynamic positive loop between gaze shift and foveal perceptual processing that is responsible for preference formation. Simion and Shimojo (2006) adapted a "gaze-contingent window" procedure in which only one facial feature, such as one of the eyes or the mouth, in the facial stimuli was allowed to see through a small peephole which moved with the participant's gaze. This procedure was meant to prohibit the initial holistic process of the stimuli, and to further decouple the orienting from the stimuli processing. As predicted in the positive loop model, the same pattern of the gaze cascade effect was found, except that it started more than 7 secs earlier than the choice response. This suggested that even the initial, local sensory samplings are already biased towards the to-be-chosen object by gaze. In other words, feature sampling, orienting, and the preference formation mutually facilitate in parallel.

Simion and Shimojo (2007) further showed that the gaze cascade effect can be observed even when the visual stimuli were terminated *before* the preference decision. In such a condition, the gaze shift and foveation would not facilitate further sensory processing of the stimulus. Yet, the gaze was biased towards the "place holder" that used to contain the to-be-preferred face. This is direct evidence against the possibility that the participant had already made the decision implicitly, and the gaze bias was only meant to gather more positive evidence (because no evidence can be gathered from the blank space). Thus, orienting, not the stimulus processing itself, is the key factor contributing to preference.

Some studies (Glaholt & Reingold, 2009a, 2009b; Glaholt, Wu & Reingold, 2010) claimed that the gaze cascade could occur not just in the preference, but some other decision tasks, and this remains controversial. In their studies, the tasks to be compared with the preference task were to judge: (a) which of the two items looked more expensive (Glaholt et al., 2010); (b) which photograph was taken more recently (Glaholt & Reingold, 2009b); or (c) which photograph looked unusual/atypical (Glaholt & Reingold, 2009a). These tasks relied on an individual's subjective evaluation, subjective recency, or subjective novelty, which may induce emotional responses. As a result, the tasks Glaholt and his colleagues used may share the same underlying mechanism as preference choice, explaining the discrepancy of results among theirs and our earlier studies.

Thus, while the specificity of the gaze cascade effect may still remain controversial, we emphasize that the preference decision-making is (and possibly other tasks which involve emotional responses are) a dynamically evolving process.

NOVELTY VERSUS FAMILIARITY

Preference/attractiveness decision-making is a dynamic process, as shown in the previous section. One implication may be that it depends on memory.

As for the effects of memory on preference/attractiveness, there is inconsistency in the literature. Some have proposed a general tendency towards choosing novel stimuli (the "novelty principle" for preference; e.g., Berlyne, 1970; Fantz, 1964; Olst, 1971), whereas others have proposed the familiarity principle (e.g., Houston-Price & Nakai, 2004; Zajonc, 1968).

Park, Shimojo and Shimojo (2010) demonstrated a segregation of the two principles for preference across object categories: novelty preference for natural scenes and familiarity preference for faces. In the experiment, two images of the same object categories (faces, natural scenes, and geometric figures as control) were presented side-by-side on the screen. Participants were required to make relative preference judgment on the paired images. One of the paired images remained the same and was repeatedly presented over trials, and the other was always a new image on each trial. The repeated image was controlled with its attractiveness as median among all the other images. Results showed that, for the face images the repeated image rated as more preferred over trial repetition, illustrating a familiarity principle for preference. On the contrary, for natural scenes, the new image instead became preferred over the repeated image, illustrating a novelty principle for preference. No strong preference bias towards novelty or familiarity was observed for geometric figures.

A follow-up study of this initial finding further investigated which aspect of the past experience is critical for the familiarity preference for faces, and novelty preference for natural scenes, respectively (Liao, Yeh & Shimojo, 2011). To be more specific, We asked participants to judge roundness on faces, overall color tones for natural scenes, and complexity for geometric figures. Later, the participants were asked to perform the relative preference judgment (5 trials) to see if the same pattern of preference can be observed or not. Results showed that familiarity preference for faces could be formed whenever the experience was passive viewing or objective judgment during the exposure. By contrast, novelty preference for natural scenes was only observed after the objective judgment, but not passive viewing. The overall results suggest that preferences for familiar faces and novel natural scenes were modulated by task-modulated memory at different processing levels or selection involvement.

HOW DOES GAZE INTERACT WITH MEMORY IN PREFERENCE DECISION?

It remains unclear how memory affects gaze/orienting and further affects preference. Animal and infant studies showed that novel objects/places attract orienting in general (e.g., Bevins & Bardo, 1999; Fantz, 1964; Klebaur & Bardo, 1999). However, Dodd, Van der Stigchel and Hollingworth (2009) showed that it depends on the task. Gaze had a stronger tendency towards novelty and inhibited the visited location (i.e., familiar location) in a searching task. By contrast, when the task was to just freeview, memorize, or judge the pleasantness of the photo, gaze tended to go back to the familiar location more efficiently (with shorter saccadic latency). Park et al. (2010) initial finding (mentioned above), together with these studies, raises a question as to whether a gaze is attracted to novelty or familiarity depending upon object category.

With regard to the main theme of this chapter that preference formation is a dynamic process, which may be governed by the positive loop generated by gaze, it is intriguing to see how memory affects the initial gaze, and how the initial gaze in turn leads the final preference decision. To address this issue, it is worth noting that saccade/gaze shift itself is already the brain's decision (albeit implicit), and the final preference decision is a mere outcome of such cumulative "micro decisions." It is thus interesting to ask whether different object categories attract gaze in different ways and contribute to final preference choices towards familiarity versus novelty differently.

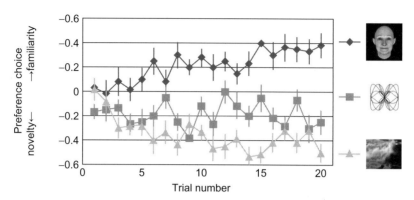

FIGURE 12.2 **Mean relative preference choices of paired images.** We code the participants' response as "1" if the repeated old image is chosen as preferred, and "−1" if the new image is chosen. At trial repeats, the familiar old faces become progressively preferred than the new faces ($R^2 = 0.74$, $p < 0.001$). In contrast, for natural scenes, it is the new natural scenes that become preferred than the repeated familiar ones ($R^2 = 0.43$, $p < 0.01$). For geometric figures, there is no systematical preference bias as the trial increases ($R^2 = 0.00$, $p > 0.8$).

Thus, we adopted the same procedure as Park et al. (2010) with modifications for further analyses on gaze behavior (Shimojo et al., 2003)[1]. Participants are required to make a 2AFC preference choice on the paired images within the three object categories (face, natural scenes, and geometric figures) while their eye movements are recorded. They were allowed to freely view the images till they make the response by pressing one of the two keys in the same way as in Shimojo et al. (2003).

In the following, we will summarize our findings one by one, together with relevant descriptions of procedures and analyses.

1   Participants' final preference decision towards familiarity versus novelty depends on stimulus type
    Final preference decision results are shown in Figure 12.2. Once the images are repeatedly presented, familiar faces are preferred rather than novel ones, whereas novel natural scenes are preferred rather than familiar ones. The results showed a segregation of memory-related preference decision (familiarity vs. novelty) by object categories, thus replicating our previous findings (Liao et al., 2011; Park et al., 2010).

---

[1] Fifteen students at Caltech participate in this experiment. All have normal or corrected-to-normal vision. Their eye movements are recorded by Eyelink II system.

Faces

Geometric figures

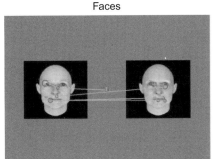

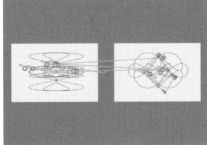

Natural scenes

FIGURE 12.3   **Examples of eye movement patterns.** Dots represent fixation allocations and lines represent the saccadic trajectories between the fixations. The original face and natural scene stimuli were full colored, whereas the geometric figures were black/white.

2 Gaze cascade effect reflects the choice type (for familiarity vs. novelty), but does not differ among object categories. This suggests that gaze reflects the underlying mechanisms of preference formations, regardless of stimulus type

Examples of eye movement patterns are shown in Figure 12.3. Participants look back and forth between the paired images until they make their decision. In order to examine whether there is a progressive gaze bias towards the chosen image before the decision is made, i.e., the gaze cascade effect, we analyzed the probability of the fixation allocation on the chosen image against the unchosen image as a function towards the button-pressing response.

Results are shown in Figure 12.4. In general, there is a gaze bias to the to-be-chosen image, which starts approximately 600 ms before the decision is made, regardless of the object category. This is qualitatively consistent with the original finding (Shimojo et al., 2003), thus suggesting that gaze reflects the general mechanism of preference formation. Most importantly, the detailed profile of the gaze cascade curve reveals the difference between the preference decision for

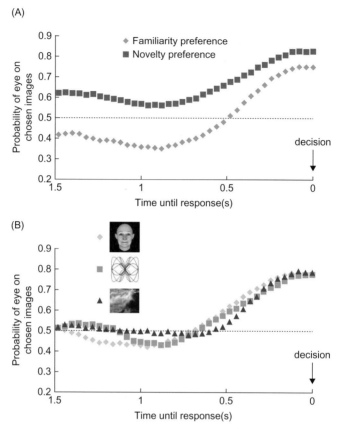

FIGURE 12.4 **Eye movement results.** The function of the probability of the participants' gaze on the **chosen** images time locked until response. The black dashed level, line represents the chance meaning no bias towards chosen or unchosen images. The data is segregated by choice type (familiarity and novelty preference) in (A), and by stimulus type (face, natural scene, and geometric figure) in (B).

familiar image versus novel image (Figure 12.4A). In contrast, the gaze bias does not differ when the chosen images are faces, natural scenes, or geometric figures[2] (Figure 12.4B).

In the following, we will show analyses on the initial gaze behavior in response to the stimulus onset, and its relationship to the final preference decision.

[2] We apply the Pairwise Kolmogorov-Smirnov Test to examine the difference between the segregated probability functions. Data from 600 ms before the decision until decision are extracted for analyses. Results show significant difference between choice types ($d = 0.50, p < 0.03$), but no difference among object categories ($ds < 0.3, ps > 0.6$).

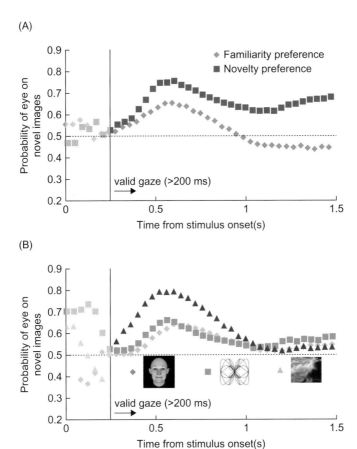

FIGURE 12.5   Eye movement results.   The function of the probability of participants' gaze on the **novel** images time locked from stimulus onset. Note that on average it takes approximate 200 ms to trigger saccade, and we thus arbitrarily define the gaze behavior to validly reflect participants' intention as >200 ms after stimulus onset. Before 200 ms, the gaze may just reflect the microsaccade or unintended eye movements. The black dashed level line represents the chances, meaning no bias towards novel or familiar images. The data curves are segregated by choice type (familiarity and novelty preference) in (A) and by stimulus type (face, natural scene, and geometric figure) in (B).

3 Novelty attracts gaze starting from stimulus onset in general, though the effect varies with stimulus type
We analyze the probability of fixation location on the novel image against the familiar image, starting from (i.e., time-locked to) the image presentation onset. Results are shown in Figure 12.5. We find a general tendency that gaze is biased towards the novel image, and this bias diminishes after approximately 1 sec. The gaze bias towards

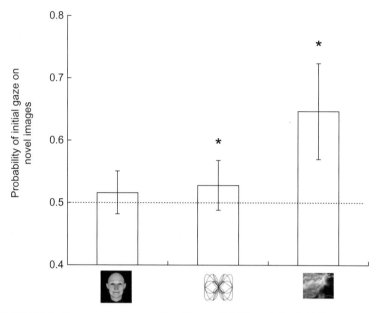

FIGURE 12.6   **Eye movement results.** The probability of the initial gaze on novel images is plotted as a function of stimulus type. The dashed line represents the chance level, meaning no initial gaze bias towards novel or familiar images. Error bars represent one standard deviation from the mean. Asterisks represent the statistically significant difference from the chance level.

novel images is more prominent when the final preference choice is for novelty rather than familiarity (Figure 12.5A). In addition, the gaze bias towards a novel image is also more prominent for natural scenes than faces or geometric figures, and there is no difference between faces and geometric figures[3] (Figure 12.5B).

4   Whether initial gaze is attracted by novelty depends on stimulus type
     Although the previous analysis shows the general gaze bias towards novelty starting from stimulus onset, the effect is accumulated by several gazes starting from the stimulus onset. It remains unclear whether the novel images attract the very first initial gaze, and whether the initial gaze bias, if it exists, differs across object categories. We have addressed this issue, and the results

[3] Data after 200 ms stimulus onset until 1 sec are extracted for the Pairwise Kolmogorov-Smirnov Test. Results show significant difference between choice types ($d = 0.43, p < 0.03$), natural scenes vs. faces ($d = 0.62, p < 0.001$), natural scenes vs. geometric figures ($d = 0.57, p < 0.001$), but no difference between faces vs. geometric figures ($d = 0.24, p < 0.5$).

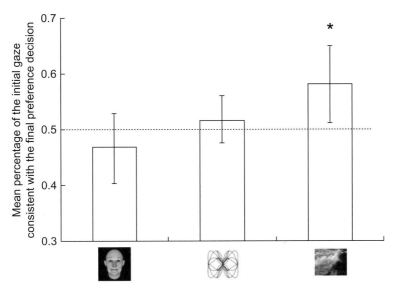

FIGURE 12.7 **Mean percentage of the initial gaze consistent with the final preference decision is plotted as a function of stimulus type.** The dashed line represents the chance level, meaning the final preference is not classified by the initial gaze. Error bars represent one standard deviation from the mean. Asterisks represent the statistically significant difference from the chance level.

are shown in Figure 12.6. The initial gaze is attracted to novel natural scenes and geometric figures, but not faces[4].

5 Whether initial gaze location predicts final preference decision depends on stimulus type

As novel natural scenes receive initial gaze bias and are preferred in later preference decisions, we wonder if initial gaze predicts final preference choice, and whether this predictive power, if it exists, differs across object categories. To address the predictive power, we examined the classification performance, i.e. to classify the final preference choices based on the initial gaze location. We have addressed this issue, and the results are shown in Figure 12.7. The initial gaze location predicts a final preference

---

[4]ANOVA shows that the initial gaze is towards novel images more prominently for natural scenes than faces or geometric figures [$F(1,14) = 24.99, p < 0.0001$]. To compare with the chance level, one-sample t-test proves the significant initial gaze bias towards novel natural scenes [$t(14) = 7.09, p < 0.001$] and geometric figures [$t(14) = 2.58, p < 0.03$], but not for novel faces [$t(14) = 1.61, p > 0.1$].

decision only for natural scenes, but not for faces or geometric figures[5].

Note that whether the initial gaze is attracted by novel images is one issue, and initial gaze predicts final preference decision is yet another, independent issue. The dissociation is well demonstrated in the geometric figures in that, while the initial gaze is attracted by novel geometric figures, it does not predict the final preference decision. For natural scenes, it is clear that our results support both hypotheses that initial gaze is attracted by novel natural scenes and predicts the final preference decision.

6 Initial gaze's dwell duration predicts final preference choice: The longer the initial gaze stays on, the more likely the image will be preferred later

As suggested by Shimojo et al. (2003), it is not just the physical exposure affecting preference (the mere exposure effect). Rather, what's important is the spontaneous gaze engagement in the specific picture to affect preference. Therefore, we conduct analysis on initial gaze duration as a potential classifier of the final preference decision, to examine whether the effects differ across object categories. The results show that initial gaze stays longer in the chosen than unchosen images, and this tendency is common across all object categories[6] . To provide more detailed information, we show frequency histograms on the dwell duration of the initial gaze, segregated by the gaze on chosen and unchosen images (Figure 12.8). As shown in the figure, the mean dwell duration is larger when the initial gaze is on the chosen rather than the unchosen images. This effect is consistently observed in all three object categories, despite the different effect sizes.

Altogether, the results from 1 to 6 above suggest that the initial gaze, not just the choice of it but also the dwell duration, has some relationship to the final preference choice, yet it also depends on object categories.

---

[5] ANOVA shows that the classification performance between the initial gaze and the final preference choice differs among object categories [$F(1,14) = 14.58$, $p < 0.02$]: better classification performance for natural scenes than faces (Tukey's test, $p < 0.05$) or geometric figures ($p < 0.01$). To compare with the chance level, the one-sample t-test shows that the final preference is classified by the initial gaze for natural scenes [$t(14) = 4.68, p < 0.001$], but not for faces [$t(14) = -1.99, p > 0.06$] or geometric figures [$t(14) = 1.71, p > 0.1$].

[6] Mean dwell durations are subjected to ANOVA with image type (chosen or unchosen) and object category (faces, natural scenes, or geometric figures) as within-subject factors. Results show the main effect of image type [$F(1,14) = 8.54, p < 0.02$] and object category [$F(2,14) = 14.24, p < 0.001$], but not the two-way interaction [$F(2,28) = 2.56, p > 0.09$].

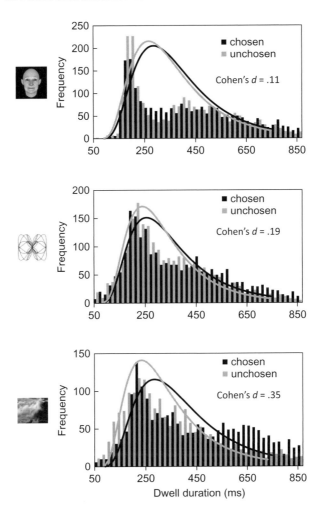

FIGURE 12.8   Frequency histogram on the dwell duration of the initial gaze. These data are fitted with log-normal distribution function (the curves; all Kolmogorov-Smirnov Goodness of Fit values < 0.13, ps > 0.4). Cohen's d represents the effect size between the two distributions. The mean dwell duration is longer when the initial gazes are on the chosen rather than unchosen images, regardless of object category.

For faces, the initial gaze's landing location does not predict the final preference decision, but once the gaze stays longer on a certain face, it would more likely be the preferred face. For natural scenes, the initial gaze, not just the choice but also the dwell duration, predicts the final preference decision.

DYNAMIC AND TIME-EVOLVING PROCESS TOWARDS PREFERENCE DECISION

To summarize, the gaze cascade effect reflects general, and implicit, precursors of conscious preference decision. With this regard, there is virtually no difference across object categories (Figure 12.4B). However, the gaze cascade profiles are indeed different between choice types (Figure 12.4A). For novelty preference, gaze is biased towards novel pictures initially, and biased towards a chosen picture with shallower accumulation. However, for familiarity preference, the gaze bias towards a chosen picture grows with the steeper accumulation.

From stimulus onset, gaze orienting is affected by memory, but differently across object categories. Natural scenes attract the initial gaze more strongly than faces (Figures 12.5 and 12.6). Results may be explained by the effect of differences in similarity. The natural scenes are less similar to each other, even within subcategories such as mountains or flowers, than faces. Dissimilarity among features may enhance the salience of the novel stimulus, and thus further attract attention and gaze.

The initial gaze predicts the preference decision in different object categories through different processes, which interacts with choice types. The preference decision in natural scenes can be predicted by initial gaze location (Figure 12.7). Considering that the initial gaze is biased towards novel natural scenes (Figure 12.6), this initial gaze bias may trigger the positive loop of gaze and preference and then lead to a final novelty preference decision. Furthermore, a preference decision in general, regardless of object category, can be predicted by initial gaze duration (Figure 12.8). The longer the image is looked at, the more likely it would be preferred. Altogether, with the results that a preference decision is biased towards familiar faces, this implies that the initial gaze stays longer in a familiar face and thus leads to a final familiarity preference decision. The findings also have an implication to the mere exposure effect (Zajonc, 1968). The exposure needs to involve gaze engagement, rather than just physical appearance, to facilitate the positive loop towards preference.

Preference decision-making is a dynamic process spreading over several seconds. During the process, active inspection itself contributes to preference decision formation. We like, so we look, and we look, so we like. This positive loop leads to preference. This idea is partly consistent with Damasio's somatic marker hypothesis (Damasio, 1996), which states that somatic marker guides decision-making. Gaze may be one of the somatic markers to the preference decision. The physiological evidence also suggests a highly overlapped neural network between preference and gaze processes (Emery, 2000; Gottfried, O'Doherty & Dolan, 2003; Kim, Adolphs, O'Doherty & Shimojo, 2007).

Meanwhile, post-hoc interviews of the participants indicate that they are mostly unaware of their own gaze bias, and even if some of them do realize, they all actively deny the causal contribution of the gaze to their preference decision (unpublished data). The same also applies to the familiarity/novelty manipulations. All participants are aware that there is always a repeated old image paired with a new image in the experimental procedure, but none of them infer their preference decision from the repetition of the images (also see Liao et al., 2011). There is also fMRI evidence that the preference decision process is implicit, mainly due to a subcortical reward system (Kim et al., 2007).

Thus in short, a preference decision is not made instantly at a moment in time, but rather it is a gradually evolving dynamic, somatic and (at least initially) implicit process.

References

Berlyne, D. E. (1970). Novelty, complexity, and hedonic value. *Percept Psychophys, 8,* 279–286.

Bevins, R. A., & Bardo, M. T. (1999). Conditioned increase in place preference by access to novel objects: Antagonism by MK-801. *Behavioural Brain Research, 99,* 53–60.

Damasio, A. R. (1996). The somatic marker hypothesis and the possible functions of the prefrontal cortex. *Philosophical Transactions of the Royal Society London, 351,* 1413–1420.

Dodd, M. D., Van der Stigchel, S., & Hollingworth, A. (2009). Novelty is not always the best policy: Inhibition of return and facilitation of return as a function of visual task. *Psychological Science, 20,* 333–339.

Emery, N. J. (2000). The eyes have it: The neuroethology, function and evolution of social gaze. *Neuroscience and Biobehavioral Reviews, 24,* 581–604.

Fantz, R. L. (1961). The origin of form perception. *Scientific American, 204,* 66–72.

Fantz, R. L. (1963). Pattern vision in newborn infant. *Science, 140,* 296–297.

Fantz, R. L. (1964). Visual experience in infants: Decreased attention to familiar patterns relative to novel ones. *Science, 146,* 668–670.

Glaholt, M. G., & Reingold, E. M. (2009). Stimulus exposure and gaze bias: A further test of the gaze cascade model. *Atten Percept Psychophys, 71,* 445–450.

Glaholt, M. G., & Reingold, E. M. (2009). The time course of gaze bias in visual decision tasks. *Visual Cognition, 17*(8), 1228–1243.

Glaholt, M. G., Wu, M. C., & Reingold, E. M. (2010). Evidence for top-down control of eye movements during visual decision making. *Journal of Vision, 10*(5), 1–10.

Gottfried, J. A., O'Doherty, J., & Dolan, R. J. (2003). Encoding predictive reward value in human amygdala and orbitofrontal cortex. *Science, 301,* 1104–1107.

Houston-Price, C., & Nakai, S. (2004). Distinguishing novelty and familiarity effects in infant preference procedures. *Infant and Child Development, 13,* 341–348.

Kim, H., Adolphs, R., O'Doherty, J. P., & Shimojo, S. (2007). Temporal isolation of neural processes underlying face preference decision. *Proceedings of the National Academy of Sciences, 104,* 18253–18258.

Klebaur, J. E., & Bardo, M. T. (1999). The effects of anxiolytic drugs on novelty-induced place preference. *Behavioual Brain Research, 101,* 51–57.

Laughlin, W. S. (1968). Hunting: An integrating biobehavior system and its evolutionary importance. In R. B. Lee & I. DeVore (Eds.), *Man the hunter.* Chicago: Aldine Publishing.

Liao, H. I., Yeh, S. L., & Shimojo, S. (2011). Novelty vs. familiarity principles in preference decisions: Task-context of past experience matters. *Frontiers in Psychology*, 2(43), 1–8.

van Olst, E. H. (1971). *The orienting reflex*. New York: Walter De Gruyter Inc.

Park, J., Shimojo, E., & Shimojo, S. (2010). Roles of familiarity and novelty in visual preference judgments are segregated across object categories. *Proceedings of the National Academy of Sciences*, 107, 14552–14555.

Shimojo, S., Simion, C., Shimojo, E., & Scheier, C. (2003). Gaze bias both reflects and influences preference. *Nature Neuroscience*, 6, 1317–1322.

Simion, C., & Shimojo, S. (2006). Early interactions between orienting, visual sampling and decision making in facial preference. *Vision Research*, 46, 3331–3335.

Simion, C., & Shimojo, S. (2007). Interrupting the cascade: Orienting contributes to decision making even in the absence of visual stimulation. *Percept Psychophys*, 69, 591–595.

Teller, D. Y. (1979). The forced-choice preferential looking procedure: A psychophysical technique for use with human infants. *Infant Behavior and Development*, 2, 135–158.

Zajonc, R. B. (1968). Attitudinal effects of mere exposure. *Journal Personality and Social Psychology, Monograph Supplement*, 9(2, Pt. 2), 1–27.

IMPLICATIONS, APPLICATION AND FUTURE DIRECTION

13

Choice Sets as Percepts

Andrew Caplin

Professor of Economics and Co-Director of the Center for Experimental
Social Science, New York University

INTRODUCTION

Two historic and important concepts, one from economics and one from psychology, are profoundly synergistic. The economic concept is utility maximization, which characterizes what a decision-maker (DM) selects from an available set of choices: the so-called "choice set." The psychological concept is the percept, corresponding to the subjective perception of external reality.

In this chapter, I highlight the potential value of research that integrates information concerning how choice sets are perceived into the utility-maximizing framework. I sketch ongoing research from both contributing disciplines that adopts this integrative approach, albeit in an unselfconscious manner. I highlight opportunities for further advance

Neuroscience of Preference and Choice
DOI: 10.1016/B978-0-12-381431-9.00023-1

based on a more conscious merging of questions and approaches, and address some of the unique intellectual challenges that such research faces. Of course, it also faces profound sociological challenges related to the compartmentalization of the academy, but the dividing walls will not hold up under the weight of successful integrative research.

The conceptual basis for the proposed research program is outlined in the section on *Samuelson meets weber*. In the section on *search and the choice process*, I present an example of ongoing research in economics in which the distinction between subjective and objective choice sets is of the essence. The section on *The Drift-Diffusion Model and Small Choice sets* does likewise for a branch of psychometrics. The section on *Next steps* lays out some intellectual challenges that confront researchers seeking to make further headway in this area, and introduces some recently initiated research that suggests how to overcome them.

SAMUELSON MEETS WEBER

Economists study the implications of individual decisions for various socially important outcomes, including the market prices of goods and services. A key ingredient of all economic analysis is therefore a theory of individual decision-making. Given the many demands on its services, such a theory has to be modular and capable of being used in a bewildering array of different settings. Economists developed the theory of utility maximization for just this reason. It fulfills its role admirably and can be used to model how choices will be made in any and all situations the DM could conceivably be faced with.

In its conceptual form, the theory of utility maximization asserts that a DM does not knowingly reject alternatives that are superior to the option that is chosen. From this intuitively reasonable starting point, it provides a coherent vision of the role of individual decisions in social affairs. It combines simplicity, generality, and intuitive content in a manner that is unmatched. No alternative vision remotely as general or as well grounded in psychological intuition is anywhere on the horizon.

With all of its positives, those who object to the theory can point to myriad behaviors that appear to contradict it, ranging from simple reversals of preference to systematic biases of various kinds. The end result is that many psychologists and even economists have come to see the utility maximizing model of decision-making as crude, and its proponents as reactionary (in the intellectual sense, at least).

The truth is otherwise. What the critics miss is that, in its conceptual form, the theory of utility maximization is extraordinarily flexible. It rests on the idea that it is rare for an individual knowingly to reject a superior alternative. At the very least, this seems like a reasonable starting point.

Indeed, it may even seem tautological. It is not easy to tell a coherent story in which a DM knowingly rejects an option perceived as superior at the moment of choice.

Given its almost circular nature, it is ironic that the theory of utility maximization has come to be regarded by so many as blatantly false. I believe that the roots of this fall from grace lie in one of the most profound breakthroughs in the history of social science. In 1938, the young Paul Samuelson posed in a precise manner the question of whether or not the theory of utility maximization is indeed tautological (Samuelson, 1938). In the process, he launched the "revealed preference" revolution which forever changed the structure of economic thought.

Revealed preference theory involves a fundamental switch of perspective. Rather than being postulated on *a priori* grounds, it identifies utility maximization with the restrictions it would impose on an idealized data set of individual choices. The almost poetic conceptualization of choice as constrained optimization is replaced with the prosaic measurement of choices from subsets of a choice set. To test the theory of utility maximization, economists were instructed to observe choices from as many choice sets as possible and judge how well they met the revealed preference restrictions.

To almost no one's surprise, rigorous tests revealed that the theory of utility maximization often failed, sometimes in spectacular fashion. One response to these failures has been to call for choice theory to start again on some new and more "behavioral" foundation. The problem is that alternative visions are hard to find, and those that have been proposed generally nest the standard theory as a special case. This makes them vulnerable to Samuelson's original question: what exactly are the restrictions that these more general theories place on choice behavior? In many cases, it turns out that the theories are close to vacuous, in the sense that they do not constrain observed choices in any manner whatever. In choice theory, as in so many other contexts, you can't beat something with nothing.

The research program described herein places the limitations of the theory of utility maximization in a different light. The launch pad is the observation that traditional revealed preference tests of utility maximization are based on an auxiliary assumption that is hard to test: that subjective and objective choice sets are one and the same. If the available choices are apples and bananas, the DM is implicitly assumed to know this. Similarly, if the objects of choice are seen by neutral experts as being monetary lotteries, it is presumed that this is precisely how they are seen by the DM.

In practice, it is unrealistic to expect DMs to fully understand all available options, particularly if they are complex or manifold. Hence there is almost surely a gap between what the DM perceives and the choice set as it would be described by a fully informed third party. The research outlined herein focuses on the extent to which the inappropriate identification of objective with subjective choice sets is responsible for the observed failings of utility theory.

The proposed focus on subjective perceptions as opposed to objective reality is in line with a tradition in psychometric research dating back to Weber (1834). Weber used a simple choice experiment to show that, while perceived weight and actual weight are not equivalent, there are lawful relations between them. Glimcher (2010) reviews the manner in which these original findings have been expanded upon in the ensuing period. In fitting with psychometric tradition, the research program outlined herein involves focusing on choice sets as DMs perceive them rather than as an all-seeing economist might.

SEARCH AND THE CHOICE PROCESS

Psychologists are not alone in their recognition of gaps between objective and subjective choice sets. Indeed the entire economic theory of imperfect information and search is premised upon the existence of such a distinction. In a foundational paper, George Stigler introduced costs of information gathering in trying to explain observed violations of the law of one price (Stigler, 1961). The basic structure of sequential search theory was set at this time: a DM faced with a large number of options and costs of search will not uncover them all. Rather, the optimal strategy is to uncover them one at a time and ultimately to settle on some object that achieves a high enough level of utility, often referred to as the "reservation" level of utility.

From this simple starting point, the theory of imperfect information has grown into one of the cornerstones of modern economics. Yet despite its centrality, economists have generally shied away from attempting to measure how much subjects know. Rather than address the obvious measurement challenge, economists have traditionally applied the theory in settings in which the incompleteness of information is clearly of the essence and in which it has a seemingly natural structure. For example, in models of asymmetric information, it is *per se* plausible that the seller of a used good knows more about its quality than do potential buyers.

In recent years, economists have begun belatedly to investigate how consumers search, in large part by importing experimental techniques from psychometric research. Examples of data that were first explored by psychologists and later picked up by leading-edge economists include: direct observation of the order of information search using MouseLab (e.g. Payne, Bettman & Johnson, 1993; Johnson, Camerer, Sen & Rymon, 2002; Gabaix, Laibson, Moloche & Weinberg, 2006); eye movements (e.g. Wang, Spezio & Camerer, 2010; Reutskaja, Nagel, Camerer & Rangel,

2011); and the time taken in arriving at a decision (e.g. Busemeyer & Townsend, 1993; Rubinstein, 2007).

One limitation of the psychometric tools employed in the above studies is that it is hard to connect them in a precise manner either with information acquisition or with the final act of choice. For example, in the case of MouseLab, while it is easy to conclude that objects that are not uncovered are not understood, it is not easy to infer what is internalized by unmasking and gazing at an object. One cannot assume that looking at a complex object is the same as understanding it, and it is also inherently difficult to know how the level of understanding is related either to the time for which each object is left uncovered or how long it is held in the gaze. So profound is the problem of inference that new observations alone are simply insufficient to identify either how the choice set is perceived, or how one chooses among the resulting percepts. One needs in addition some theoretical construct.

It is here that the theory of utility maximization again shows its worth. In an effort to develop a simpler data set with which to understand the subjective perception of the choice set, Caplin and Dean (2011) consider "choice process" data. This data conveys not only the final option that the DM selects, but also how provisional choices change during the prior period of contemplation (see also Campbell, 1978).

One advantage of choice process data over less familiar forms of data is that it comes in the standard form of choices, and hence is readily modeled using traditional tools of the trade, in particular the theory of constrained optimization. In the search theoretic context, this enables choice process data to provide a test bed for Stigler's simple theory of search, in which the DM searches sequentially through the available options, comparing searched options in full according to a fixed utility function. This form of search has no implications for final choices *per se*, since choice of one option over another can be rationalized by ignorance of the alternative. However, Caplin and Dean show that sequential search has a distinctive signature in choice process data. So too does a "reservation-based" search in which the DM searches until an object is identified with utility above a fixed reservation level. Not only is this form of search optimal in standard search models, but it also characterizes the "satisficing" model of Simon (1955). Simon posited a process of item-by-item search, and the existence of a level of reservation utility, attainment of which would induce the DM to curtail further search.

Caplin, Dean and Martin (Forthcoming) introduce an experimental design that enables data on the choice process to be gathered in an appropriately incentivized manner. Subjects are presented with a collection of objects from which they must choose. They are given a fixed period to consider which of these items to select, and are told that they can select any option at any time during the consideration period by

V. IMPLICATIONS, APPLICATION AND FUTURE DIRECTION

clicking on it, changing their selection as many times as they like. The key to the experimental design is that subjects are informed up front that they will be paid based not on their final choice, but rather on their choice at a randomly selected point during the period of contemplation. They are also informed that failure to make a selection results in the worst possible outcome. Hence they are incentivized to make a quick first choice, and thereafter always to switch to their most preferred option as the period of contemplation unfolds.

To provide a simple measure of the quality of decisions, subjects in the Caplin, Dean and Martin experiments selected among "dollar objects": monetary prizes presented as sequences of addition and subtraction operations. Given that the choice problem is nontrivial, subjects regularly failed to find the best option. However, they broadly obeyed the restrictions associated with sequential search and with Simon's satisficing model. Most subjects stopped searching when an environmentally-determined level of reservation utility was realized. The estimated level of reservation utility was found to vary with object complexity and with set size: the former is in fitting with the theory of sequential search, while the latter is not.

To set the above research in the larger context of exploring choice sets as percepts, the key feature is that Stigler's theory of ABS involves a particularly simple set of possible percepts. Specifically, all percepts are subsets of the objective choice set. The simplicity of this vision is that it calls for no changes in the underlying model of choice, which is the standard theory of utility maximization. All that is needed are experimental enrichments that enable one to identify signature features of the various modes of search, and that in so doing provide information on how the DM perceives the choice set.

There are interesting connections between the search theoretic approach to decision-making and various noted behavioral anomalies, such as status quo bias: DMs observed tendency to stick with a default option (Samuelson & Zeckhauser, 1988). Current research work by Geng (2010) relying both on MouseLab and on the choice process experimental design suggests that a significant portion of such status quo bias may result from incomplete search. The search theoretic viewpoint may have much to offer those wishing to deepen understanding of otherwise anomalous choice behavior.

THE DRIFT-DIFFUSION MODEL AND SMALL CHOICE SETS

From its inception, the theory of sequential search was designed for settings in which there are many options that are potentially available, each of which is relatively simple to evaluate. This makes it well-suited to searching for the cheapest source of a particular good, yet ill-suited to considering choice among a small set of options, each of which is complex. In such settings it is unlikely that any of the available options is ever perfectly understood, so that choice must be based on a comparison of the goods in some rather than in all aspects (e.g. Payne, Bettman & Johnson, 1992).

An important psychometric paradigm designed for modeling choice among a small number of complex objects is the drift-diffusion model (Ratcliff, 1978). In its basic form, the model applies to choices between two and only two options that are made under a certain amount of time pressure. The DM is assumed to start with imperfect information on how the options compare, which information gets successively refined as the choice is considered for longer and longer periods of time. At some point in the process of contemplation, information improves to the point that one or the other of the options is seen as having a clear edge. At a critical threshold, it is assumed that this edge is sufficiently large to induce the DM to make a selection in favor of the option that is judged at that time to be superior.

In any given experimental context, the underlying model parameters (e.g. the rate of information acquisition and the decision threshold) are identified based on how well they fit the observed joint distribution of choices and of response times. The resulting fit data appears close in a wide variety of different decision-making domains, despite the small number of model parameters and the implied restrictions on this joint distribution (see Ratcliff & McKoon, 2008).

In addition to providing a reasonable fit with a large body of behavioral data, the drift-diffusion model has gained great currency in recent years due to apparent connections with neuro-physiological data. Among the most powerful such connections was uncovered by Shadlen and Newsome (2001) based on neurons in the lateral intraparietal area (LIP) of rhesus monkeys. They trained monkeys (based on a juice reward) to decide the direction of motion of a field of dots that had a certain coherence of motion. The selection was made by means of an eye movement to a target in the relevant direction. They arranged the task so that they could separately identify neurons in the response fields related to the distinct target motions. They found strong connections between the direction and coherence of motion and the firing rate in the corresponding LIP neurons. They also found neural responses to be largely consistent with the drift-diffusion model, with a threshold level of differential neuronal activity triggering eye motion in the corresponding direction.

At first, it may seem that the drift-diffusion model represents an entirely mechanistic approach to choice that has no relationship with the theory of utility maximization. Yet this distinction is only skin deep: the drift-diffusion model is intimately connected with models of signal processing and choice in fitting with the economic tradition. To make the analogy precise, consider a DM trying to identify whether a good prize is in the box to their left or the box to their right, given that either location is initially equally likely. While contemplating this decision, the DM observes a sequence of independent and identically distributed binary signals, with one such signal indicating that the good prize is more likely to be on the left, the other that it is more likely to be on the right, and updates in standard Bayesian fashion in the face of this information. Finally, suppose that the DM's rule for selecting an object and thereby stopping the flow of information is to continue until the absolute difference between the number of left and right signals hit a critical threshold. The behavior of this expected utility maximizing Bayesian would perfectly mirror the prescriptions of the drift-diffusion model.

To be sure, the connection between the drift-diffusion model and optimal signal processing remains controversial. In particular, there is a debate concerning whether the model is best seen as describing hardware that constrains decision-making or as an approximation to a Bayesian model of optimal signal processing. However, even if it is seen in terms of hardware, the model remains strongly tied with utility maximization. After all, in all versions of the model the final choice is made to maximize the probability of receiving the good prize, precisely as in standard expected utility theory. In that sense, the model can be seen as combining expected utility maximization (the decision is always made to maximize the expected utility of the final prize) with a specific model of updating and of terminating search.

NEXT STEPS

Search theory is designed for choice among many options, each of which is simple. The drift-diffusion model is designed for choice between two complex options. In the former case, it is assumed implicitly that the gap between subjective and objective perception is that only a subset of the options is ever identified by the DM. In the latter case, the gap revolves around the incomplete examination of the small number of available options, with the degree of incompleteness depending on the time for which the choice is considered. A high priority in future research that explores how choice sets are perceived is to consider hybrid settings in which there are more than two options, each of which is complex. The hope is that what will be identified is a corresponding hybrid approach to understanding these decisions that somehow combines the theory of sequential search and signal-processing theory.

A barrier to this form of hybrid research is that testing percept-based theories is profoundly difficult. It is not coincidental that in the above examples there are experimental measures that go beyond just the final choice: provisional choices in the case of a sequential search, and decision times in the case of the drift-diffusion model. I believe that these and other data enrichments will constitute a defining feature of the proposed research program. Until and unless choice-based data sets can be suitably enriched, the challenge posed by Samuelson will remain open: once one allows for subjective interpretations of the choice set, the theory of utility maximization is rendered vacuous. Hence, for the proposed research to prosper, great attention will have to be given to measurement issues and to the associated challenges of engineering and experimental design.

While new measurements are vital, they are not sufficient to make the proposed research thrive. In fact, non-standard data raises deep theoretical challenges. How does one develop a theory that has implications not only for choices but also for new and far less familiar data? I believe that the key to progress is ensuring a tight match between experimental design and the corresponding theories. There must be conceptual innovation in formulating how various choice environments impact the generation of percepts. Simultaneously, one needs a well-grounded model of choice among the resulting percepts.

Recently initiated research illustrates the potential of this form of integrative research. Caplin and Martin (2011) develop a theory of imperfect perception that has testable implications for observed choices. The "rational expectations perception based representations" (RE-PREP) that they analyze involve DMs who have fully internalized how their perceptions, whatever they may be, relate to choices and to their consequences. The signature of the resulting theory is that stochastic choices must be "unimprovable," with the precise improvements that can be ruled out depending on the data available for model testing. The model allows both the layout of the prizes (e.g. order in a list or position on a screen) and the information content of the environment (e.g. the extent to which the first in the list has turned out to be the best) impact choice behavior. Hence it allows rich study of how informational "nudges" impact attention and choice.

More broadly, there are three stages to the proposed integrative research on choice sets as percepts. First, a vision of the class of objects that may be perceived by the DM in considering a particular choice set. Second, a vision of how the mode of presentation of the choice set translates into this space of perceptions. Third, a theory of choice from the corresponding percepts is needed. In research of this kind, theory and experiments will co-evolve. What will be needed are complementary tools of a theoretical and experimental nature that are as tightly integrated as is possible. Not only will the dividing lines between economists and psychologists need to further blur, but also those between theorists and experimentalists.

V. IMPLICATIONS, APPLICATION AND FUTURE DIRECTION

Acknowledgments

I thank Mark Dean, Paul Glimcher, Daniel Martin, and Antonio Rangel for their guidance and encouragement in relation to this article.

References

Busemeyer, J., & Townsend, J. (1993). Decision field theory: A dynamic-cognitive approach to decision making in an uncertain environment. *Psychological Review, 100*(3), 432–459.

Campbell, D. (1978). Realization of choice functions. *Econometrica, 46,* 171–180.

Caplin, A., & Dean, M. (2011). Search, choice, and revealed preference. *Theoretical Economics, 6,* 19–48.

Caplin, A., Dean, M., & Martin, D. (Forthcoming). Search and satisficing. *American Economic Review.*

Caplin, A., & Martin, D. (2011). A testable theory of imperfect perception (Forthcoming).

Gabaix, X., Laibson, D., Moloche, G., & Weinberg, S. (2006). Costly information acquisition: experimental analysis of a boundedly rational model. *American Economic Review, 96*(4), 1043–1068.

Geng, S. (2010). Asymmetric search, asymmetric choice errors, and status quo bias. *Mimeo.* New York University.

Glimcher, P. (2010). *Foundations of neuroeconomic analysis.* Oxford University Press.

Johnson, E. J., Camerer, V., Sen, S., & Rymon, T. (2002). Detecting failures of backward induction: Monitoring information search in sequential bargaining. *Journal of Economic Theory, 104,* 16–47.

Payne, J. W., Bettman, J. R., & Johnson, E. J. (1988). Adaptive strategy selection in decision making. *Journal of Experimental Psychology. Learning, Memory, and Cognition, 14,* 534–552.

Ratcliff, R. (1978, March). A theory of memory retrieval. *Psychological Review, 85*(2), 59–108.

Ratcliff, R., & McKoon, G. (2008). The diffusion decision model: theory and data for two-choice decision tasks. *Neural Computation, 20,* 873–922.

Reutskaja, E., Nagel, R., Camerer, C., & Rangel, A. (2011). Search dynamics in consumer choice under time pressure: An eyetracking study. *American Economic Review, 101,* 900–926.

Rubinstein, A. (2007). Instinctive and cognitive reasoning: A study of response times. *Economic Journal, 117,* 1243–1259.

Samuelson, W., & Zeckhauser, R. (1988). *Status quo bias* in decision making. *Journal of Risk and Uncertainty, 1*(1), 7–59.

Shadlen, M. N., & Newsome, W. T. (2001). Neural basis of a perceptual decision in the parietal cortex (area LIP) of the rhesus monkey. *Journal of Neurophysiology, 86,* 1916–1936.

Simon, H. (1955). A behavioral model of rational choice. *The Quarterly Journal of Economics, 69*(1), 99–118.

Stigler, G. (1961). The economics of information. *The Journal of Political Economy, 69*(3), 213–225.

Wang, J., Spezio, M., & Camerer, C. (2010). Pinocchio's pupil: using eyetracking and pupil dilation to understand truth telling and deception in sender-receiver games. *American Economic Review, 100*(3), 984–1007.

Weber, E. H. (1834; 1996). *On the Tactile Senses,* H. E. Ross & D. J. Murray Trans. and Ed. New York: Experimental Psychology Series.

14

Preferences and Their Implication for Policy, Health and Wellbeing

Ivo Vlaev[1] and Ara Darzi[2]

[1]Centre for Health Policy, Imperial College London, London
[2]Division of Surgery, Imperial College London, London

INTRODUCTION

Many domains of everyday life require concepts and institutions that aspire to realize a behavioral goal of individual and social well-being. Examples include encouraging people not to smoke, to improve their

Neuroscience of Preference and Choice
DOI: 10.1016/B978-0-12-381431-9.00024-3

diet, to exercise more, to practice safe sex, to use seat belts, to follow speed limits, to protect the environment, and so on. In addition, many important public policy questions for the twenty-first century are likely to be concerned with how individuals respond to information and incentives aimed to prompt changes in individual health-related behavior. A better understanding of how best to bring about desired behavior change is vital if health is to be improved and environment preserved.

Over the last 50 years, disciplines of behavioral medicine and behavioral epidemiology have evolved to identify, explain, and address personal risk factors (Davidson, Goldstein, Kaplan et al., 2003; Heller & Page, 2002; Rychetnik, Frommer, Howe & Shiell, 2002). For example, a massive accumulation of evidence supports the premise that sedentary lifestyles are a primary cause of cardiovascular disease, cancer and numerous other morbidities (Blair, Kohl, Barlow et al., 1995; Broman, 1995; Pate, Pratt, Blair et al., 1995). Similar evidence has accumulated for other risk factors like dieting, smoking, alcohol consumption, sexual hygiene, and medical self-examination. Therefore, substantial health losses are attributable to lifestyle, particularly amongst the least well-off (Barr, 1987; Uitenbroek, Kerekovska & Festchieva, 1996) while significant gains in health can, in principle, be realized through relatively small changes in the choices people make (Department of Health, 2004). It remains true that policy makers and healthcare professionals are still faced with a dearth of generalizable, effective, and sustainable interventions that can be translated into effective health promotion practice (Glasgow, Kleges, Dzewaltowski et al., 2004).

New models of effective behavior change are needed for two reasons. First, existing theories and methods leave a substantial proportion of the variance in behavior unexplained, beyond the effect of informed intentions – several recent meta-analyses imply that changing intentions can account for up to 28% of the variance in behavior change (Sheeran, 2002), but estimates based only on experimental studies report explained variance as low as 3% (Webb & Sheeran, 2006). Second, evidence in the behavioral sciences amply testifies to the fact that human behavior is highly susceptible to subtle change in the environment (Ariely, 2008; Thaler & Sunstein, 2008). Such "contextual" influence on human choices are often beyond intentional control, an observation that might explain why they are neglected by traditional policy making and public health institutions, which focus mainly on intention as a route to behavior change. Webb and Sheeran (2006, p. 259) provide evidence that standard intervention models based on changing cognitions, such as beliefs and attitudes, can produce effects comparable to contextual change interventions. Traditional models of behavior change have not fully integrated this evidence yet, even though it promises to improve the effectiveness of population-wide interventions. This Chapter is a step in this direction and we propose a conceptual framework that integrates several research streams including traditional models of behavior change, behavioral economics, and neuroscience. As a result, we can now begin to comprehend why and how changing the "choice architecture" (Thaler & Sunstein, 2008) can lead people to more desirable behaviors.

ROUTES TO BEHAVIOR CHANGE

Two general paradigms for population-wide behavior change have emerged in recent years – models and interventions that aim to change cognitions such as beliefs and attitudes, and models that change the context or environment within which a person acts. For example, most persuasion and education campaigns aim to change attitudes by relying on reflective processing of information (see Norman, Abraham & Conner, 2000; Shumaker et al., 2008). A second route relies mostly on contextual changes without necessitating change in the underlying cognitions. This route has received relatively less attention in research on population-level behavior change – even though the distinction between behaviors resulting from internally cued, reflective, and intentional changes compared to those arising from externally cued, automatic and reactive change is well-understood in theories of self-regulation (Bargh & Chartrand, 1999).

To describe how the two paradigms accomplish behavior change, we first need to consider the underlying psychological mechanisms causing overt change in behavior. We propose that the two routes accomplish behavioral change by relying, to various degrees, on distinct types of psychological and neural mechanism. A common account is that behavior emerges from the operation of two distinct control mechanisms. An evolutionarily old "System 1" process is described as *automatic, uncontrolled, effortless, associative, fast, unconscious* and *affective*; a more recently evolved, human "System 2" process is described as *reflective, controlled, effortful, rule-based, slow, conscious* and *rational* (see Chaiken & Trope, 1999; Evans, 2008; Slovic, Finucane, Peters & McGregor, 2002, for surveys of the research on dual-process theories in psychology). For example, conscious information processing is effortful and capacity limited, whereas automatic processing can deal with simultaneous inputs independent of each other, often, but not necessarily, outside the window of awareness (e.g., walking and eating a sandwich is automatic, while having a conversation with somebody is conscious and reflective). Another distinction is that the latter is often considered "superficial" and heuristic, while the former provides a more systematic and "deeper" analysis of the world.

The historical foundations of dual-process theories can be traced back to William James who proposed two distinct modes of thinking: associative and true reasoning (Sloman, 1996). Nowadays dual-process theories abound in social, personality, cognitive, and clinical psychology. For example, in cognitive psychology attention and working memory are conceptualized as relying on two distinct processes (Barrett, Tugade & Engle, 2004). Dual-process models are also common in the study of social psychological phenomena related to population behavior change, such as persuasion and attitude change. Social psychologists explain the fact that people behave differently at different times by positing two inner processes activated by different stimuli (e.g., general versus specific, symbolic versus pictorial, physical versus social, and so on). For example, the *heuristic-systematic model* (Chaiken, Liberman & Eagly, 1989), the *elaboration likelihood model* (Petty & Cacioppo, 1986), the *mood as information model* (Schwarz, 1990), and the *self-evaluation maintenance model* (Tesser, 1986) are among a few well-known social-psychological models that invoke dual-process frameworks to explain the effects of persuasive messages on attitudes and behavior. In general, people are assumed to behave in a different way on different occasions insofar as they are inhabited by two qualitative distinct processes, which are simultaneously activated by any situation. Thus, behavior change is understood as the joint function of both processes, but on some occasions, a given stimulus may activate only one process while the other is dormant.

The evolution of the dual system approach has also resulted in the recent development of formal or computational models that can more precisely account for known and puzzling phenomena across several domains in psychology. In cognitive neuroscience, for example, the distinction between propositional and associative processes is known as *model-based* and *model-free* reinforcement learning (see Chapter 2 by Peter Dayan). The former learns an explicit representation model of the environment, while the latter learns the values of different actions in distinct states of the environment without explicitly representing outcomes (Dayan & Niv, 2008).

Our approach takes a similar stance although we do not develop formal aspects in detail here, but instead borrow insights from recent developments using computational models. We propose a more detailed account of how information is represented and processed in each system, which in turn is critical to understanding a mapping between psychological processes and behavior change interventions. Therefore, we assume that the crucial difference between the two systems is the nature of their computations. In particular, the core of our framework is an extension of a dual-process model of implicit and explicit attitude change (Gawronski & Bodenhausen, 2006). We postulate that the external world and inner states are represented in different ways in the two systems, because elements in the two systems are connected by different types of relations – *propositional reasoning* and *associative activation*.

Propositional reasoning temporarily connects perceptual concepts or abstract semantic concepts through semantic relations to which a *truth value* is assigned – this includes logical relations, such as *"is a"*, *"is not"*, or *"implies"*, as well as abstract relations, such as *"causes"*, and social relations such as *"friend "*, *"enemy"*, or *"mother"*. Associative activation is assumed to represent long-term memory, in which elements of sensory, conceptual, affective, and motor representations can be interconnected (see Strack & Deutsch, 2004). Such relations are associative links formed according to principles of contiguity and similarity (i.e., links are created or strengthened if stimuli are presented or activated in close temporal or spatial proximity). Thus, links in the automatic system are not assumed to have any semantic meaning by themselves (i.e., do not carry a truth value and do not reflect declarative knowledge about objects in the world) and the relation between two or more elements is that of a mutual activation. Such associative links reflect correlations between aspects of the environment and cognitive, affective, or motor reactions, without representing the causes of such multimodal or multidimensional correlations. As a result, structures emerge in the associative systems that bind together frequently co-occurring features and form associative clusters.

Dual-process paradigms serve as a unified framework for behavior change, accounting for numerous theories and intervention methods. In essence, the propositional reasoning processes change behavior through reflective, conscious changes in cognitions, as a response to new information – e.g. in the form of arguments for and against a specific behavior, after receiving an advice, or education and skill training; while automatic processes change behavioral responses as a result of an automatic reaction to incentives and cues in the context within which action choices are made. Thus, an automatic route to behavior change holds that people are often more influenced by external, contextual cues (i.e., situational, "bottom-up" factors) than by internal, "top-down" factors like cognitions or motivations.

In order to explain behavior change, Vlaev and Dolan (2010) proposed a new framework that further elaborates the models proposed by Gawronski and Bodenhausen (2006) by providing a more elaborated account of how the associative systems controls behavior (see Figure 14.1). In particular, on the basis of animal and human behavioral evidence, and a recent integrative approach to decision-making in cognitive neuroscience (Glimcher, Camerer, Fehr & Poldrack, 2009; Rangel, Camerer & Montague, 2008), we introduced two distinct systems for automatic behavioral control: *impulsive (Pavlovian) systems*, based on evolutionarily acquired affective responses to particular environmental stimuli, where assignment of value elicits a set of "prepared" behaviors such as approach, consummatory responses to a reward, avoidance, defensive, retaliatory, offensive and fighting responses; and *habit systems*, learning

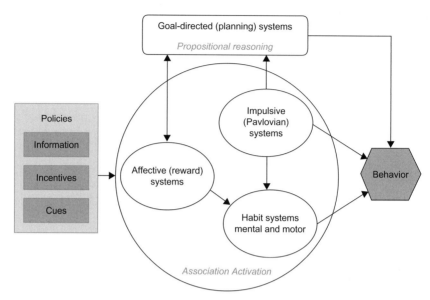

FIGURE 14.1 A dual-process multi-system framework for population behavior change.

through repeated practice in a stable environment, to flexibly assign values to actions and even mental operations. We also distinguish separate *affective systems*, which do not directly control action (see Berridge, Robinson & Aldridge, 2009), but they can influence the controllers and hence are important targets for behavior change interventions.

Indeed, the idea that the brain contains such multiple, separate decision systems is ubiquitous in psychology, neuroscience, and even economics (see Balleine, 2005, for a systematic review of evidence pointing to multiple valuation systems being involved in value-based decision-making). These routes to behavior change are also motivated by knowledge about the existence of separate brain structures, or modules, for automatic processing of different kinds information like perceptual processing, mechanisms for emotional reactions, memory retrieval, and motor actions (Anderson, Bothell, Byrne et al., 2004; Meyer & Kieras, 1997). What uniquely differentiates impulsive systems and habit systems from any other learning is that they readily engage a rich network of value in the brain (see Seymour, Singer & Dolan, 2007), while simple associative learning can be embodied in any neural circuitry or network. Also, much evidence points to competition between a cognitive planning system and impulsive and habitual controllers associated with dopamine and the basal ganglia (see Glimcher et al., 2009; Rolls, 2005). The consensus is that all three systems are subject to mutual interactions and therefore each system must be assigned a "value" so that it can compete with actions favored by the other systems. Such competition has often been implicated in self-control issues such as dieting or drug addiction (see Daw, Niv & Dayan, 2005). In summary, the three systems can independently influence behavior and contribute their own valuations of each option.

Our framework proposes three classes of interventions: *information*, *incentives* and *cues*, which differently engage these control systems. We identify these three intervention types for two reasons. First, these interventions have distinct affects on the three processing systems and behavior change. The second reason is historical, as each intervention type has been predominantly used by only one of the three behavior change paradigms that are integrated into our framework (behavior change theories, behavioral economics, and persuasion models, respectively).

Next, we discuss how the different systems specifically relate to behavior change. We provide evidence about how intervention types influence these control systems and thus cause behavior change (see Vlaev & Dolan, 2010, for a more complete review of the literature). In particular, we argue that the most influential models of population behavior change predominantly rely on information to "persuade" the propositional system to change behavior by inferring new beliefs. We then illustrate how incentives have been used to change behavior via the three control systems and show how redesigning the physical and social environment creates opportunities for new behaviors. Finally, we present evidence about how cues can bring about sustainable behavior change solely via the automatic route.

SYSTEMS FOR SELF-CONTROL

Goal-Directed Systems

Goal-directed or planning systems are centered prefrontal cortex (Rolls, 2005), but may also subsume other mechanisms localized to hippocampus and dorsomedial striatum (see Glimcher et al., 2009). Goal-directed systems in humans represent models of the world or the organism in a propositional format, and engage propositional model-based reasoning to simulate future *outcomes* and calculate sequences of actions to achieve the most valuable goals. Such representations can be flexibly generated and changed, but to do necessitates attentional and computational resources. Dual-process models in social psychology assume that propositional representations are generated by retrieving elements of the proposition, such as concepts and the relation applied to them, from the associative systems and assigning truth values to the relation. Once knowledge has been generated, syllogistic rules (used for the transfer of truth from the premises to the conclusion) are applied to draw inferences that go beyond the information given (Bruner, 1973). For example, during a persuasion campaign against smoking, the perceivable features of the campaign will spread activation to the *cancer* concept in the associative systems. Then the relational schema of *causal* relationship will be retrieved and combined with the labels *smoking* and *cancer* (the labels can also be combined with a visual input such as the image of a smoking person). Thus, the propositional representation "*smoking causes cancer*" is generated.

As a result, the same person may wonder how often he or she smokes and whether they are a passive smoker. Categorical knowledge or *new beliefs* about oneself may be derived from the categorization of oneself as an active or passive smoker, and the proposition that being a smoker is associated with bad health. The propositional belief "*I will become ill*" may then be inferred based on this categorization. Propositional reasoning (e.g. in the form of modus tollens – denying the consequent) may also lead to the conclusion that "*smoking cessation will help avoid illness*", leading a person to infer new behavioral goals (e.g., avoiding being amongst smokers). In this way, propositional reasoning generates new knowledge about the person at risk from illness, and to a conclusion about the desirability of a particular action and goal. This behavioral decision should then activate behavioral schemata (e.g., walking out of a smoking venue) that are appropriate in the situation (Strack & Deutsch, 2004).

In summary, propositional inferences based on premises with evaluative content can be made about what is the case, and about what is good or bad, because syllogistic rules also apply to evaluative judgments, such as attitudes (see Schwarz & Bohner, 2001). Human language is a symbolic implementation of goal-directed thinking that is embodied in propositional reasoning (Rangel et al., 2008). Therefore, providing verbal information tends to trigger goal-directed systems as persuasion information is often about state–action–outcome sequences.

Habit Systems

Control over decisions often transfer from goal-directed mechanisms to a habit-based system that controls both action habits and mental habits. This distinction is encouraged by recent evidence supporting these two constructs and also by the specific purposes of designing behavior change interventions. According to Bargh (1996, p. 28) "*any skill, be it perceptual, motor, or cognitive, requires less and less conscious attention the more frequently and consistently it is engaged.*" Habitual control is mediated by instrumental learning, whereby an individual learns to associate a particular action or cognitive strategy with its value in a given state without an explicit representation of the specific outcome or goal (which is a privilege of the goal-directed system). Consequently, actions and mental operations that lead to a reward are executed more frequently in the specific state/ environment, whereas those that lead to aversive events are executed less often. The dissociation between habit systems and planning systems is supported by an important characteristic of habits, namely their resistance to change when they are in opposition to intentions (Neal, Wood & Quinn, 2006; Oullette & Wood, 1998).

Action Habits

Habit systems are responsible for instrumental and adaptive state-action associations (also known as operant conditioning), and thus avoiding the need to compute the expected outcomes. Conventionally, habits have been defined as "behavioral dispositions to repeat well-practiced actions given recurring circumstances" (Wood, Tam & Witt, 2005). They are assumed to develop through repetition of the behavior (e.g., smoking, or reading the news) in the presence of consistent stimuli (e.g., coffee, home), environment, or context (Neal et al., 2006), which leads to habits being automatically cued by environment and easier to perform over time. Wood and Neal (2007) argue that habits are usually the consequence of past goal pursuit and they arise when people repeatedly use a particular behavior (e.g., drinking) in particular contexts (e.g., dinner table) to pursue their goals (e.g., relaxation and socializing). However, once acquired, habits are performed without mediation of a goal and they can be triggered just by relevant context cues (e.g., sight and taste of food provokes desire to drink without the purpose of social relaxation). Wood, Quinn and Kashy (2002) estimate that about 50% of our everyday activities are performed habitually. For example, most behaviors of relevance to public health (and behavior change) tend to be habitual, such as eating, exercising, drinking, driving and hygiene practices.

At a neural level, habitual controllers are implemented in the dorsolateral striatum under the dopamine neurons into this area (utilizing a dopamine-dependent circuit via the substantia nigra and ventral tegmental area) are important for learning the value of actions; but stimulus–response representations might also be encoded in cortico-thalamic loops and the infralimbic (medial) prefrontal cortex. At a computational level, associative clusters can bind together frequently co-occurring motor representations with their conditions and their consequences or reward values. Research on the connection between perception and behavior suggests that semantic concepts can also be automatically connected to motor programs (e.g., Bargh, Chen & Burrows, 1996; Dijksterhuis & Bargh, 2001). Such sensory-motor and conceptual-motor clusters are called *behavioral schemata*, which have important implications for habit learning and behavior change interventions, as environmental cues can be used in interventions to trigger such schemata.

Mental Habits

A widely held assumption is that automaticity of cognitive procedures is achieved through frequent execution (e.g., Anderson, 1993). Mental habits can be classified as "reflexive reasoning" as distinct from "reflective reasoning" – the former described as a more automatic and effortless process that, like memory retrieval, must rely strictly on existing connections in long-term memory (see Hummel & Holyoak, 2003, for a more formal treatment of these two reasoning types in cognitive sciences). Several recent studies demonstrate the importance of cues in the automatic operation of mental habits, which as a concept is relatively new in psychology. Recognizing that it is the automaticity of repeated behavior that makes it a habit, Verplanken, Friborg, Wang et al. (2007) applied this habit concept to mental events and investigated negative self-thinking as a mental habit – with a key distinction between mental content (negative self-thoughts) and mental process (negative self-thinking habit). The negative self-thinking habit was assessed with a metacognitive instrument (Habit Index of Negative Thinking; HINT) measuring whether negative self-thoughts occur often, are unintended, are initiated without awareness, are difficult to control, and are self-descriptive. Verplanken and Velsvik (2008) demonstrate habitual negative body image thinking as psychological risk factor in adolescents. In behavior change setting, Orbell and Verplanken (2010) report related work on mental habits in health behavior – situational habit cues eliciting wanted and unwanted habit responses (smoking when drinking alcohol in a pub, and dental flossing when forming an implementation intention to floss in response to a specified situational cue).

Thus, mental habits are within the realm of automatic processes, but we also extend this concept to well-documented mental automatisms known as *heuristics* (or mental shortcuts), such as the tendency to make certain kinds of judgments that may depart from rationality. For example, "lexicographic heuristics" (such as "satisficing" and one-reason decision-making) describe the tendency to make decisions by sequentially using only one choice dimension at a time, starting with the most important or salient attribute, and continuing until all unsatisfactory alternatives are eliminated (see Gigerenzer & Goldstein, 1996, for a review). Heuristics are often considered to be adaptive cognitive strategies that exploit the informational and statistical structures in the environment, and hence they are often described as "ecologically rational" and part of the human "adaptive toolbox" (Gigerenzer & Selten, 2001; Gigerenzer, Todd & the ABC Research Group, 1999). Some heuristics have evolved by building upon other psychological principles and capacities (Gigerenzer, Hoffrage &

Goldstein, 2008). For example, the "recognition heuristic" (Goldstein & Gigerenzer, 2002) makes inferences from patterns of missing knowledge by exploiting a fundamental adaptation of many organisms: the capacity for recognition. Many other heuristics are triggered only by specific environments and decision contexts, for example, when deciding whom to befriend and whom to compete against, which is an adaptive problem for humans and other social animals (Cosmides & Tooby, 1992).

Impulsive (Pavlovian) Systems

Motivation is often defined as activation of goal-oriented behavior (Elliott & Dweck, 1988) even though it may not involve an explicit model of the expected outcomes as in a goal-directed system. Impulsive or instinctive controllers emit evolutionarily appropriate, innate behaviors to specific predetermined stimuli, but state-based associative learning allows organisms to deploy such behavior in response to other stimuli (such behaviors are also known as "unconditioned" and "conditioned" Pavlovian responses). Therefore, purely impulsive Pavlovian actions (e.g., approach or avoidance, fight or flight, consume or repel) are stimulus-triggered responses, ones that provide only a restricted set of options for action. Nonetheless, impulsive systems can control a surprisingly wide range of human behaviors with important consequences, such as overeating, addiction, obsessive–compulsive behaviors and opting for immediate smaller rewards at the expense of delayed larger rewards (see Dayan, Niv, Seymour & Daw, 2006). This follows from the character of the Pavlovian responses – resulting from classically conditioned value predictions, which are myopic and directed toward immediate reinforcers and their predictors.

In terms of computations, impulsive systems can learn to associate conditioned with unconditioned stimuli (also known as classical/Pavlovian conditioning), thus enabling the organism to predict whether certain environmental *states* or cues are likely to be followed by (un)desirable *outcomes*. Pavlovian systems include the amygdala which, through its connections to brainstem nuclei and core of the nucleus accumbens, orchestrates hard-wired responses while more specific responses are controlled through amygdala's connections to hypothalamus and periaqueductal grey (e.g., responses to negative stimuli have specific and spatial organizations along an axis of the dorsal periaqueductal grey). More broadly, learning in impulsive systems is implemented in basolateral amygdala, the ventral striatum and orbitofrontal cortex.

Direct effects of impulsive systems on habitual responses (see Figure 14.1) is demonstrated in the phenomenon known as *conditioned suppression* (Estes & Skinner, 1941), whereby stimuli predicting aversive Pavlovian outcomes suppress appetitive instrumental responding leading to inappropriate actions such as withdrawal. Niv, Joel and Dayan (2006) discuss and model related phenomenon known as *Pavlovian-instrumental transfer* (PIT) demonstrating (under extinction conditions) that the vigor with which people perform an habitual response for a positive outcome is influenced by the mere presentation of Pavlovian stimuli that are associated with either appetitive or aversive outcomes. Talmi, Seymour, Dayan and Dolan (2008) have gone on to show that enhanced motivational vigor is correlated with enhanced activity in the nucleus accumbens, a brain area viewed as a "limbic–motor interface" that enables associative information, such as Pavlovian incentive value, to influence habitual responding. This unidirectional effect of motivation systems on the habit systems is supported by related findings that vigor increases when habituation is stronger (i.e., a direct interaction effect). PIT has been used to explain aspects of drug addiction, and particularly the effects of environmental drug-associated cues in eliciting drug-seeking behavior and relapse in recovering addicts (see Everitt, Dickinson & Robbins, 2001).

Goal-directed control can also use inputs from the impulsive systems for hedonic valuation of future states (Gray, Braver & Raichle, 2002; Ochsner & Gross, 2005), although these systems can still compete when controlling behavior (as shown in Figure 14.1). Dayan and Seymour (2009) suggest effects of the impulsive system on the goal-directed (planning) system, which are relevant for behavior change interventions. Such effects emerge when predictions of negative affective states lead to Pavlovian withdrawal. For example, when people entertain chains of thought or plans about the future, any chain of thought leading towards a negative outcome would trigger a Pavlovian withdrawal response, which may lead to this decision tree being terminated or pruned (i.e., this would block out branches of the decision tree during a tree-based search). Therefore, if individuals contemplate the future, they tend to favor branches with more positive outcomes. Thus, they may, for example, ignore possibilities that relate to negative health outcomes and focus only on the positive consequences of current actions; this might explain the unsatisfactory effects of health promotion campaigns targeting the goal-directed systems. Similarly, people also exhibit related behaviors, such as not collecting information if it is likely to provide bad news (e.g., when receiving persuasion messages to participate in regular health screening tests, which might explain the low uptake of cancer screening).

Affective Systems

Affective systems generate psychological states known as *drives* and *emotions*, and human behavior is strongly influenced by the attendant evoked feelings (Aunger & Curtis, 2008). Zajonc (1980) argued that affective reactions are faster and more automatic than cognitive reactions and showed that people can experience an affective reaction to a stimulus before they realize what it is they are reacting to; which suggests that affective and motivational reactions are automatic. For example, sudden and unexpected noises can cause fear before people figure out the source of the noise. Recently, there has been an increase in research on the role of emotions in decision-making. Cohen, Pham, and Andrade (2008) argue that judgments that are evoked by subjective feelings and moods (for example, sadness or disgust) are influenced by an *affect heuristic*. Slovic et al. (2002) also consider the affect heuristic at work if subconscious emotional evaluations are used as the basis of decisions, although they occur even before cognitive evaluation takes place (Kahneman, 2003). Loewenstein (2000, p. 427) suggested that affects like negative emotions (e.g., anger, fear), drive states (e.g., hunger, thirst, sexual desire), and feeling states (e.g., pain) are more important in individual daily lives than higher level cognitive processes (that are often presumed by researchers to underlie decision-making).

A more elaborate understanding of the affective responses triggering the three control systems is likely to be highly informative for the design of behavior change interventions, which require deconstructing "drives" and "emotions" into more specific affective or motivational states. In a systematic meta-analysis of interventions to promote hygiene behavior in eleven developing countries, Curtis, Danquah and Aunger (2009) subdivided motivated behavior into several categories of drives and emotions that tend to trigger hand-washing behaviors: *affiliation* (seek to conform so as to reap the benefits of social living), *attraction* (be attracted to, and want to attract, high-value mates), *comfort* (place one's body in optimal physical, chemical conditions), *disgust* (avoid objects and situations carrying disease risk), *fear* (avoid objects and situations carrying risk of injury or death), *nurture* (want to care for offspring), *status* (seek to optimize social rank). Note that even though the local beliefs and the social and physical environments were quite varied across the eleven countries, the specific motivations, however, represented a common universal set. Judah, Aunger, Schmidt et al. (2009) utilized this framework to develop intervention messages aimed at increasing hand washing in a developed western society, which proved that the most effective messages are based on automatic motivational mechanisms. In our discussion of the literature, we employ similar classification of specific affective states, or primary rewards (e.g., *belonging, disgust, status, fear*, etc.), which can be interpreted as targeting specific self-control systems. We also realize that this list is not complete as other evolved states might yet be proven effective in behavior change interventions (e.g., *control, greed, curiosity, boredom, guilt, hope*) (see Fiske, 2010). Thus, our review does not aim to catalog all possible affective states. Instead, we focus on the basic underlying mechanisms triggered by behavior change interventions.

V. IMPLICATIONS, APPLICATION AND FUTURE DIRECTION

Brain structures commonly linked to affect include the amygdala, the cingulate cortex, and the insular cortex (LeDoux, 2000). The amygdala is implicated in a range of emotional and non-emotional processes (e.g., fear learning and enhanced attention to threat stimuli – Phelps, 2006). Anterior cingulate cortex is implicated in assessing the salience of emotion and motivational information (Allman et al., 2001). Insular cortex, represents somatic information, particularly as it relates to arousal and feelings (Critchley, Wiens, Rotshtein et al., 2004; Damasio, 2000), as well as disgust (Phillips, Young, Senior et al., 1997) and empathy (Singer, Seymour, O'Doherty et al., 2004). The anterior insula in particular, is involved in mapping physiological states of the body, including pain, touch and visceral sensations of autonomic arousal (Critchley, 2005). The right anterior insula is considered a cortical center for interoception that may be involved in decision-making by representing and computing valenced subjective feeling states (Critchley et al., 2004; Damasio, 1994). Investigations of the neural systems mediating valence (e.g. in amygdala) and arousal (e.g. in anterior cingulate cortex) have shown some clear dissociations between regions sensitive to valence versus arousal (e.g., Anderson & Sobel, 2003), but there may also be some brain regions most sensitive to specific arousal–valence combinations (Cunningham, van Bavel & Johnsen, 2008). In summary, the accumulated evidence suggests that a wide-spread network of brain regions is involved in processing motivational relevance.

Differentiation between the components of affective systems and action control systems (as in Figure 14.1) also resonate with recent neurobiological evidence dissecting three dissociable psychological components of reward: "liking" (hedonic impact), "wanting" (incentive salience), and "learning" (predictive associations and cognitions) (see Berridge et al., 2009, for a review). In particular, hedonic hotspots for opioid enhancement of sensory pleasure, i.e. "liking," are located in the nucleus accumbens' medial shell (part of ventral striatum) and the target for its outputs – the ventral pallidum (posterior half). Suppression/activation of endogenous dopamine neurotransmission reduces/amplifies "wanting" but not "liking."

Translated in terms of our framework, "liking" overlaps with affective systems, "wanting" with impulsive (Pavlovian) or motivational systems, and learning signifies the processes within and between the associative systems (see Figure 14.1). The reason we attribute "wanting" with impulsive (Pavlovian) systems is because "wanting" can apply to innate incentive stimuli (unconditioned stimuli) or to learned stimuli that were originally neutral but now predict the availability of reward (Pavlovian conditioned stimuli). Such stimuli can increase motivation to approach or search for rewards, and increase the vigor with which they are sought – these are properties characteristic of Pavlovian systems

(Talmi et al., 2008). Thus, "wanting" is distinguishable from more cognitive forms of desire that involve declarative goals or explicit expectations of future outcomes, which are mainly mediated by cortical circuits (i.e. the goal-directed systems).

Goal-directed control can use direct inputs from the affective systems for the hedonic valuation of future states (Ochsner & Gross, 2005), and neuroimaging data demonstrate integration of emotion and cognition in the lateral prefrontal cortex (Gray et al., 2002). The goal-directed system also influences the affective system, which is supported by evidence that dorsolateral prefrontal cortex (DLPFC) is involved in cognitive control over emotions (Miller & Cohen, 2001). For example, when DLPFC activation exceeds anterior insula activation, people are more likely to accept unfair offers in bargaining games (Sanfey, Rilling, Aronson et al., 2003); and DLPFC is also involved in overriding emotional biases when delaying gratification (McClure, Laibson, Loewenstein & Cohen, 2004).

BEHAVIOR CHANGE INTERVENTIONS

Information

Many behavior change interventions focus on the way people think by providing information (in the form of messages, education, advice) that, assuming rationality, uses persuasion to adopt a specific behavior and to train the skills needed to adopt this new behavior (see Abraham & Michie, 2008). For example, most research on health-related choices involves using messages that contain arguments describing either the benefits of adopting a healthy behavior (e.g., physical wellness) or costs (risks) of unhealthy behavior (e.g., heart disease) (Gray, 2008). Information usually contains arguments, by definition propositional statements. For the purposes of population behavior change, we do not assume that informative arguments can affect behavior via automatic/ associative systems – not because we render it impossible (e.g., some arguments can easily trigger various affective automatisms), but because traditional approaches assume such arguments are aimed at changing beliefs, attitudes and goals through propositional reasoning.

Arguments vary in strength, and researchers have put forward guidelines on how to develop such arguments in persuasion campaigns and behavior change interventions (Petty & Cacioppo, 1986, pp. 31–33). Ruiter, Kok, Verplanken and Brug (2001) applied this approach in health promotion to demonstrate that respondents reported a more positive attitude toward breast self-examination (BSE) after reading persuasive messages containing stronger (e.g., *"By performing BSE you are able to detect breast cancer in an earlier and therefore more treatable stage"*) as opposed to weaker arguments (e.g., *"Performing BSE is a nice way to be intimate with yourself"*). In the context of a health threat, a typical communication of health risk contains a proposition stating the perceived cause of a threat and a related effective coping procedure: *"IF high blood pressure is caused by being unfit THEN exercise will reduce it"* (Marteau & Weinman, 2006).

Interventionists also try to uncover the characteristics of risk information (such as states, actions and probable outcomes) likely to motivate behavior change (e.g. information about unhealthy behaviors, or DNA risk regarding an inherited predisposition to certain diseases). Such interventions identify the cognitions to target (e.g., cognitive representations of threat) so as to optimize the motivational impact of risk information (see Marteau & Weinman, 2006). Marteau and Lerman (2001) demonstrate that providing information about genetic risk (associations between *states, actions* and adverse *outcomes*) may not be sufficient to increase motivation to change behavior, but change is more likely if people are persuaded that changing their behavior can indeed reduce the risk of an adverse health outcome – i.e. information about associations between states, actions and positive outcomes. The latter type of information (e.g., presented as bar-charts showing changes in health-related risks before-and-after behavioral change) is needed to alter beliefs about the target behavior and beliefs about one's ability to modify this behavior.

The explicit focus on outcomes is also evident in interventions showing that people are more likely to change behavior (e.g., quitting smoking) when they generate more avoidance goals or outcomes during propositional reasoning (e.g., getting rid of a hacking cough, not developing cancer and heart diseases, getting rid of the smell of smoke on clothes and belongings) (see Worth, Sullivan, Hertel et al., 2005). It is also essential to infer short-term goals that are vivid and detectable, as people can easily monitor their progress (e.g., get rid of the cough), as well as to infer long-term goals (Rothman, Hertel, Baldwin & Bartels, 2007). In summary, this evidence shows that people need to maintain explicit and detailed knowledge of the decision tree (including its short-term and long-term states and outcomes) that guides their behavior change.

In line with models assuming that health beliefs underlie propositional inferences about appropriate courses of action, Curtis et al. (2009) found that flaws in the chain of belief linking hand-washing to long-term and uncertain beneficial outcomes (preventing child diarrhea and hence the possible serious illness or loss of the child) explains why mothers in developing countries do not adopt hand-washing with soap. For example, mothers know about germ theory but they also have prior beliefs that diarrhea is a benign, non-life-threatening, disease, and hence unlikely to lead to adverse outcomes. In such circumstances, the causal chain for belief, or the decision tree, about diarrhea is possibly too long, and without immediate value, to motivate change in current behavior.

This also illustrates that intervention should try to elicit the naive beliefs and decision trees in the target population.

In summary, almost five decades of research, on whether changes in cognitions engender population-level behavior change, have been embodied in dozens of theories and documented in hundreds of publications. The domains of application cover most maladaptive and problematic behaviors, which have been the focus of public policy concern. Such topical behaviors include (just to name a few) condom use, use of dental tablets, testicular self-examination, parent–child communication, smoking, skin examination, course enrolment, sunscreen use, visiting an internet site, low-fat diets, contraceptive use, exercise, indoor tanning, donating behavior, sun protection, sexual behavior, breast self-examination, seat belt use, cycle helmet use, study behavior, AIDS-risk behavior, smoking, HIV-preventive behavior, calcium intake, and others (see Webb & Sheeran, 2006, for a review of traditional psychological models and their application). Such "traditional" interventions differ in terms of the theoretical basis of the intervention, but all rely *mostly* on providing people with information that should engage a reflective cognitive system.

In terms of theoretical models that usually underlie such interventions, the richest comprises numerous theories in social and health psychology, which assume that providing information that changes various beliefs, also produces intentions to change one's behavior, and those intentions ultimately cause the altered behavior. Most prominent behavior change models based on attitude change are the *health belief model* (Rosenstock, 1966), *protection motivation theory* (Rogers, 1983), the *theory of reasoned action* (Fishbein & Ajzen, 1975), the *theory of planned behavior* (Ajzen, 1991), the *model of interpersonal behavior* (Triandis, 1977), and the *social-cognitive theory* (Bandura, 1977). All these models postulate that behavior is changed by change in beliefs linking the behavior of interest to expected outcomes (which is the subjective probability that the behavior will produce a given outcome). It is assumed that these beliefs in combination with the subjective values of the expected outcomes determine the *attitude* toward the behavior. Other beliefs (not only attitudinal ones) are also proposed to play role in these theories, such as beliefs about normative approval, self-efficacy, and behavioral control (e.g., see Ajzen, 1991).

Incentives

Behavioral economics has convincingly demonstrated that people respond to incentives, which usually tend to activate both reflective and automatic processing – for example, people rationally respond to change in prices and costs, while they still overreact to losses relative to gains of equal magnitude (Kahneman & Tversky, 1979, 1991, 2000). Thus, incentives may require complex propositional reasoning, such as computing the value of the incentive, making trade-offs, and updating beliefs and goals as a result; or incentives also provoke simple automatic association between the incentive and a specific action. This dichotomy is embodied in the distinction between short-run and long-run patience. Evidence from neuroscience shows that the former is driven by an affective system that responds preferentially to immediate rewards and is less sensitive to the value of future rewards, as opposed to a more patient system that evaluates trade-offs between abstract rewards, including rewards in the more distant future (see Daw et al., 2005, for a review). Given such evidence, our approach is to utilize these biases to "lead" people into making behavioral choices that maximize their welfare (e.g., Thaler & Sunstein, 2008). Next, we consider three types of incentive-based interventions depending on the decision system that is mostly affected in provoking behavior change.

Incentives for Goal-Directed Control

People are sensitive to prices and costs – the economic *law of demand* (Pearce, 1986). Most demand curves (depicting the relationship between the price of a certain commodity and the amount of it that consumers are willing to purchase at that given price) slope downwards from left to right (Kreps, 1990). In other words, higher cost reduces our desire to consume (e.g., for health-related consumption of alcohol and cigarettes, see Chaloupka, Grossman & Saffer, 2002; Colman, Grossman & Joyce, 2003; Manning, Blumberg & Moulton, 1995). Traditional incentives-based interventions also involve introducing rules (legislation, regulation) involving sanctions (e.g., smoking ban, compulsory seatbelts), which have proven to be very effective in behavior change (Chaloupka & Grossman, 1996; Chaloupka et al., 2002).

In the health domain, for example, drug use decreases as constraints on drug consumption increase – in terms of the price and effort required to obtain and use the substance, or in terms of the quality and magnitude of the alternatives forfeited as a consequence of drug use. Similarly, as constraints on drug use decrease (e.g., drugs are readily available at low cost or there is little in the way of alternatives to be forfeited by using drugs), consumption increases (Chaloupka, Grossman, Bickel & Saffer, 1999). An alternative way to foster healthier lifestyles is to offer incentives that encourage people to eat healthier foods, take more exercise, drink less alcohol, and give up smoking (Marteau, Ashcroft & Oliver, 2009). Interest in using incentives to promote behavior change has increased as the economic and social costs of unhealthy behaviors have become apparent (Sindelar, 2008).

Incentives for Habitual Control

Learned action habits in the domain of incentives relate to a behavioral legacy in the application of incentives to provoke reinforcement learning in individual-level behavior change (see Wolpe, 1990). The reinforcement principle has also been successfully used at a population-level to develop interventions to treat drug addiction and substance misuse (including smoking and alcohol consumption) and improve medication compliance (see Higgins, Heil & Plenbani-Lussier, 2004, for a review). Such interventions usually include earning points contingent on outpatients submitting drug/substance-negative urine specimens. Points usually begin at a low value (e.g., $2.50) and increase with each consecutive negative test result. Points can then be exchanged for retail items kept onsite at the clinic, or clinic staff can make purchases on behalf of the patient. A drug-positive result or failure to provide a scheduled specimen resets the voucher value back to an initial low value from which it could increment again (e.g., see Roll, Higgins & Badger, 1996).

In general, such contingency management, which involves the systematic delivery/withholding of reinforcing consequences contingent on the occurrence/absence of a target response, can be used to improve outcomes across a range of different behaviors and populations. This research also shows that drug use is a form of operant behavior that is sensitive to environmental consequences, and that the degree of control that drugs exert over behavior as reinforcers is malleable and dependent on an environmental context offering alternative reinforcers.

Incentives for Impulsive Control

The presentation or removal of money has been shown to elicit physiological responses indicative of an emotional reaction (see Delgado, Gillis & Phelps, 2008), which can result in Pavlovian responses. Many important questions in public policy and behavior change relate to how individuals evolved to respond to various kinds of incentives like price (cost) and benefits. Thornton (2008) reports an example of economic incentives at work in a field study conducted in Malawi where the problem was the high frequency of people tested for HIV/AIDS who did not bother to subsequently pick up their results. Thornton demonstrated that offering small incentives can circumvent this failure where the key observation was the biggest jump between zero and a very small incentive (10–20 cents which is about one-tenth of a day's wage) – sufficient to increase take up by 50%. Offering more money further affects behavior but to a much lesser degree. Here, the interesting result is a change in behavior when the financial reward increases from zero to a very small amount, which seems like a salient, "attention grabbing" change that commands an impulsive action to obtain this (insignificant in economic terms) immediate reward.

Similar effects of small incentives are reported by Duflo, Kremer and Robinson (2010) in a field study in Kenya, where a small time-limited reduction in the cost of purchasing fertilizer at harvest induces a substantial increase in fertilizer use, equivalent to larger price cuts later in the season. This intervention was based on a modeling of farmers' decisions as present-biased (over-discounting their future utility) and partially naive (underestimating the odds that they will be impatient in the future), which leads to procrastination and postponing purchasing fertilizer until proceeds from the harvest are spent. Thus, human welfare can be improved by exploiting known effects of incentives on an impulsive system that drives a hyperbolic discounting that overvalues a proximal/immediate stimulus (Frederick, Loewenstein & O'Donoghue, 2001). McClure et al. (2004) report neurobiological evidence that competing neural valuation systems, one with a low discount rate and one with a high discount rate, play a part in decisions that involve choosing between immediate small monetary payoffs and larger but delayed payoffs.

Impulsive or Pavlovian automatisms elicited by commitment devices have been utilized in interventions aimed to change financial behaviors. Thaler and Benartzi (2004) discuss the results of a commitment-based behavior change intervention known as the *save more tomorrow plan*, which allows employees to pre-commit to increase their contribution (savings) rate in a 401(k) pension plan whenever they get a pay raise. In this plan, various Pavlovian automatisms either play a role or are taken into account in the design, such as delaying the salient immediate cost of foregoing current consumption, and discounting the future loss in consumption (hyperbolic agents procrastinate and fall into inertia, because they feel that later activities will not be as important as the current activity). As a result, people are happy to pre-commit by agreeing to incur the future cost of the action (e.g. saving or paying a fine if not losing weight). In this case, the average saving rates for the participants increased from 3.5% to 11.6% over the course of 28 months. In a study conducted in the developing world, Ashraf, Karlan and Yin (2006) report that individuals identified as having hyperbolic preferences also desire commitment devices – a saving product offered by a Philippine bank, intended for individuals who want to commit now to restrict access to their savings.

Similar mechanisms play a role in *defaults*, which are options that are assumed as preselected if the individual does not make an active choice of another available alternative (see Thaler & Sunstein, 2008, for a discussion). Individuals procrastinate in their decision to opt out of the default, because they avoid current costs but are happier to incur future opt-out costs. Powerful effects of defaults on behavior (and wellbeing) have been observed in organ donation decisions (Johnson & Goldstein, 2003; Abadie & Gay, 2004), employees' contributions to healthcare flexible-spending accounts (Schweitzer, Hershey & Asch, 1996). Halpern, Ubel and Asch (2007) discuss various ways in which default options can improve the quality of healthcare such as an opt-out policy of various routine vaccinations and tests for patients and staff.

Cues

According to dual-process models of persuasion (e.g., see Chaiken & Trope, 1999), people respond to contextual *cues* that tend to trigger an automatic processing system, cues that are not mediated by propositional thinking and goal-directed planning. Recent evidence suggests that human behavior departs from cognitive rationality, because it is very susceptible to subtle cues in the environment within which choices are made, or the *choice architecture* (Thaler & Sunstein, 2008). This evidence suggests that the situation within which an individual acts and makes choices, triggers automatic processes that influence action. Therefore, we suggest that interventionists can gain from a more systematic approach to behavior change that takes greater consideration of contextual cues in behavior. Here, we present evidence for several types of motivational cues and habit cues, some of which have not yet been proposed by existing models of persuasion or behavior change. We demonstrate that when developed as methods for behavior change, such cues can have a powerful effect on behavioral decisions. This is not an exhaustive list of cues – instead we intend to illustrate the basic principles; our suggestion is that, in order to trigger behavior change, these cues tend to engage affective, impulsive, or habit systems.

Cues for Habitual Control

Abundant evidence demonstrates that salient cues trigger *action habits*. For example, prolonged exposure to salient stimuli increases the likelihood of succumbing to (healthy) temptation. Some interventions have used salient cues to provoke healthy eating by positioning the healthy foods at the beginning of the queue in canteens and at the most visible place, while less healthy foods are positioned last and at the least visible places (Thaler & Sunstein, 2008, pp. 1–3, report a successful application of this principle). Lally, Chipperfield and Wardle (2008) successfully used an action habit-based approach in an intervention to enable participants to control their weight, delivered as a leaflet containing advice on habit formation – simple recommendations such as eating roughly at the same times and incorporating target activity behaviors into daily routines. Lally, van Jaarsveld, Potts and Wardle (2010) apply this approach to developing healthy habit formation across a variety of healthy eating, drinking and activity behaviors. In another health domain, interventions that alter subtle cues in eating situations have been shown to successfully control eating habits. Sobal and Wansink (2007) demonstrated that the amounts of food and drink people serve and consume decreases when smaller plates, spoons, and glasses are used. Smaller plates lead to less food intake, because people habitually consume around 92% of what they serve themselves (Wansink & van Ittersum, 2011). Thus, the amount of food on a plate is a contextual cue that prompts habitual control of food intake. Therefore, by using small plates and utensils, behavioral change interventions could take advantage of the fact that people's choices for serving size and food intake are habitually cued in relation to container size (Wansink & Cheney, 2005). For example, if a 12 inch plate is reduced to a 10 inch plate, this would result in a 22% reduction of calories per serving, which would cause a weight loss of around 8 kilograms per year for an average sized adult.

Mental habits are also triggered by salient cues that focus respondents' attention on the relevant choice attributes (e.g., calorie content, alcohol units, or costs). Habitual, heuristic use of one cue at a time (Gigerenzer & Goldstein, 1996) is well exemplified in interventions that change behavior by focusing people's attention on important dimensions. This mechanism is illustrated in field intervention testing whether information on HIV risks can change sexual behavior among teenagers in Kenya (Dupas, 2011). Providing information on the relative risk of HIV infection by partner's age group led to a 28% decrease in teen pregnancy and 61% decrease in the incidence of pregnancies with older, riskier partners. In contrast, there was no statistically significant decrease in teen pregnancies after the introduction of the national HIV education curriculum, which provided only general information about the risk of HIV and did not focus the message on the risk distribution in the population. These results suggest that teenagers are responsive to risk information that would enable them to reduce their exposure to risk. Deciding whether to have sex with a particular partner is a complex choice problem based on various social and situational factors, which often leads to inaction or wrong choices. By making the age of partner salient, the intervention reduced a multi-attribute choice dilemma to a habitual (heuristic) decision based on a single attribute or cue, which enabled the teenagers to select behaviors that improved their welfare.

Cues for Impulsive Control

Our review organizes the evidence according to the primary rewards or motivational states that we interpret as being triggered by the interventions in question. For example, prominent behavioral effects in the literature are usually due to the affective and/or impulsive systems when interventions elicit states such as *belonging, attraction, greed, disgust, and fear*. We do not catalog all possible impulsive and affective responses enlisted in the literature (e.g., see Curtis et al., 2009; Fiske, 2010). Note that some examples here are interpreted as interactions between affective and impulsive systems, but that does not imply that the affective system cannot interact with habit and goal-directed systems in other interventions. With these caveats in mind, our key message is that people's in-build predispositions to specific emotional and impulsive responses can be affective "drivers" in behavior change interventions.

Humans "want" to *belong* and they impulsively seek *affiliation* with groups and similar others. This resonates with Fiske's (2010) suggestion that this is the most fundamental motive, because *"as a foundation, people first and foremost need to belong (to relationships and groups) to survive; the environment to which people adapt is the social group, and culture codifies survival rules in different groups."* The instinct for belonging or affiliation is usually triggered by cues indicating social *norms*, which causes people to conform. Using norms as cues for behavior change is reported in several papers by Cialdini and his colleagues. Goldstein, Cialdini and Griskevicius (2008) tested whether a hotel-towel reuse sign conveying information about social norms might be more persuasive than a sign widely adopted throughout the hotel industry. One sign was designed to reflect the basic environmental-protection message asking guests to help to save the environment and show their respect for nature by participating in the program, which resulted in 35.1% participation. A second sign utilized the social-norm information with the truthful message that most guests at the hotel recycled their towels at least once during their stay, which increased the percentage of participation to 44.1%. Note that this response may well be explained by goal-directed thinking (e.g., *"I want to be seen to belong to the social class of the customers using this hotel, and hence should be environmentally conscious"*), or by habitual imitation of local customs and rules that has been frequently rewarded in the past. However, a third type of message conveyed that most previous occupants of the room had reused towels at some point during their stay, which significantly increased the towel reuse rate to 49.3%. Obviously, a logical chain of thought could not lead to the conclusion that the inhabitants of this particular room are representative of all customers; nor could habitual imitation of specific "hotel room norms" have been reinforced. Most likely, people often impulsively and automatically want to conform to what others, with whom they have something in common (e.g. inhabit the same hotel room), usually do.

Cialdini (2003) reports a related intervention to reduce antisocial behavior, which placed a pair of signs in different areas of a national park. The first sign urged visitors not to take wood, and depicted a scene showing three thieves stealing wood, while the second sign depicted a single thief – indicating that stealing is definitely not a social/collective norm. The measure of "message" effectiveness was the percentage of marked pieces of wood stolen over five weeks. As expected, the second message resulted in significantly less theft than the first message (7.92% versus 1.67%), which is likely to be driven by impulsive avoidance/suppression of the harmful actions. Schultz, Nolan, Cialdini et al. (2007) demonstrate that this lesson applies to population-level interventions aiming to prevent environmentally harmful behavior such as energy wastage – a descriptive normative message describing average energy usage in the neighborhood produced either a desirable energy saving when households were already consuming at a low rate, or an undesirable increase in consumption when households were consuming at a low rate. Also adding an injunctive norm conveying social approval or disapproval – by showing smiling or sad faces, respectively – eliminated the undesirable increase. This example does not exclude the possibility that habitual imitation can work in the same direction as an impulsive desire to conform.

In the health domain, norms are very affective in interventions promoting hygiene behaviors in a natural setting. Judah et al. (2009) tested the effectiveness of messages aimed at increasing hand washing with soap in highway service station restrooms in the UK. The intervention messages were displayed on an electronic dot matrix screen over the entryway to the two restrooms, and wireless devices recorded entry and soap use. Both genders responded well to messages based on norms. Indeed, the only message that was effective for both genders was the norms message *"Is the person next to you washing with soap?,"* which was the most effective message for men and the second most effective for women (other effective messages aimed to provoke disgust or activated knowledge about the dangers of failing to wash hands). A related study by Ybarra and Trafimow (1998) demonstrated that priming the collective self – by asking respondents what they have in common with their family and friends – makes subjective norms the strongest predictor of intentions towards using a condom during sex. Rivis and Sheeran (2004) report a meta-analysis of twenty-one studies, which found a medium to strong correlation between descriptive norms and intention for health-risk behaviors (smoking, drug use, binge drinking, condom use, extradyadic sex, gambling) and health-promoting behaviors (healthy eating, dieting, physical exercise).

Cues provoking the affective systems have also been used to motivate Pavlovian avoidance responses (see Figure 14.1), which have evolved to evade objects and situations carrying disease risk. *Disgust* can have powerful effects on behavior change compared to traditional behavior change models relying on providing health information alone, as shown by Curtis, Garbrah-Aidoo and Scott (2007) in the context of promoting hygiene related behaviors like hand-washing with soap around African countries. In particular, mothers in Ghana tend to use most soap for cleaning clothes, washing dishes, and bathing instead of hand washing, but only 3% wash hands with soap after toilet use. Studying hundreds of mothers and their children revealed that previous health campaigns had failed to change behavior, because Ghanaians used soap when they felt that their hands were dirty (e.g., after cooking or traveling) and hand-washing is provoked by feelings of disgust (parents also felt deep concerns about exposing their children to anything disgusting). As a result, these authors designed an intervention campaign, television commercials, focusing on provoking disgust rather than promoting soap use. For example, soapy hand-washing was shown only for 4 seconds in one 55-second video clip, but there was a clear message that toilet prompts worries of contamination, disgust, and requires soap. This campaign was a completely different approach from most public health campaigns, because it did not mention sickness and hence did not try to educate and change health-related beliefs. Instead, the campaign just provoked an innate emotional reaction of disgust, which resulted in a 13% increase in the use of soap after the toilet and 41% increase in reported soap use before eating. Translating this evidence to the language of the impulsive (Pavlovian) system, an encounter with a conditioned stimulus for a punishment (toilets) also triggers conditioned reflexes or "wanting" to avoid or remove its own associated unconditioned stimulus (disgusting, disease-carrying substances), presumably via transfer of incentive salience to associatively linked representations of the absent punisher. In interventions, this can also be manifest as cue-triggered increases in instrumental behavior, such as washing with soap, which we described before as Pavlovian-instrumental transfer.

Other interventions have tried to provoke *fear* by using gain versus loss framed messages in terms of benefits versus detriments. This intervention strategy appears to rely on an amygdala-based fear response provoking Pavlovian approach/engagement or avoidance, triggered by framing the choice information in terms of gains or losses relative to some psychological reference point such as current health or wealth (De Martino, Kumaran, Seymour & Dolan, 2006). As an example, Banks, Salovey, Greener et al. (1995) demonstrated that due to loss-aversion, loss-framed persuasive messages emphasizing the risks of not obtaining mammography have a stronger impact on behavior promotion (opting for mammography) than gain-framed messages emphasizing the benefits of obtaining mammography, even though both messages are factually equivalent. Webb and Sheeran (2006) review similar interventions and report that loss-framed messages have been found to elicit greater behavioral intention for health-detecting behaviors such as breast self-examination and cancer self-examination (see McCormick & McElroy, 2009).

CONCLUSION

We describe a conceptual framework for population behavior change, which offers a novel perspective of how various constructs from diverse models and research domains link together. The key concepts are borrowed from cognitive neuroscience, which now offers insights into how the human brain implements high level psychological functions, including decision-making. Such knowledge when combined with insights from other disciplines has spawned new disciplines, a pertinent example being neuroeconomics (Glimcher et al., 2009). This new field has already generated remarkable findings into the distinct neural systems that underlie how people learn in an optimal fashion, how human preferences are formed and the mechanisms that explain common deviations from rationality in human choice behavior. Emergent findings suggest a revision in how we construe the architecture of the human mind and the challenge is now to understand how these systems interact during the expression of behavior including how they impact on, for example, self-control.

The framework presented here explicitly sets behavioral models at the center of the intervention and policy planning process – in helping to understand specific behaviors and by identifying the underlying factors which influence them. The behavioral model outlined here can be combined with existing frameworks for designing and developing interventions. For example, see Bartholomew, Parcel and Kok's (1998) *intervention mapping* approach, which can lead to more theoretically grounded and effective interventions and policies.

The interventions that we describe pertain mainly to the domains of behavior economics and social and cognitive psychology, which have relied mostly on experimental evidence and to a lesser degree on field studies (but see DellaVigna, 2009, for a recent review of field evidence). In this respect, there are some limitations of applying results of behavioral economics to population behavior change (e.g., in health) as there are, as yet, few long-term studies and randomized controlled trials. Therefore, future research should target these limitations to provide more solid evidence for population-wide interventions and public policy that addresses the relative effectiveness, and cost-effectiveness, of cues and control systems in different wellbeing domains (e.g., health, environment, finance) and specific behaviors (e.g., smoking, dieting, saving).

Acknowledgments

The authors would like to thank Paul Dolan and Peter Dayan for valuable comments and suggestions.

References

Abadie, A., & Gay, S. (2004). The impact of presumed consent legislation on cadaveric organ donation: A cross country study. *Journal of Health Economics*, 25, 599–620.
Abraham, C., & Michie, S. (2008). A taxonomy of behavior change techniques used in interventions. *Health Risk & Society*, 27, 379–387.
Ajzen, I. (1991). The theory of planned behavior. *Organizational Behavior and Human Decision Processes*, 50, 179–211.

Allman, J. M., Hakeem, A., Erwin, J. M., Nimchinsky, E., & Hof, P. (2001). The anterior cingulate cortex. The evolution of an interface between emotion and cognition. *Annals of the New York Academy of Sciences, 935*, 107–117.

Anderson, J. R. (1993). *The architecture of cognition*. Cambridge, MA: Harvard University Press.

Anderson, A., & Sobel, N. (2003). Dissociating intensity from valence as sensory inputs to emotion. *Neuron, 39*, 581–583.

Anderson, J. R., Bothell, D., Byrne, M. D., Douglass, S., Lebiere, C., & Qin, Y. (2004). An integrated theory of the mind. *Psychological Review, 111*, 1036–1060.

Ariely, D. (2008). *Predictably irrational: The hidden forces that shape our decisions*. New York, NY: Harper Collins.

Ashraf, N., Karlan, D., & Yin, W. (2006). Tying Odysseus to the mast: Evidence from a commitment savings product in the Philippines. *Quarterly Journal of Economics, 121*, 635–671.

Aunger, R., & Curtis, V. (2008). Kinds of behavior. *Biology & Philosophy, 23*, 317–345.

Balleine, B. W. (2005). Neural bases of food-seeking: affect, arousal and reward in corticostriatolimbic circuits. *Physiology & Behavior, 86*, 717–730.

Bandura, A. (1977). Self-efficacy: toward a unifying theory of behavioral change. *Psychological Review, 84*, 191–215.

Banks, S. M., Salovey, P., Greener, S., Rothman, A. J., Moyer, A., Beauvais, J., & Epel, E. (1995). The effects of message framing on mammography utilization. *Health Risk & Society, 14*, 178–184.

Bargh, J. A. (1996). The automaticity of everyday life. In R. S. Wyer, Jr. (Ed.), *Advances in social cognition* (Vol. 10). New Jersey: Lawrence Erlbaum Associates.

Bargh, J. A., Chen, M., & Burrows, L. (1996). Automaticity of social behavior: Direct effects of trait construct and stereotype activation on action. *Journal of Personality and Social Psychology, 71*, 230–244.

Bargh, J. A., & Chartrand, T. L. (1999). The unbearable automaticity of being. *American Psychologist, 54*, 462–479.

Barr, N. A. (1987). *The economics of the welfare state*. London: Weidenfeld & Nicolson.

Barrett, L. F., Tugade, M. M., & Engle, R. W. (2004). Individual differences in working memory capacity and dual-process theories of the mind. *Psychological Bulletin, 130*, 553–573.

Berridge, K. C., Robinson, T. E., & Aldridge, J. W. (2009). Dissecting components of reward: "liking", "wanting", and learning. *Current Opinion in Pharmacology, 9*, 65–73.

Blair, S. N., Kohl, H. W., III, Barlow, C. E., Paffenbarger, R. S., Jr, Gibbons, L. W., & Macera, C. A. (1995). Changes in physical fitness and all-cause mortality: A prospective study of healthy and unhealthy men. *Journal of the American Medical Association, 273*, 1093–1098.

Broman, C. L. (1995). Leisure-time physical activity in an African–American population. *Journal of Behavioral Medicine, 18*, 341–353.

Bruner, J. S. (1973). *Beyond the information given: Studies in the psychology of knowing*. Oxford, UK: Norton.

Chaiken, S., Liberman, A., & Eagly, A. H. (1989). Heuristic and systematic information processing within and beyond the persuasion context. In J. S. Uleman & J. A. Bargh (Eds.), *Unintended thought* (pp. 212–252). New York, NY: Guilford.

Chaiken, S., & Trope, Y. (Eds.), (1999). *Dual-process theories in social psychology*. New York, NY: Guilford Press.

Chaloupka, F. J., & Grossman, M. (1996). Price, tobacco control policies and youth smoking. National Bureau of Economic Research Working Paper No. W5740.

Chaloupka, F. J., Grossman, M., & Saffer, H. (2002). The effects of price on alcohol consumption and alcohol-related problems. *Alcohol Research & Health, 26*, 22–34.

Chaloupka, F. J., Grossman, M., Bickel, W. K., & Saffer, H. (Eds.), (1999). Introduction. In *The economic analysis of substance use and abuse: An integration of economic and behavioral economic research* (pp. 1–12). Chicago, IL: University of Chicago Press.

Cialdini, R. B. (2003). Crafting normative messages to protect the environment. *Current Directions in Psychological Science, 12*, 105–109.

Cohen, J. B., Pham, M. T., & Andrade, E. B. (2008). The nature and role of affect in consumer behavior. In C. P. Haugtvedt, P. Herr & F. Kardes (Eds.), *Handbook of consumer psychology* (pp. 297–348). New York, NY: Erlbaum.

Colman, G., Grossman, M., & Joyce, T. (2003). The effect of cigarette excise taxes on smoking before, during and after pregnancy. *Journal of Health Economics, 22*, 1053–1072.

Critchley, H. D. (2005). Neural mechanisms of autonomic, affective, and cognitive integration. *Journal of Comparative Neurology, 493*, 154–166.

Critchley, H. D., Wiens, S., Rotshtein, P., Ohman, A., & Dolan, R. J. (2004). Neural systems supporting interoceptive awareness. *Nature Neuroscience, 7*, 189–192.

Cunningham, W. A., van Bavel, J. J., & Johnsen, I. R. (2008). Affective flexibility: evaluative processing goals shape amygdala activity. *Psychological Science, 19*, 152–160.

Curtis, V. A., Danquah, L. O., & Aunger, R. V. (2009). Planned, motivated and habitual hygiene behavior: An eleven country review. *Health Expectations, 24*, 655–673.

Curtis, V., Garbrah-Aidoo, N., & Scott, B. (2007). Masters of marketing: Bringing private sector skills to public health partnerships. *American Journal of Public Health, 97*, 634–641.

Damasio, A. R. (1994). *Descartes' error: Emotion, reason, and the human brain.* New York, GP: Putnam.

Damasio, A. R. (2000). *The feeling of what happens.* New York: Harcourt Brace & Company.

Davidson, K., Goldstein, M., Kaplan, R., Kaufmann, R. M., Knatterud, G. L., & Orleans, C. T., et al. (2003). Evidence-based behavioral medicine: What is it and how do we achieve it?. *Annals of Behavioral Medicine, 26*, 161–171.

Daw, N. D., Niv, Y., & Dayan, P. (2005). Uncertainty-based competition between prefrontal and dorsolateral striatal systems for behavioral control. *Nature Neuroscience, 8*, 1704–1711.

Dayan, P., & Niv, Y. (2008). Reinforcement learning: The good, the bad and the ugly. *Current Opinion in Neurobiology, 18*, 185–196.

Dayan, P., Niv, Y., Seymour, B. J., & Daw, N. D. (2006). The misbehavior of value and the discipline of the will. *Neural Networks, 19*, 1153–1160.

Dayan, P., & Seymour, B. (2009). Values and actions in aversion. In P. W. Glimcher, C. F. Camerer, E. Fehr & R. A. Poldrack (Eds.), *Neuroeconomics: Decision-making and the brain.* New York, NY: Academic Press.

Delgado, M. R., Gillis, M. M., & Phelps, E. A. (2008). Regulating the expectation of reward via cognitive strategies. *Nature Neuroscience, 11*, 880–881.

DellaVigna, S. (2009). Psychology and economics: Evidence from the field. *Journal of Economic Literature, 47*, 315–372.

De Martino, B., Kumaran, D., Seymour, B., & Dolan, R. (2006). Frames, biases, and rational decision-making in the human brain. *Science, 313*, 684–687.

Department of Health. (2004). Choosing health: Making healthy choices easier. Public Health White Paper (The Stationery Office, Department of Health).

Dijksterhuis, A., & Bargh, J. A. (2001). The perception–behavior expressway: Automatic effects of social perception on social behavior. In M. P. Zanna (Ed.), *Advances in experimental social psychology* (Vol. 33, pp. 1–40). San Diego: Academic.

Duflo, E., Kremer, M., & Robinson, J. (2010). Nudging farmers to use fertilizer: Evidence from Kenya. *American Economic Review* (forthcoming).

Dupas, P. (2011). Do teenagers respond to HIV risk information? Evidence from a field experiment in Kenya. *American Economic Journal: Applied Economics, 3*, 1–36.

Elliott, E. S., & Dweck, C. S. (1988). Goals: an approach to motivation and achievement. *Journal of Personality and Social Psychology, 53*, 5–12.

Estes, W. K., & Skinner, B. F. (1941). Some quantitative properties of anxiety. *Journal of Experimental Psychology, 29*, 390–400.

Evans, J. St. B. T. (2008). Dual-processing accounts of reasoning, judgment, and social cognition. *Annual Review of Psychology, 59,* 255–278.

Everitt, B. J., Dickinson, A., & Robbins, T. W. (2001). The neuropsychological basis of addictive behavior. *Brain Research Reviews, 36,* 129–138.

Fishbein, M., & Ajzen, I. (1975). *Belief, attitude, intention and behavior: An introduction to theory and research.* Reading, MA: Addison Wesley.

Fiske, S. T. (2010). *Social beings: Core motives in social psychology.* New York, NY: Wiley.

Frederick, S., Loewenstein, G., & O'Donoghue, T. (2001). Time discounting: A critical review. *Journal of Economic Literature, 40,* 351–401.

Gawronski, B., & Bodenhausen, G. V. (2006). Associative and propositional processes in evaluation: An integrated review of implicit and explicit attitude change. *Psychological Bulletin, 132,* 692–731.

Gigerenzer, G., & Goldstein, D. G. (1996). Reasoning the fast and frugal way: Models of bounded rationality. *Psychological Review, 103,* 650–669.

Gigerenzer, G., Hoffrage, U., & Goldstein, D. G. (2008). Fast and frugal heuristics are plausible models of cognition: Reply to Dougherty, Franco-Watkins, and Thomas (2008). *Psychological Review, 115,* 230–237.

Gigerenzer, G., & Selten, R. (Eds.), (2001). *Bounded rationality: The adaptive toolbox.* Cambridge, MA: MIT Press.

Glasgow, R. E., Kleges, L. M., Dzewaltowski, D. A., Bull, S. S., & Estabrooks, P. (2004). The future of health behavior change research: what is needed to improve translation of research into health promotion practice?. *Annals of Behavioral Medicine, 27,* 3–12.

Glimcher, P. W., Camerer, C. F., Fehr, E., & Poldrack, R. A. (Eds.), (2009). *Neuroeconomics: Decision-making and the brain.* New York, NY: Academic Press.

Goldstein, N. J., Cialdini, R. B., & Griskevicius, V. (2008). A room with a viewpoint: Using social norms to motivate environmental conservation in hotels. *Journal of Consumer Research, 35,* 472–482.

Goldstein, D. G., & Gigerenzer, G. (2002). Models of ecological rationality: The recognition heuristic. *Psychological Review, 109,* 75–90.

Gray, J. B. (2008). Framing: A communication strategy for the medical encounter. *Journal of Community Health, 1,* 422–430.

Gray, J. R., Braver, T. S., & Raichle, M. E. (2002). Integration of emotion and cognition in the lateral prefrontal cortex. *Proceedings of the National Academy of Sciences of the United States of America, 99,* 4115–4120.

Halpern, S. D., Ubel, P. A., & Asch, D. A. (2007). Harnessing the power of default options to improve health care. *New England Journal of Medicine, 357,* 1340–1344.

Heller, R. F., & Page, J. (2002). A population perspective to evidence based medicine: Evidence for population health. *Journal of Epidemiology and Community Health, 56,* 45–47.

Higgins, S. T., Heil, S. H., & Plenbani-Lussier, J. (2004). Clinical implications of reinforcement as a determinant of substance use disorders. *Annual Review of Psychology, 55,* 431–461.

Hummel, J. E., & Holyoak, K. J. (2003). A symbolic-connectionist theory of relational inference and generalization. *Psychological Review, 110,* 220–264.

Johnson, E. J., & Goldstein, D. G. (2003). Do defaults save lives? *Science, 302,* 1338–1339.

Judah, G., Aunger, R., Schmidt, W. P., Michie, S., Granger, S., & Curtis, V. (2009). Experimental pretesting of hand-washing interventions in a natural setting. *American Journal of Public Health, 99,* S405–S411.

Kahneman, D. (2003). A perspective on judgment and choice – mapping bounded rationality. *American Psychologist, 58,* 697–720.

Kahneman, D., & Tversky, A. (1979). Prospect theory: An analysis of decision under risk. *Econometrica, 47,* 263–292.

Kahneman, D., & Tversky, A. (1991). Loss aversion in riskless choice: A reference-dependent model. *Quarterly Journal of Economics, 106*, 1039–1061.

Kahneman, D., & Tversky, A. (Eds.). (2000). *Choices, values and frames*. New York, NY: Cambridge University Press and the Russell Sage Foundation.

Kreps, D. M. (1990). *A course in microeconomic theory*. New York, NY: Harvester Wheatsheaf.

Lally, P., Chipperfield, A., & Wardle, J. (2008). Healthy habits: Efficacy of simple advice on weight control. *International Journal of Obstetric Anesthesia, 32*, 700–707.

Lally, P., van Jaarsveld, C. H. M., Potts, H. W. W., & Wardle, J. (2010). How are habits formed: Modelling habit formation in the real world. *European Journal of Social Psychology, 40*, 998–1009.

LeDoux, J. E. (2000). Emotion circuits in the brain. *Annual Review of Neuroscience, 23*, 155–184.

Loewenstein, G. (2000). Emotions in economic theory and economic behavior. *American Economic Review, 90*, 426–432.

Manning, W. G., Blumberg, L., & Moulton, L. (1995). The demand for alcohol: The differential response to price. *Journal of Health Economics, 14*, 123–148.

Marteau, T. M., Ashcroft, R. E., & Oliver, A. (2009). Using financial incentives to achieve healthy behavior. *British Medical Journal, 338*, 983–985.

Marteau, T. M., & Lerman, C. (2001). Genetic risk and behavior change. *British Medical Journal, 322*, 1056–1059.

Marteau, T. M., & Weinman, J. A. (2006). Self-regulation and the behavioral response to DNA risk information: A theoretical analysis and framework for current research and future practice. *Social Science & Medicine, 62*, 1360–1368.

McClure, S. M., Laibson, D. I., Loewenstein, G., & Cohen, J. D. (2004). Separate neural systems value immediate and delayed monetary rewards. *Science, 306*, 503–507.

McCormick, M., & McElroy, T. (2009). Healthy choices in context: How contextual cues can influence the persuasiveness of framed health messages. *Judgment and Decision-Making, 4*, 248–255.

Meyer, D. E., & Kieras, D. E. (1997). A computational theory of executive cognitive processes and multiple-task performance (Part 1). Basic mechanisms. *Psychological Review, 104*, 2–65.

Miller, E., & Cohen, J. (2001). An integrative theory of prefrontal cortex function. *Annual Review of Neuroscience, 24*, 167–202.

Neal, D. T., Wood, W., & Quinn, J. M. (2006). Habits – A repeated performance. *Current Directions in Psychological Science, 15*, 198–202.

Niv, Y., Joel, D., & Dayan, P. (2006). A normative perspective on motivation. *Trends in Cognitive Sciences, 8*, 375–381.

Norman, P., Abraham, C., & Conner, M. (Eds.), (2000). *Understanding and changing health behavior: From health beliefs to self-regulation* (2nd ed.). Amsterdam: Harwood Academic.

Ochsner, K. N., & Gross, J. J. (2005). The cognitive control of emotion. *Trends in Cognitive Sciences, 9*, 242–249.

Orbell, S., & Verplanken, B. (2010). The automatic component of habit in health behavior: habit as cue-contingent automaticity. *Health Risk & Society, 29*, 374–383.

Oullette, J. A., & Wood, W. (1998). Habit and intention in everyday life: the multiple processes by which past behavior predicts future behavior. *Psychological Bulletin, 124*, 54–74.

Pate, R. R., Pratt, M., Blair, S. N., Haskell, W. L., Macera, C. A., & Bouchard, C., et al. (1995). Physical activity and public health. A recommendation from the Centers for Disease Control and Prevention and the American College of Sports Medicine. *Journal of the American Medical Association, 273*, 402–407.

Pearce, D. W. (Ed.). (1986). (3rd ed.). Cambridge, MA: MIT Press.

Petty, R. E., & Cacioppo, J. T. (1986). *Communication and persuasion: Central and peripheral routes to attitude change*. New York, NY: Springer-Verlag.

Phelps, E. A. (2006). Emotion and cognition: insights from studies of the human amygdala. *Annual Review of Neuroscience, 57*, 27–53.

Phillips, M. L., Young, A. W., Senior, C., Brammer, M., Andrew, C., & Calder, A. J., et al. (1997). A specific neural substrate for perceiving facial expressions of disgust. *Nature, 389*, 495–498.

Rivis, A. J., & Sheeran, P. (2004). Descriptive norms as an additional predictor in the theory of planned behavior: A meta-analysis. *Current Psychology, 22*, 264–280.

Rogers, R. W. (1983). Cognitive and physiological processes in fear appeals and attitude change: A revised theory of protection motivation. In J. T. Cacioppo & R. E. Petty (Eds.), *Social psychophysiology: A sourcebook*. London: Guilford Press.

Roll, J. M., Higgins, S. T., & Badger, G. J. (1996). An experimental comparison of three different schedules of reinforcement of drug abstinence using cigarette smoking as an exemplar. *Journal of Applied Behavior Analysis, 29*, 495–505.

Rolls, E. T. (2005). *Emotion explained*. Oxford: Oxford University Press.

Rothman, A. J., Hertel, A. W., Baldwin, A. S., & Bartels, R. (2007). Understanding the determinants of health behavior change: Integrating theory and practice. In J. Shah & W. Gardner (Eds.), *Handbook of motivation science* (pp. 494–507). New York, NY: Guilford Press.

Rosenstock, I. (1966). Why people use health service. *Milbank Memorial Fund Quarterly, 44*, 94–123.

Ruiter, R. A. C., Kok, G., Verplanken, B., & Brug, J. (2001). Evoked fear and effects of appeals on attitudes to performing breast self-examination: An information processing perspective. *Health Expectations, 16*, 307–319.

Rychetnik, L., Frommer, M., Howe, P., & Shiell, A. (2002). Criteria for evaluating evidence on public health interventions. *Journal of Epidemiology and Community Health, 56*, 119–127.

Sanfey, A. G., Rilling, J. K., Aronson, J. A., Nystrom, L. E., & Cohen, J. D. (2003). The neural basis of economic decision-making in the Ultimatum Game. *Science, 300*, 1755–1758.

Schultz, P. W., Nolan, J. M., Cialdini, R. B., Goldstein, N. J., & Griskevicius, V. (2007). The constructive, destructive, and reconstructive power of social norms. *Psychological Science, 18*, 429–434.

Schwarz, N. (1990). Feelings as information: Informational and motivational functions of effective states. In E. T. Higgins & R. M. Sorrentino (Eds.), *Handbook of motivation and cognition: Foundations of social behavior* (Vol. 2, pp. 527–561). New York: Guilford Press.

Schwarz, N., & Bohner, G. (2001). The construction of attitudes. In A. Tesser & N. Schwarz (Eds.), *Blackwell handbook of social psychology: Intrapersonal processes* (pp. 436–457). Oxford, UK: Blackwell.

Schweitzer, M., Hershey, J. C., & Asch, D. A. (1996). Individual choice in spending accounts: Can we rely on employees to choose well?. *Medical Care, 34*, 583–593.

Seymour, B., Singer, T., & Dolan, R. (2007). The neurobiology of punishment. *Nature Reviews Neuroscience, 8*, 300–311.

Sheeran, P. (2002). Intention–behavior relations: A conceptual and empirical review. In W. Stroebe & M. Hewstone (Eds.), *European review of social psychology* (Vol. 12, pp. 1–36). London: Wiley.

Shumaker, S. A., Schron, E., Ockene, J., & McBee, W. L. (Eds.), (2008). *The handbook of health behavior change* (2nd ed.). New York, NY: Springer.

Sindelar, J. L. (2008). Paying for performance: the power of incentives over habits. *Health Education & Behavior, 17*, 449–451.

Singer, T., Seymour, B., O'Doherty, F., Kaube, H., Dolan, R. J., & Frith, C. D. (2004). Empathy for pain involves the affective but not sensory components of pain. *Science, 303*, 1157–1162.

Sloman, S. A. (1996). The empirical case for two systems of reasoning. *Psychological Bulletin, 119*, 3–22.

Slovic, P., Finucane, M., Peters, E., & McGregor, D. G. (2002). The affect heuristic. In T. Gilovich, D. Griffin & D. Kahneman (Eds.), *Heuristics and biases: The psychology of intuitive judgement* (pp. 397–420). New York, NY: Cambridge University Press.

Sobal, J., & Wansink, B. (2007). Kitchenscapes, tablescapes, platescapes, and foodscapes: Influence of microscale built environments on food intake. *Environment and Behavior, 39,* 124–142.

Strack, F., & Deutsch, R. (2004). Reflective and impulsive determinants of social behavior. *Personality and Social Psychology Review, 8,* 220–247.

Talmi, D., Seymour, B., Dayan, P., & Dolan, R. (2008). Human Pavlovian-instrumental transfer. *The Journal of Neuroscience, 28,* 360–368.

Tesser, A. (1986). Some effects of self-evaluation maintenance on cognition and action. In R. M. Sorrentino & E. T. Higgins (Eds.), *Handbook of motivation and cognition: Foundations of social behavior* (pp. 435–464). New York, NY: Guilford Press.

Thaler, R. H., & Benartzi, S. (2004). Save more tomorrow: Using behavioral economics to increase employee savings. *Journal of Political Economy, 112,* 164–187.

Thaler, R. H., & Sunstein, C. R. (2008). *Nudge: Improving decisions about health, wealth, and happiness.* New Haven, CT: Yale University Press.

Thornton, R. L. (2008). The demand for, and impact of, learning HIV status. *American Economic Review, 98,* 1829–1863.

Triandis, H. C. (1977). *Interpersonal behavior.* Monterey, CA: Brooks/Cole.

Uitenbroek, D. G., Kerekovska, A., & Festchieva, N. (1996). Health lifestyle behavior and socio-demographic characteristics. A study of Varna, Glasgow and Edinburgh. *Social Science & Medicine, 43,* 367–377.

Verplanken, B., Friborg, O., Wang, C. E., Trafimow, D., & Woolf, K. (2007). Mental habits: Metacognitive reflection on negative self-thinking. *Journal of Personality and Social Psychology, 92,* 526–541.

Verplanken, B., & Velsvik, R. (2008). Habitual negative body image thinking as psychological risk factor in adolescents. *Body Image, 5,* 133–140.

Vlaev, I., & Dolan, P. (2010). From changing cognitions to changing the context: A dual-process framework for population behavior change. Manuscript submitted for publication.

Wansink, B., & Cheney, M. M. (2005). Super bowls: Serving bowl size and food consumption. *Journal of the American Medical Association, 293,* 1727–1728.

Wansink, B., & van Ittersum, K. (2011). The perils of plate size: Waist, waste, and wallet. *Journal of Marketing* (in press).

Webb, T. L., & Sheeran, P. (2006). Does changing behavioral intentions engender behavior change? A meta-analysis of the experimental evidence. *Psychological Bulletin, 132,* 249–268.

Wolpe, J. (1990). *The practice of behavior therapy* (4th ed.). New York, NY: Pergamon Press.

Wood, W., & Neal, D. T. (2007). A new look at habits and the habit–goal interface. *Psychological Review, 114,* 843–863.

Wood, W., Quinn, J. M., & Kashy, D. (2002). Habits in everyday life: thought, emotion, and action. *Journal of Personality and Social Psychology, 83,* 1281–1297.

Wood, W., Tam, L., & Witt, M. G. (2005). Changing circumstances, disrupting habits. *Journal of Personality and Social Psychology, 88,* 918–933.

Worth, K., Sullivan, H., Hertel, A. W., Jeffery, R. W., & Rothman, A. J. (2005). Are there times when avoidance goals can be beneficial? A look at smoking cessation. *Basic and Applied Social Psychology, 27,* 107–116.

Ybarra, O., & Trafimow, D. (1998). How priming the private self or the collective self affects the relative weights of attitudes and subjective norms. *Personality and Social Psychology Bulletin, 24,* 362–370.

Zajonc, R. B. (1980). Feeling and thinking: Preferences need no inferences. *American Psychologist, 35,* 151–175.

Index